高等学校计算机科学与技术项目驱动案例实践系列教材

Java高级框架应用开发与项目案例教程

——Spring+SpringMVC+MyBatis

梁立新　编著

清华大学出版社
北京

内容简介

本书应用"项目驱动"(Project-Driven)最新教学模式，通过完整的项目案例系统地介绍了使用Spring+Spring MVC+MyBatis高级框架进行应用开发的方法和技术。本书内容包括Spring+Spring MVC+MyBatis概述；eGov电子政务项目概述、Spring技术和Spring控制反转(Inversion of Control，IoC)、Spring面向方面编程(Aspect-Oriented Programming，AOP)和事务处理、Spring MVC基础、Spring MVC高级特性、MyBatis基础、MyBatis实体关系映射、MyBatis高级特性以及Spring+Spring MVC+MyBatis集成等内容。

本书注重理论与实践相结合，内容详尽，提供了大量实例，突出应用能力的培养，将一个实际项目的知识点分解在各章作为案例讲解，是一本实用性突出的教材。本书可作为普通高等院校计算机专业本、专科生Spring+Spring MVC+MyBatis高级框架应用开发课程的教材，也可供设计开发人员参考。

图书在版编目(CIP)数据

Java高级框架应用开发与项目案例教程：Spring+Spring MVC+MyBatis/梁立新编著．—北京：清华大学出版社，2021.5(2024.7重印)
高等学校计算机科学与技术项目驱动案例实践系列教材
ISBN 978-7-302-58064-5

Ⅰ．①J…　Ⅱ．①梁…　Ⅲ．①JAVA语言－程序设计－高等学校－教材　Ⅳ．①TP312.8

中国版本图书馆CIP数据核字(2021)第075691号

责任编辑：张瑞庆　常建丽
封面设计：常雪影
责任校对：胡伟民
责任印制：刘海龙

出版发行：清华大学出版社
网　　址：https://www.tup.com.cn，https://www.wqxuetang.com
地　　址：北京清华大学学研大厦A座　　**邮　　编**：100084
社 总 机：010-83470000　　**邮　　购**：010-62786544
投稿与读者服务：010-62776969，c-service@tup.tsinghua.edu.cn
质量反馈：010-62772015，zhiliang@tup.tsinghua.edu.cn
课件下载：https://www.tup.com.cn，010-83470236
印 装 者：三河市铭诚印务有限公司
经　　销：全国新华书店
开　　本：185mm×260mm　　**印　　张**：26.5　　**字　　数**：656千字
版　　次：2021年7月第1版　　**印　　次**：2024年7月第3次印刷
定　　价：69.99元

产品编号：087985-01

高等学校计算机科学与技术项目驱动案例实践系列教材

编写指导委员会

主　任

李晓明

委　员

（按姓氏笔画排序）

卢先和　杨　波

梁立新　蒋宗礼

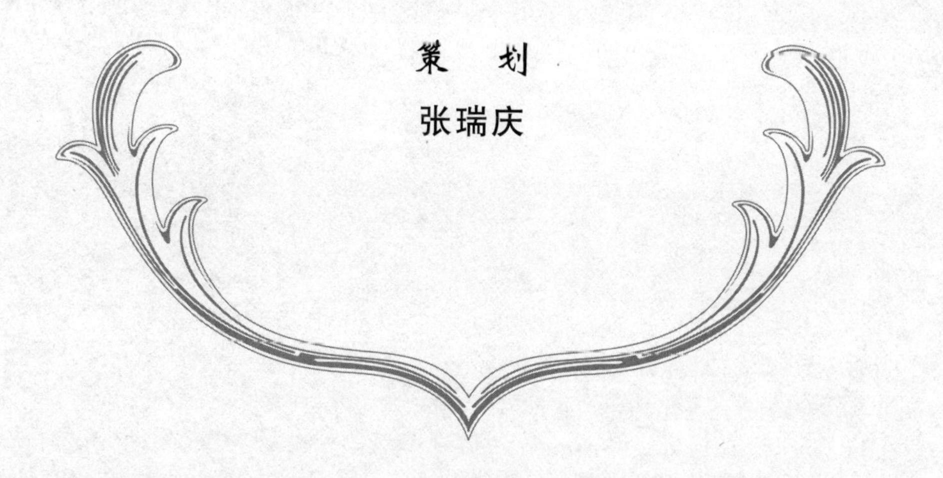

策　划

张瑞庆

FOREWORD

序 言

作为教育部高等学校计算机科学与技术教学指导委员会的工作内容之一，自从2003年参与清华大学出版社的“21世纪大学本科计算机专业系列教材”的组织工作以来，陆续参加或见证了多个出版社的多套教材的出版，但是现在读者看到的这一套“高等学校计算机科学与技术项目驱动案例实践系列教材”有着特殊的意义。

这个特殊性在于其内容。这是第一套我所涉及的以项目驱动教学为特色，实践性极强的规划教材。如何培养符合国家信息产业发展要求的计算机专业人才，一直是这些年人们十分关心的问题。加强学生的实践能力的培养，是人们达成的重要共识之一。为此，高等学校计算机科学与技术教学指导委员会专门编写了《高等学校计算机科学与技术专业实践教学体系与规范》(清华大学出版社出版)。但是，如何加强学生的实践能力培养，在现实中依然遇到种种困难。困难之一，就是合适教材的缺乏。以往的系列教材，大都比较“传统”，没有跳出固有的框框。而这一套教材，在设计上采用软件行业中卓有成效的项目驱动教学思想，突出“做中学”的理念，突出案例(而不是“练习作业”)的作用，为高校计算机专业教材的繁荣带来了一股新风。

这个特殊性在于其作者。本套教材目前规划了十余本，其主要编写人不是我们常见的知名大学教授，而是知名软件人才培训机构或者企业的骨干人员，以及在该机构或者企业得到过培训的并且在高校教学一线有多年教学经验的大学教师。我以为这样一种作者组合很有意义，他们既对发展中的软件行业有具体的认识，对实践中的软件技术有深刻的理解，对大型软件系统的开发有丰富的经验，也有在大学教书的经历和体会，他们能在一起合作编写教材本身就是一件了不起的事情，没有这样的作者组合是难以想象这种教材的规划编写的。我一直感到中国的大学计算机教材尽管繁荣，但也比较“单一”，作者群的同质化是这种风格单一的主要原因。对比国外英文教材，除了 Addison Wesley 和 Morgan Kaufmann 等出版的经典教材长盛不衰外，我们也看到 O'Reilly“动物教材”等的异军突起——这些教材的作者，大都是实战经验丰富的资深专业人士。

这个特殊性还在于其产生的背景。也许是由于我在计算机技术方面的动手能力相对比较弱，其实也不太懂如何教学生提高动手能力，因此一直希望有一个机会实际地了解所谓“实训”到底是怎么回事，也希望能有一种安排让现在

FOREWORD

教学岗位的一些青年教师得到相关的培训和体会。于是作为 2006—2010 年教育部高等学校计算机科学与技术教学指导委员会的一项工作，我们和教育部软件工程专业大学生实习实训基地（亚思晟）合作，举办了 6 期"高等学校青年教师软件工程设计开发高级研修班"，时间虽然只是短短的 1~2 周，但是对于大多数参加研修的青年教师来说都是很有收获的一段时光，在对他们的结业问卷中充分反映了这一点。从这种研修班得到的认识之一，就是目前市场上缺乏相应的教材。于是，这套"高等学校计算机科学与技术项目驱动案例实践系列教材"应运而生。

当然，这样一套教材，由于"新"，难免有风险。从内容程度的把握、知识点的提炼与铺陈，到与其他教学内容的结合，都需要在实践中逐步磨合。同时，这样一套教材对我们的高校教师也是一种挑战，只能按传统方式讲软件课程的人可能会觉得有些障碍。相信清华大学出版社今后将和作者以及高等学校计算机科学与技术教学指导委员会一起，举办一些相应的培训活动。总之，我认为编写这样的教材本身就是一种很有意义的实践，祝愿成功。也希望看到更多业界资深技术人员加入到大学教材编写的行列中来，和高校一线教师密切合作，将学科、行业的新知识、新技术、新成果写入教材，开发适用性和实践性强的优秀教材，共同为提高高等教育教学质量和人才培养质量做出贡献。

李晓明

原教育部高等学校计算机科学与技术教学指导委员会副主任、北京大学教授

PREFACE

前　言

21世纪,什么技术将影响人类的生活?什么产业将决定国家的发展?信息技术与信息产业是首选的答案。高等学校学生是企业和政府的后备军,国家教育行政部门计划在高校中普及信息技术与软件工程教育。经过多所高校的实践,信息技术与软件工程教育受到学生的普遍欢迎,取得了很好的教学效果,但也存在一些不容忽视的共性问题,其中突出的是教材问题。

从近两年信息技术与软件工程教育研究看,许多任课教师提出目前教材不合适。具体体现在:第一,来自信息技术与软件工程专业的术语很多,没有这些知识背景的学生学习起来具有一定难度;第二,书中案例比较匮乏,与企业的实际情况相差太远,致使案例可参考性差;第三,缺乏具体的课程实践指导和真实项目。因此,针对高等学校信息技术与软件工程课程的教学特点与需求,编写适用的规范化教材已刻不容缓。

本书就是针对以上问题编写的,它是一本融合项目实践与开发思想于一体的书。它的特色是以项目实践作为主线贯穿其中。本书提供了一个完整的电子政务项目案例,通过该项目的学习,读者能够快速掌握使用 Spring+Spring MVC+MyBatis 高级框架进行应用开发的方法和技术,具体内容包括 Spring+Spring MVC+MyBatis 概述、eGov 电子政务项目概述、Spring 技术和 Spring 控制反转(Inversion of Control, IoC)、Spring 面向方面编程(Aspect-Oriented Programming,AOP)和事务处理、Spring MVC 基础、Spring MVC 高级特性、MyBatis 基础、MyBatis 实体关系映射、MyBatis 高级特性以及 Spring+Spring MVC+MyBatis 集成等内容。

本书特点如下。

(1) 重项目实践。

我们多年的经验体会是"IT 是做出来的,不是想出来的",理论虽然重要,但一定要为实践服务。以项目为主线,带动理论的学习是最好、最快、最有效的方法。通过此书,希望读者对项目开发流程有个整体了解,减少对项目实践的盲目感,能够根据本书的体系循序渐进地动手做出自己的真实项目。

(2) 重理论要点。

本书以项目实践为主线,着重介绍 Spring+Spring MVC+MyBatis 高级框架理论中最重要、最精华的部分,融会贯通,这是本书的特色。读者首先通过项目把握整体概貌;其次深入局部细节,系统学习理论;最后不断优化和扩展细节,

完善整体框架和改进项目。

为了便于教学，本书配有教学课件，读者可从清华大学出版社的网站下载。

本书由深圳技术大学梁立新编著，本教材获得深圳技术大学的大力支持和教材出版资助，在此特别感谢。

鉴于编者的水平有限，书中难免有不足之处，敬请广大读者批评指正。

梁立新

2021年2月

CONTENTS

目　录

CONTENTS

C O N T E N T S

CONTENTS

第1章 Spring+Spring MVC+MyBatis 概述

学习目的与学习要求

学习目的：了解软件开发中框架的概述，简单了解Spring、Spring MVC 和 MyBatis 框架，准备好学习3个框架需要的开发工具及环境配置。

学习要求：按照开发工具及配置章节认真搭建开发工具及配置环境，是练习后面章节理论实例及案例的重要前提。

本章主要内容

本章主要内容包括框架的概述，Spring、Spring MVC 和 MyBatis 框架的基本原理，开发工具的安装与配置，其中包括集成开发工具 MyEclipse、服务器 Tomcat 和数据库 MySQL。

目前，国内外信息化建设已经进入以 Web 应用为基础核心的阶段。Java 语言应该算是开发 Web 应用的最佳语言。然而，就算用 Java 建造一个不是很烦琐的 Web 应用系统，也不是一件轻松的事情。有很多东西需要仔细考虑，如要考虑怎样建立用户接口？在哪里处理业务逻辑？怎样持久化数据？幸运的是，Web 应用面临的一些问题已经由曾遇到过这类问题的开发者建立起相应的框架（Framework）解决了。事实上，企业开发中直接采用的往往并不是某些具体的技术（如大家熟悉的 Core Java、JDBC、Servlet、JSP 等），而是基于这些技术之上的应用框架，Spring、Spring MVC 和 MyBatis 就是其中最常用的几种。

1.1 框架概述

在介绍软件框架之前，首先要明确什么是框架和为什么要使用框架。这要从企业面临的挑战谈起，如图 1-1 所示。

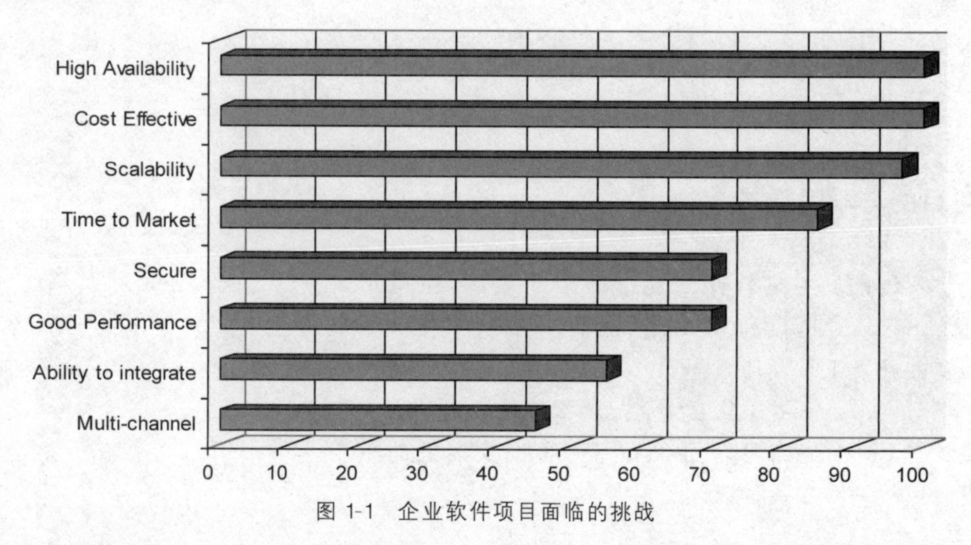

图 1-1 企业软件项目面临的挑战

可以看到，随着项目规模的扩大和复杂性的提高，企业会面临前所未有的各个方面的挑战。根据优先级排序，主要包括高可靠性(High Availability)、低成本(Cost Effective)、可扩展性(Scalability)、投放市场快速性(Time to Market)、安全性(Secure)、性能(Good Performance)、可集成性(Ability to integrate)以及多平台支持(Multi-channel)等。那么，如何面对并且解决这些挑战呢？这需要采用通用的、灵活的、开放的、可扩展的软件框架，由框架帮助解决这些挑战，之后再在框架基础之上开发具体的应用系统，如图 1-2 所示。

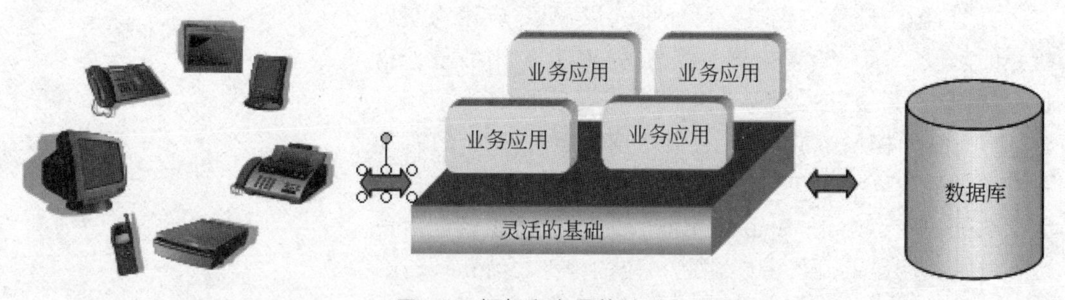

图 1-2 框架和应用的关系

这种基于框架的软件开发方式和传统的汽车生产方式类似，如图 1-3 所示。

那么，到底什么是软件框架呢？框架的定义如下：

- 是应用系统的骨架，将软件开发中反复出现的任务标准化，以可重用的形式提供使用。
- 大多提供了可执行的具体程序代码，支持迅速地开发出可执行的应用；但也可以是

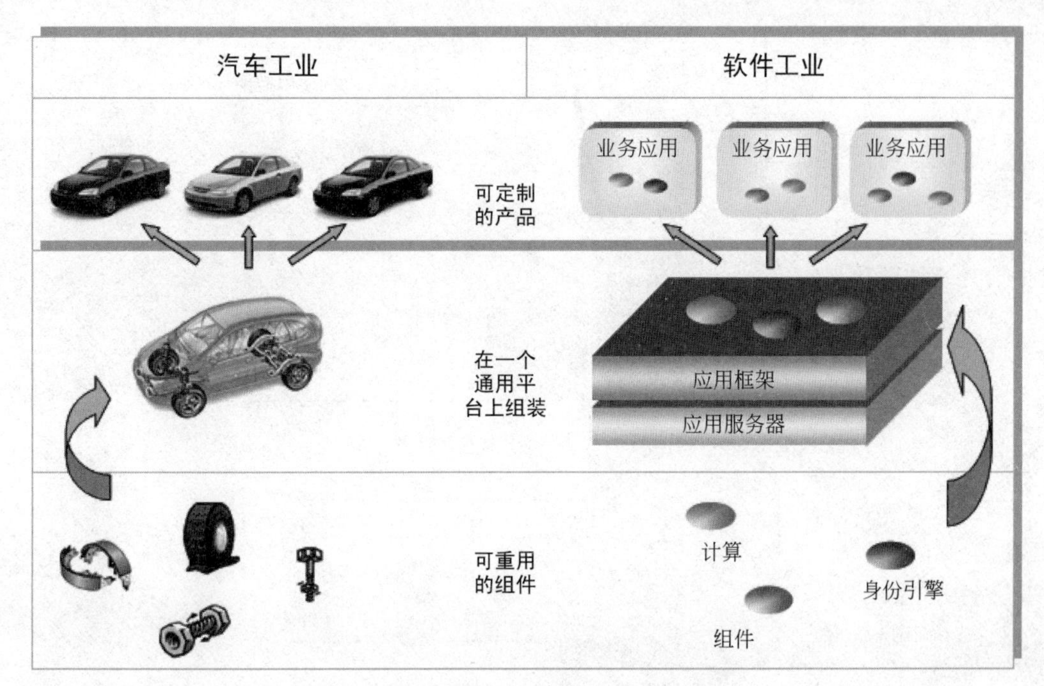

图 1-3 软件开发方式和传统的汽车生产方式

抽象的设计框架,帮助开发出健壮的设计模型

- 好的抽象、设计成功的框架,能够大大缩短开发应用系统的周期。
- 在预制框架上加入定制的构件,可以大量减少编码量,并容易测试。
- 分别用于垂直和水平应用。

框架具有以下特点:

- 框架具有很强(大粒度)的可重用性,远远超过单个类;它是一个功能连贯的类集合,通过相互协作为应用系统提供服务和预制行为。
- 框架中的不变部分,定义了接口、对象的交互和其他不变量。
- 框架中的变化部分即应用中的个性。

一个好的框架定义了开发和集成组件的标准。为了利用、定制或扩展框架服务,通常需要框架的使用者从已有框架类继承相应的子类;以及通过执行子类的重载方法,用户定义的类将会从预定义的框架类获得需要的消息。这会给我们带来很多好处,包括代码重用性和一致性,对变化的适应性,特别是它能够让开发人员专注于业务逻辑,从而大大减少开发时间。图 1-4 对有没有使用框架对项目开发所需工作量(以人 * 月衡量)的影响进行了对比。

从图 1-4 中不难看出,对于没有使用框架的项目而言,开发所需工作量(以 Man days,即人 * 月衡量)会随着项目复杂性的提高(以 Business Functions,即业务功能衡量)以几何级数递增,而对于使用框架的项目而言,开发所需工作量会随着项目复杂性的提高以代数级数递增。假定开发团队人数一样,一个没有使用框架的项目所需的周期为 6~9 个月,那么同样的项目如果使用框架则只需要 3~5 个月。

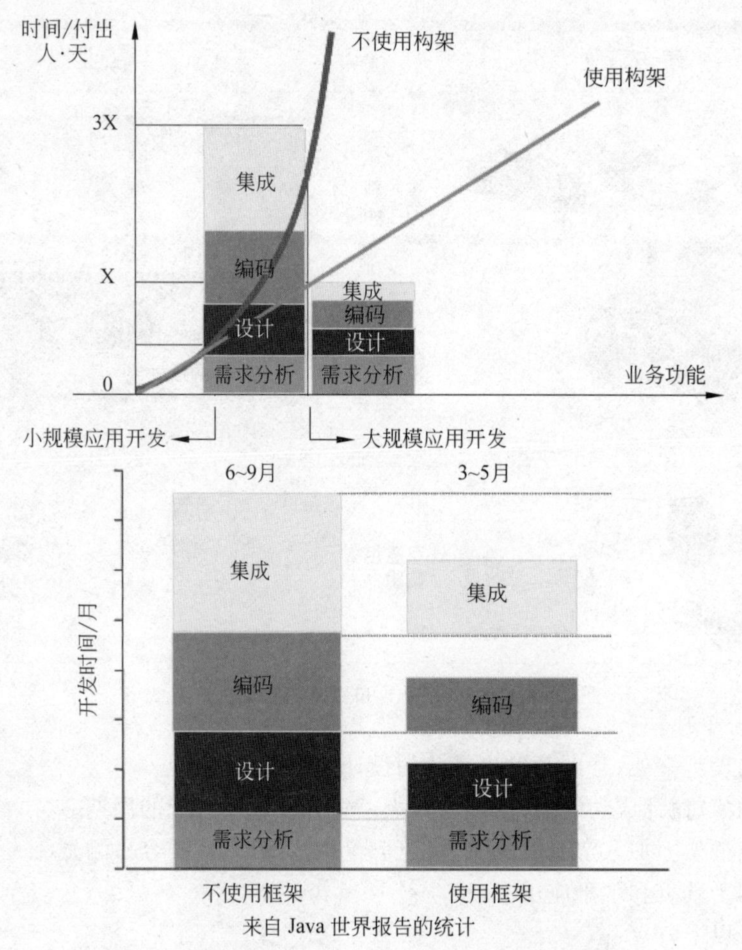

图 1-4　有没有使用框架对项目开发所需工作量的比较

1.2　Spring、Spring MVC 和 MyBatis 简介

在本书中，我们会具体讨论如何使用 3 种著名的框架 Spring、Spring MVC 和 MyBatis 使应用程序在保证质量前提下得以快速开发。

在软件架构设计中建立软件系统的高层结构，常常用到分层架构模式。

- 分层模式是一种将系统的行为或功能以层为首要的组织单位进行分配(划分)的结构模式。

  —通常在逻辑上进行垂直的层次 Layer 划分。

  —在物理上发明则进行水平的层级 Tier 划分。

- 分层要求

  —层内的元素只信赖于当前层和之下的相邻层中的其他元素。

注意，这并非绝对的要求。

大部分的 Web 应用在职责上至少能被分成 3 层：表示层(Presentation Layer)、业务层(Business Layer)和持久层(Persistence Layer)。每个层在功能上都应该十分明确，而不应该与其他层混合。每个层要相互独立，通过一个通信接口而相互联系。

这里讨论一个使用 3 种开源框架的策略：表示层用 Spring MVC；业务层用 Spring；而数据库访问层则用 MyBatis，如图 1-5 所示。

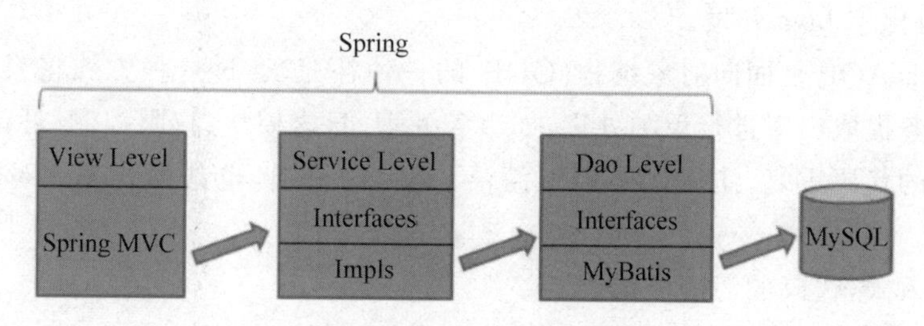

图 1-5 Spring+ Spring MVC+ MyBatis

Spring＋Spring MVC＋MyBatis 简称 SSM 框架，这是继 Struts＋Spring＋Hibernate(SSH)之后主流的 Java EE 企业级框架，它适用于搭建各种类型的企业级应用系统。SSM 这套技术是绝大部分公司明确要求掌握的技术，而 SSM 的组合搭配已经渐渐成为 JavaWeb 开发者必备的技能，虽然不是全部，但仍是目前的主要趋势，掌握 SSM 技术刻不容缓。

1. Spring

Spring 是一个功能强大并且十分完整的轻量级框架，是为了解决企业应用开发的复杂性而创建的。Spring 于 2004 年由 Rod Johnson 发布了 1.0 版本，经过多年的更新迭代，已经逐渐成为 Java 开源世界的第一框架。Spring 框架是 Java EE 应用的一站式解决方案，与各个优秀的 MVC(Model-View-Controller)框架(如 Spring MVC、Struts 2、JSF 等)可以无缝整合，与各个 ORM(Object-Relationship Mapping)框架(如 Hibernate、MyBatis、JPA 等)也可以无缝衔接，其他各种技术也因为 Spring 的存在而被很容易地整合进项目开发中，如 Redis、Log4J、Quartz 和 Thymeleaf 等。只要开发中需要使用到的技术，Spring 都提供了极好的封装和整合体验，这也是 Spring 生命力如此强大的原因，目前还没有出现能够替代 Spring 的框架。

Spring 最重要的两大核心技术，包括 IoC(Inversion of Control，控制反转，使用 Java 反射机制实现)和 AOP(Aspect-Oriented Programming，面向方面编程，使用动态代理机制实现)。简单来说，Spring 是一个轻量级的 IoC 和 AOP 的容器框架。Spring 在项目中的核心功能是解耦，也就是实现高内聚、低耦合。

(1) IoC：由 Spring 负责控制对象的生命周期和对象间的关系。所有类都会在 Spring 容器中注册，告诉 Spring 你是什么对象，需要什么对象，然后 Spring 会在系统运行到适当的时候，把你需要的对象给你，同时也把你交给其他需要你的对象。所有类的创建、销毁都由 Spring 控制，也就是说，控制对象生存周期的不再是引用它的对象，而是 Spring。对于某个具体的对象而言，以前是它控制其他对象，现在是所有对象都被 Spring 容器控制，从而降低了对象之间的耦合度，new 的使用不再频繁出现。所以这叫 IoC。

(2) AOP：在编译期间、装载期间或运行期间实现在不修改源代码的情况下给程序动态添加功能的一种技术。通常采用动态代理技术，利用截取消息的方式，对该消息进行装饰新加入的功能，以取代原有对象行为的执行；通俗点说就是把可重用的功能提取出来，然后将这些通用功能在合适的时候织入应用程序中。

Spring 框架的优点如下：

- Spring 的 IoC 容器将对象之间的依赖关系交由 Spring 控制，提高了组件之间的解耦，简化了 Java 开发。
- Spring AOP 是面向对象编程（OOP）的一种补充，通过这种方式将系统中的一些通用任务提取出来进行单独处理，如事务处理、日志模块、权限控制、性能监控等，避免大量的代码重复，使得代码更加简洁，复用性更强，也帮助开发人员更加关注业务逻辑。
- 非侵入式，代码的污染极低。
- 无与伦比的兼容性，与其他优秀的第三方框架无缝整合。
- 高度的开放性。
- 开源社区十分活跃，文档齐全，学习成本不高。

2. Spring MVC

Spring MVC 是一个典型的 Web 框架，它功能强大，且简单易学。Spring MVC 是 Spring 中的一个模块，它是基于 MVC 的思想建立起来的专门用于处理请求和响应的框架。Spring MVC 框架围绕 DispatcherServlet（分发器）这个核心类展开，DispatcherServlet 是 Spring MVC 的总策划，它负责截获请求并将其分派给相应的处理器（Controller 层）处理。它可以调用 Spring 容器进行工作，将系统更详细地划分为 Controller（控制）层、Service（服务）层、DAO（数据持久）层、POJO（实体）层。

Spring MVC 的基本原理：

（1）客户端（如 JSP 页面等）发送请求到 DispacherServlet（分发器）。

（2）由 DispacherServlet 控制器查询 HanderMapping，找到处理请求的 Controller。

（3）Controller 调用业务逻辑处理层（Service），Service 调用 DAO 层和 POJO 层与数据库交互返回给 Service 执行结果，之后把结果返回给 Controller，再交给 DispacherServlet。

（4）DispacherServlet 查询视图解析器，找到 ModelAndView 指定的视图。

（5）视图负责将结果显示到客户端。

Spring MVC 框架的优点如下：

- Spring 和 Spring MVC 无须复杂的操作即可整合，灵活度高，非侵入性。
- 配置简单，学习成本低。
- 设计合理，各模块分工明确，功能强大。
- 非常容易与第三方视图技术集成整合。
- 能够进行 Web 层的单元测试。
- 支持灵活的 URL 到页面控制器的映射。
- 灵活且强大的数据验证、格式化和数据绑定机制。

3. MyBatis

MyBatis 是一款优秀的持久层 SQL 操作封装框架，它支持定制化 SQL、存储过程以及高级映射。MyBatis 避免了几乎所有的 JDBC 代码和手工设置参数以及抽取结果集。它使用简单的 XML 或注解配置和映射实体,将接口和 Java 的 POJOs(Plain Old Java Objects,普通的 Java 对象)映射成数据库中的记录。MyBatis 是基于 SQL 配置的持久层框架,关注如何方便地通过配置 SQL 语句访问数据库,而不需要将 SQL 语句嵌入 Java 代码中(也是解耦操作),相比 Hibernate 更轻量级,更简便。

MyBatis 框架的优点如下:

- 封装了 JDBC 大部分操作,减少了开发人员的工作量。
- 相比一些自动化的 ORM 框架,"半自动化"使得开发人员可以自由编写 SQL 语句,灵活度更高。
- Java 代码与 SQL 语句分离,降低了维护的难度。
- 自动映射结果集,减少了重复的编码工作。
- 开源社区十分活跃,文档齐全,学习成本不高。

这里只简单介绍了 SSM 技术,后面章节将会详细介绍。

1.3 开发工具与配置

1.3.1 开发工具与环境

1. 集成开发环境(IDE):MyEclipse

对于 Java 应用开发人员来讲,好的集成开发环境(Integrated Development Environment,IDE)非常重要。目前在市场上占主导地位的 Java 集成开发平台就是基于 Eclipse 之上的 MyEclipse 工具。下面使用 MyEclipse 开发 Java Web 应用,在这里选择 MyEclipse 2017 版本作为开发工具,读者可以到 https://www.myeclipsecn.com 下载并安装。

MyEclipse 2017 集成(内置)了 Java Development Kits 和 Tomcat 8.5,这样就不需要单独安装它们了。(当然,为了测试和部署的方便性,也可以单独安装 Tomcat。)

2. 服务器:Tomcat

Tomcat 是一个免费的、开源的 Servlet 容器,它是 Apache 基金会的 Jakarta 项目中的一个核心项目,由 Apache 其他一些公司及个人共同开发而成。由于有企业的参与和支持,最新的 Servlet 和 JSP 规范总能在 Tomcat 中得到体现。

Tomcat 提供了各种平台的版本供下载,建议使用 Tomcat 8.5 版,可以从 https://tomcat.apache.org/index.html 下载其源代码版或者二进制版。由于 Java 的跨平台特性,基于 Java 的 Tomcat 也具有跨平台性。

3. 数据库:MySQL

MySQL 是一个多用户、多线程的 SQL(Structured Query Language,结构化查询语言)数据库,是一个客户机/服务器结构的应用,它由一个服务器守护程序 mysqld 和很多不同的客户程序和库组成。它是目前市场上运行最快的 SQL 数据库之一,可以从 http://dev.mysql.com/downloads/下载 MySQL Community Server 安装软件包。它提供了其他数据库少有的编程工具,而且 MySQL Community Server 对于个人用户是免费的。建议安装

MySQL 5.5 之后的版本，这里使用的是 MySQL 8.0.19。

MySQL 的功能特点如下：可以同时处理几乎不限数量的用户；可以处理多达 50 000 000 条以上的记录；命令执行速度快，也许是现今最快的；简单有效的用户特权系统。

1.3.2 工具集成步骤

1. MyEclipse 连 Tomcat

MyEclipse 2017 里面已经内置了 Tomcat 8.5，不需要再单独安装。当然，也可以选择外部独立的 Tomcat。

（1）首先确定自己下载并安装了 Tomcat 8.5，假定安装目录是 C:\apache-tomcat-8.5.34，之后打开 MyEclipse 2017。

（2）选择 Window→Preferences...，进入如图 1-6 所示的界面。

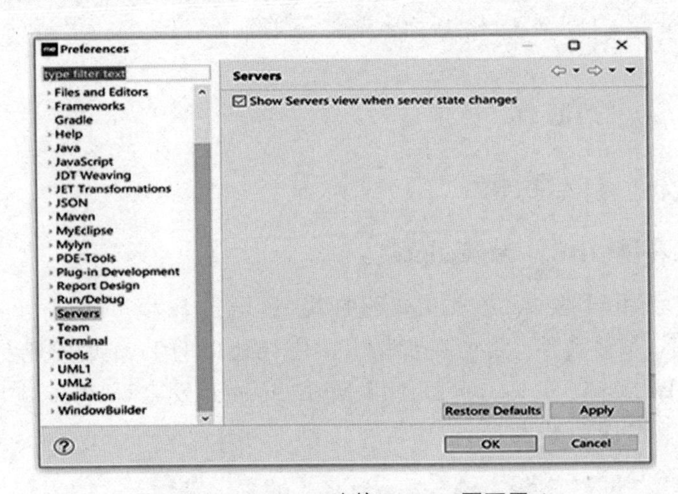

图 1-6　MyEclipse 连接 Tomcat 页面图 1

（3）选择 Servers→Runtime Environments，如图 1-7 所示。

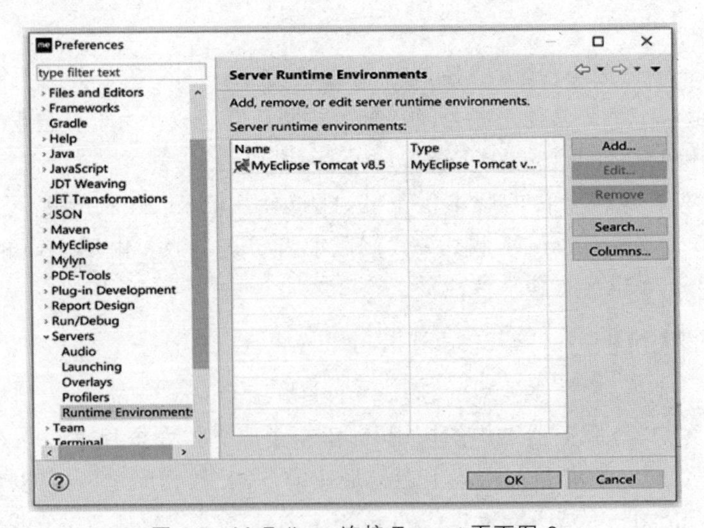

图 1-7　MyEclipse 连接 Tomcat 页面图 2

(4) 选择 Add→Tomcat→Apache Tomcat v8.5,如图 1-8 所示。

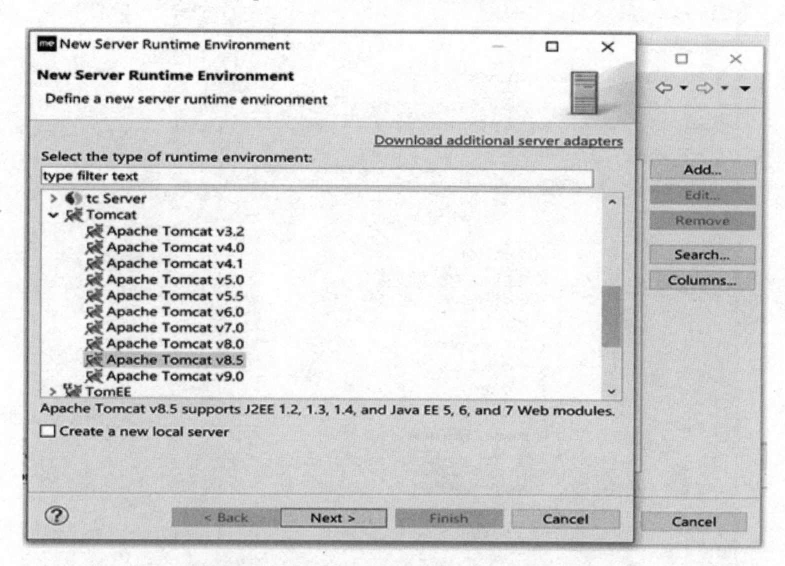

图 1-8 MyEclipse 连接 Tomcat 页面图 3

(5) 单击 Next 按钮,如图 1-9 所示。

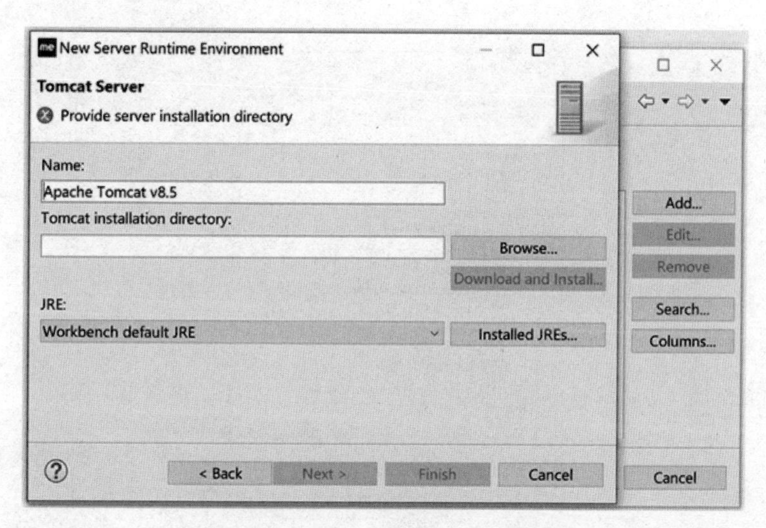

图 1-9 MyEclipse 连接 Tomcat 页面图 4

(6) 单击 Browse 按钮,找到 Tomcat 在本机的安装目录,这里是 C:\apache-tomcat-8.5.34,如图 1-10 所示。

(7) 单击"确定"按钮,最后单击 Finish 按钮即可。

2. MyEclipse 连 MySQL 数据库

(1) 单击 Window→Show View,打开如图 1-11 所示的对话框。

(2) 选择 Database→DB Browser,单击 OK 按钮,出现图 1-12。

(3) 上面操作打开了 DB Browser 视图,在该视图空白区右击,如图 1-13 所示。

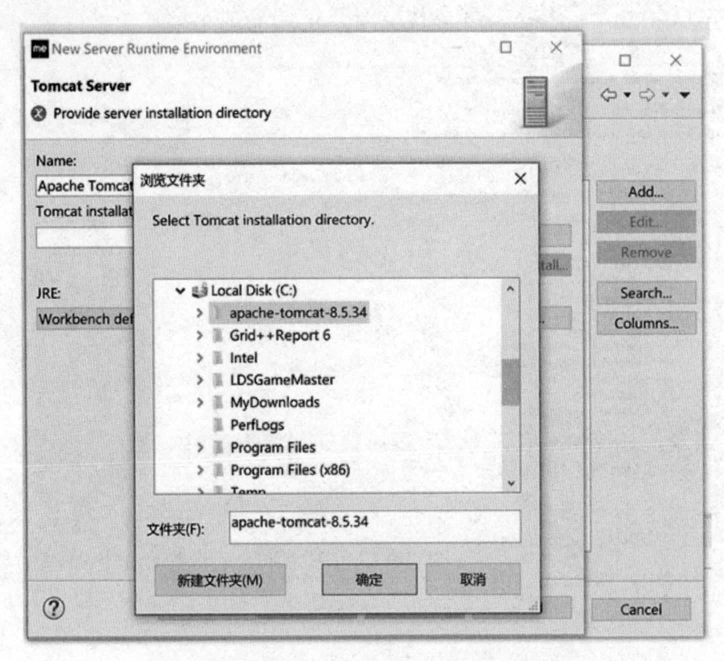

图 1-10　MyEclipse 连接 Tomcat 页面图 5

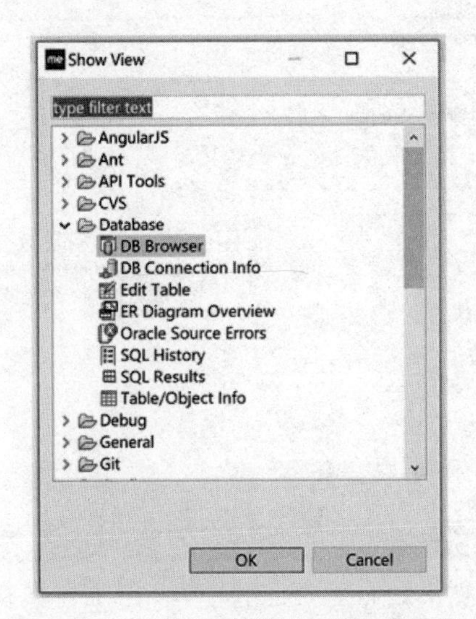

图 1-11　MyEclipse 连接 MySQL 页面图 1

(4) 单击 New...后出现配置连接页面，如图 1-14 所示。

(5) Driver template：选择 MySQL Connector/J；Driver name：自己随意取名，这里命名为 llx；Connection URL：配置自己的数据库 url，这里是 jdbc:mysql://localhost:3306/test；User name、Password 为 MySQL 用户名、密码，这里是 root、root；Driver JARs：选择 MySQL 驱动包，这里用的是 D:\Tools\mysql-connector-java-5.1.46.jar；Driver classname：选择 com.mysql.jdbc.Driver；单击 Test Driver 按钮确保连接成功，可以勾选 Save password 以便以后每次连接不用再输入密码，之后单击 Next 按钮，如图 1-15 所示。

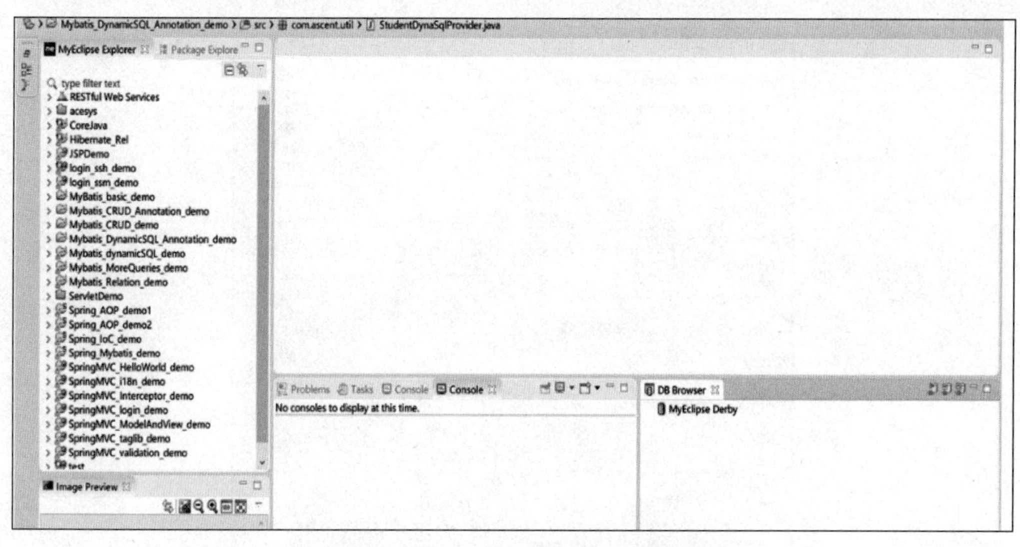

图 1-12　MyEclipse 连接 MySQL 页面图 2

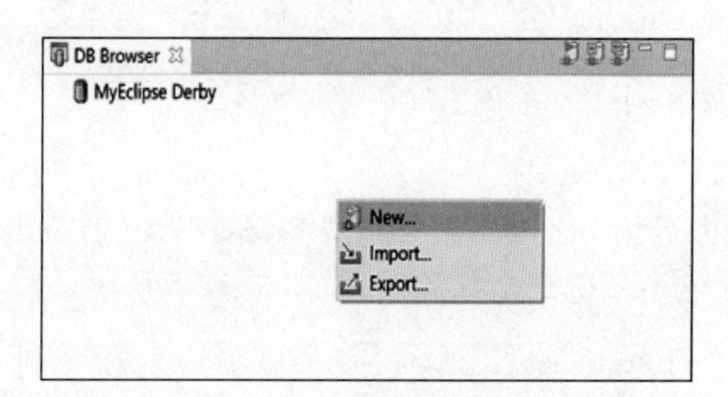

图 1-13　MyEclipse 连接 MySQL 页面图 3

Database Driver

New Database Connection Driver

Create a new connection driver

Driver template: Axion

Driver name:

Connection URL: jdbc:axiondb:<database-name>[:<database-directory>]

User name:

Password:

Driver JARs

Add JARs

Remove

Driver classname:

Test Driver

Connect to database on MyEclipse startup

Save password

Saved passwords are stored on your computer in a file that's difficult, but not impossible, for an intruder to read.

< Back　Next >　Finish　Cancel

图 1-14　MyEclipse 连接 MySQL 页面图 4

图 1-15　MyEclipse 连接 MySQL 页面图 5

最后单击 Finish 按钮配置完成。

(6) 成功设置后，DB Browser 视图区会出现刚设置的连接，右击连接，从弹出的快捷菜单中选择 Open Connection，正确连接到数据库，如图 1-16 所示。

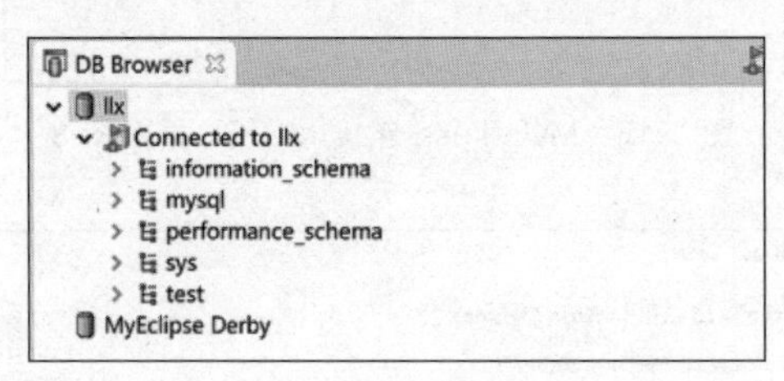

图 1-16　MyEclipse 连接 MySQL 页面图 6

上述操作已经正确设置了 MySQL 连接，在进行应用开发时，可以通过设置好的连接查看和操作数据。

习题

1. 什么是软件框架？
2. 框架具有什么特点？
3. Spring、Spring MVC 和 MyBatis 框架分别有哪些优点？

第2章 eGov电子政务项目概述

学习目的与学习要求

学习目的：了解本书项目案例——eGov电子政务系统的项目需求分析，项目系统分析与设计以及运行指南等。

学习要求：认真了解整个项目案例的需求分析与设计，为后面学习各个章节的案例开发作准备。

本章主要内容

本章主要内容包括eGov电子政务系统的需求分析和系统分析与设计。

本书采用先进的“项目驱动式”教学法，通过一个完整的eGov电子政务项目，贯穿Spring＋Spring MVC＋MyBatis 3个框架的学习。首先介绍eGov电子政务系统项目的背景知识，包括需求分析、系统设计和运行指南等。这个项目的开发过程将会贯穿在之后各个章节中的案例部分中，结合相关知识点详细讲解和实现。

2.1 项目需求分析

eGov电子政务系统是基于互联网的应用软件。从研究中心的网上能了解到已公开发布的不同栏目（如新闻、通知等）的内容，各部门可以发表栏目内容（如新闻、通知等），有关负责人对需要发布的内容进行审批。其中，有的栏目（如新闻）必须经过审批才能发布，有的栏目（如通知）则不需要审批就能发布。系统管理人员对用户及其权限进行管理。

整体功能用例图（Use-Case Diagram），如图2-1所示。

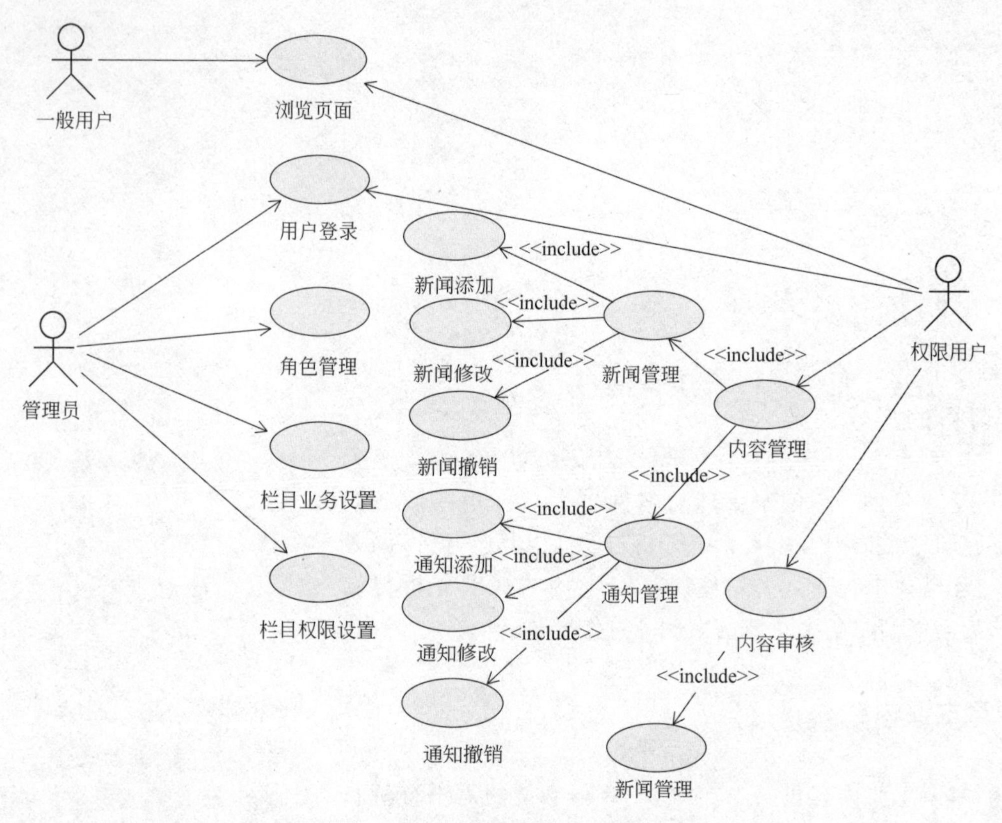

图 2-1　eGov 电子政务系统用例图

2.1.1　一般用户浏览的内容管理：首页显示及其他页面

首页显示是数据量最大的一页，是为所有模块展示内容的部分。从该页还可以登录进入管理等后端功能模块，如图 2-2 所示。

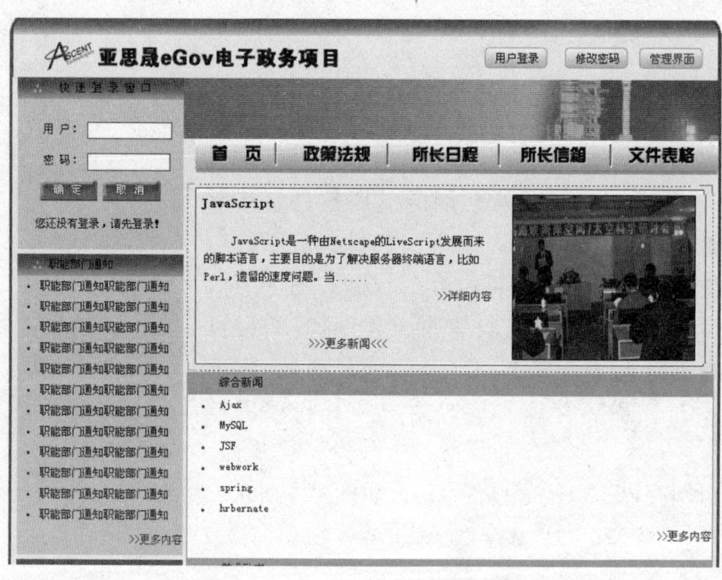

图 2-2　eGov 电子政务系统首页

首页最上面为头版头条栏目，左栏为职能部门通知，下面是综合新闻等。左上部分为快速登录窗口。

2.1.2 系统管理

系统管理是给系统管理人员使用的，主要包括以下功能模块：登录，栏目业务设置，栏目权限设置，用户管理设置。

1. 登录

1）用例描述

（1）角色：注册用户（用户和管理员）。

（2）前提条件：无。

（3）主事件流：

① 用户登录该网站的登录页面（E1）。

② 显示登录页面信息，如用户名、密码。

③ 输入用户名和密码，单击“登录”按钮（E2）。

④ 验证登录信息。

⑤ 加载用户拥有的权限信息，并显示在页面中。

（4）异常事件流

E1：输入非法的标识符，指明错误。

E2：用户账号被管理员屏蔽，无法登录。

2）用户界面图

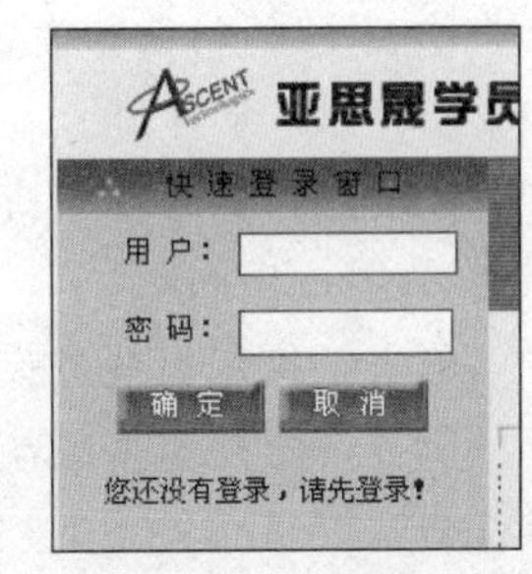

图 2-3 eGov 用户登录

用户在首页登录，如图 2-3 所示。

输入正确的用户名和密码后进入系统管理的入口页面，如图 2-4 所示。

图 2-4 系统管理入口页面

2. 栏目业务设置

1）用例描述

（1）角色：管理员。

（2）前提条件：用户必须完成登录的用例。

（3）主事件流：

① 用户登录该网页(E1)，单击“栏目业务设置”。

② 进入栏目业务设置页面。

③ 设置每个栏目的内容管理(S1）和内容审核(S2)（单击内容管理图标会更改)。

（4）分支事件流。

S1：设置内容管理。

① 单击“内容管理”链接。

② 内容管理和内容审核的权限改变。

③ 返回栏目业务设置页面。

S2：设置内容审核。

① 单击“内容审核”链接。

② 内容审核的权限改变。

③ 返回栏目业务设置页面。

（5）异常事件流。

E1：用户账号被管理员屏蔽或删除，无法设置，提示重新激活账号。

2）用户界面图

单击链接栏目业务设置，进入该模块，设定栏目是否具有内容管理和内容审核的权限。栏目业务设置是整个系统管理模块的最高级权限设置，它的操作影响到栏目权限设置以及所有与本栏目有关系的权限设置，如图 2-5 所示。

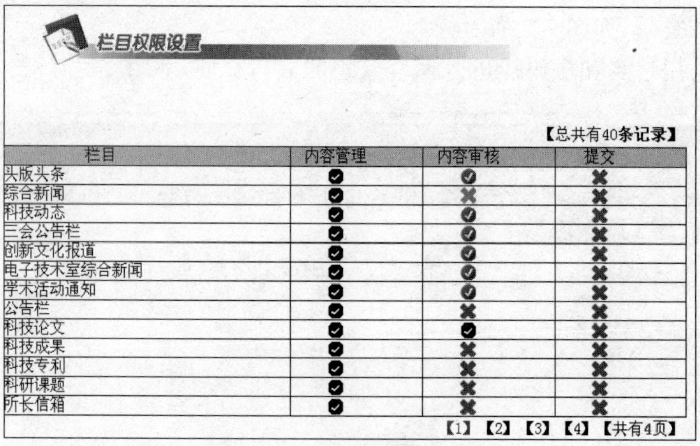

栏目权限设置

【总共有40条记录】

栏目	内容管理	内容审核	提交
头版头条	✔	✔	✖
综合新闻	✔	✖	✖
科技动态	✔	✔	✖
三会公告栏	✔	✔	✖
创新文化报道	✔	✔	✖
电子技术室综合新闻	✔	✔	✖
学术活动通知	✔	✔	✖
公告栏	✔	✖	✖
科技论文	✔	✔	✖
科技成果	✔	✖	✖
科技专利	✔	✖	✖
科研课题	✔	✖	✖
所长信箱	✔	✖	✖

【1】 【2】 【3】 【4】 【共有4页】

图 2-5　栏目权限设置

每个栏目可以设定是否具有内容管理和内容审核的权限，对于某些栏目(如新闻)，二者都有，因为新闻必须经过有关领导审核批准，才可以在网上发布。对于某些栏目(如通知)，只需要内容管理，不需要内容审核就可以在网上发布。

3. 栏目权限设置

1）用例描述

（1）角色：管理员。

(2) 前提条件：用户必须完成登录的用例。

(3) 主事件流：

① 用户登录该网站，单击“栏目权限设置”。

② 进入栏目权限设置页面。

③ 单击“设置”按钮。

④ 进入栏目权限设置页面。

⑤ 选中用户名，单击添加(S1)或删除(S2)，然后保存修改。

⑥ 该栏目的用户被添加或删除。

⑦ 返回权限栏目设置页面。

(4) 分支事件流。

S1：添加用户。

① 选中用户，单击“删除”按钮。

② 添加用户。

③ 单击“返回”按钮。

④ 返回权限栏目设置页面。

S2：删除用户。

① 选中用户，单击“删除”按钮。

② 删除用户。

③ 单击“返回”按钮。

④ 返回权限栏目设置页面。

2) 用户界面图

单击链接权限设置，进入该模块，这里主要是为用户分配栏目的管理权限，这个业务也是这个项目的核心，需要在所有部门里选取用户分配权限，如图 2-6 所示。

栏目	内容管理	内容审核	设置
头版头条	列出3333 11 99 44 测试用户11	无	设置
综合新闻	11	无	设置
科技动态	11	22	设置
三会公告栏	11	22	设置
创新文化报道	11	22	设置
电子技术室综合新闻	11	22	设置
学术活动通知	11	22	设置
公告栏	11	无	设置
科技论文	11	22	设置

图 2-6 栏目权限分配

单击设置，打开如图 2-7 所示的页面。

左面是用户过滤，也是备选用户，右面为管理权限和审核权限。选择不同部门时，该部门的所有人员应该显示在备选用户列表里。单击上面的“添加”按钮时，用户会被放入管理权限列表里。

单击下面的“添加”按钮时，用户会被放入审核权限列表里。这里有一个业务大家要记住：一个用户不可以既分配到管理权限，又分配到审核权限。

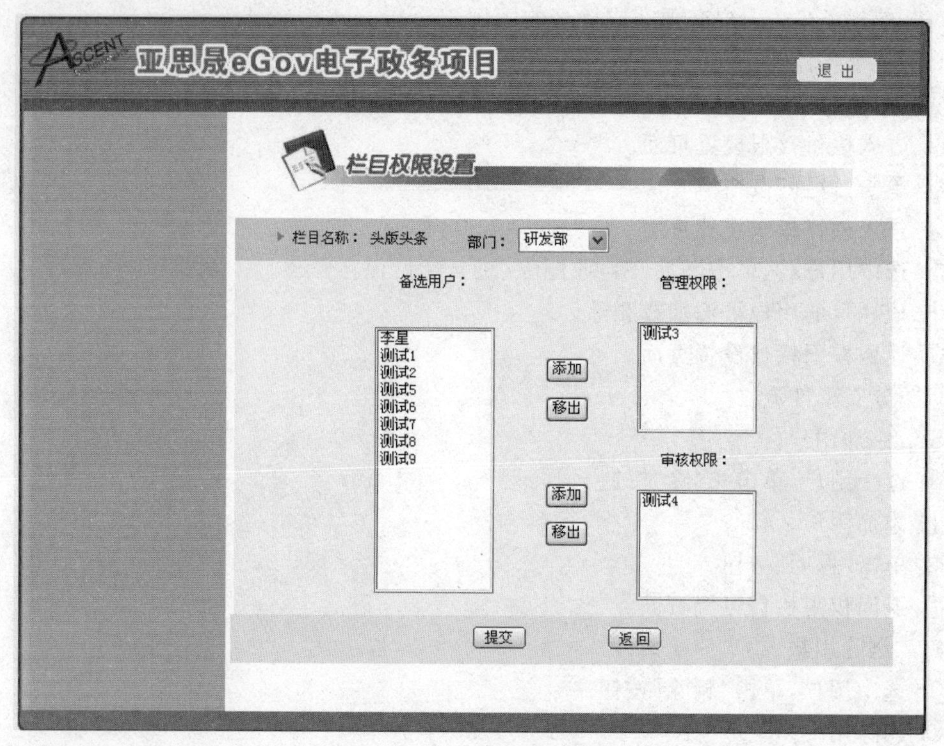

图 2-7　栏目权限分配管理

4. 用户管理设置

1）用例描述

（1）角色：管理员。

（2）前提条件：用户必须完成登录的用例。

（3）主事件流：

① 用户登录该网站，单击“用户管理设置”。

② 进入用户管理设置页面。

③ 单击“新增”按钮（S1）、“修改”按钮（S2）和“删除”按钮（S3）。

（4）分支事件流：

S1：单击“新增”按钮。

① 单击“新增”按钮。

② 进入添加新用户页面。

③ 添加用户基本信息，之后单击“添加”按钮（E1）。

④ 保存用户信息。

⑤ 返回用户管理设置页面。

S2：单击“修改”按钮。

① 单击某条用户信息的“修改”按钮。

② 进入修改用户页面。

③ 修改用户资料并单击“修改”按钮。

④ 更新用户信息。

⑤ 返回用户管理设置页面。

S3：单击“删除”按钮。

① 单击某用户的“删除”按钮。

② 删除该用户。

③ 返回用户管理设置页面。

(5) 异常事件流。

E1：输入非法的标识符，指明错误。

2) 用户界面图

用户管理设置显示用户、添加用户、修改用户、删除用户。

单击链接“用户管理设置”，进入该模块，如图 2-8 所示。

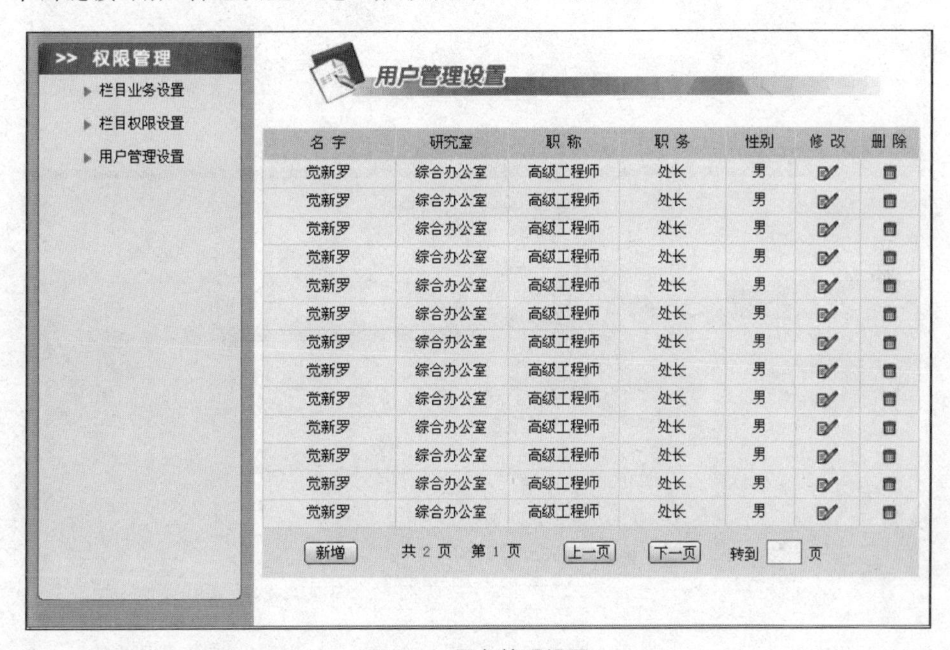

图 2-8 用户管理设置

添加用户：单击“新增”按钮，如图 2-9 所示。

输入新的用户信息，单击“提交”按钮。

修改用户：单击“修改”按钮，如图 2-10 所示。

删除用户：单击“删除”按钮，用户信息被删除。

2.1.3 内容管理和审核

内容管理和审核主要包括以下功能模块：登录；新闻管理（新闻的编辑、修改、屏蔽、删除）；通知管理（通知的编辑、修改、屏蔽、删除），新闻内容的审核等。

1. 登录

1) 用例描述

(1) 角色：注册用户（用户和管理员）。

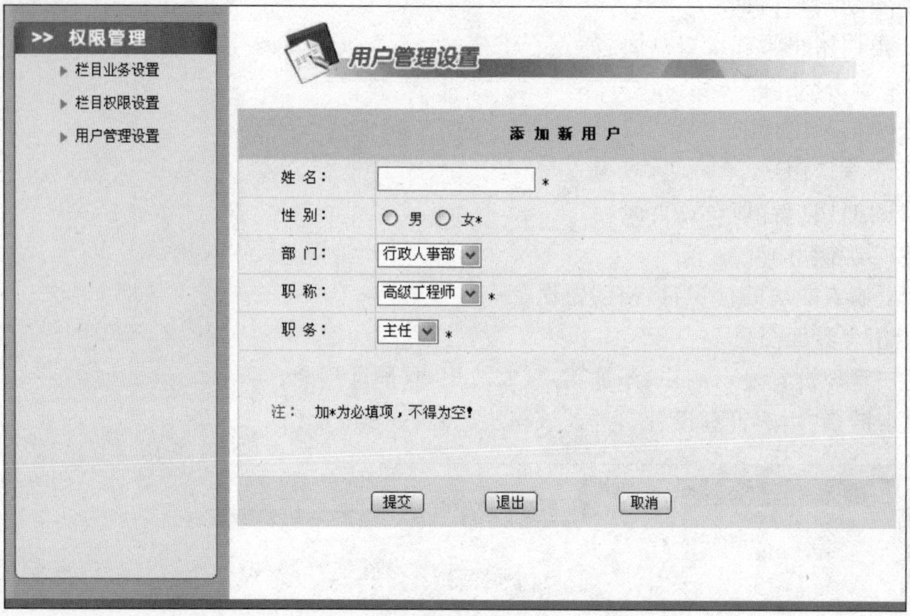

图 2-9　添加用户

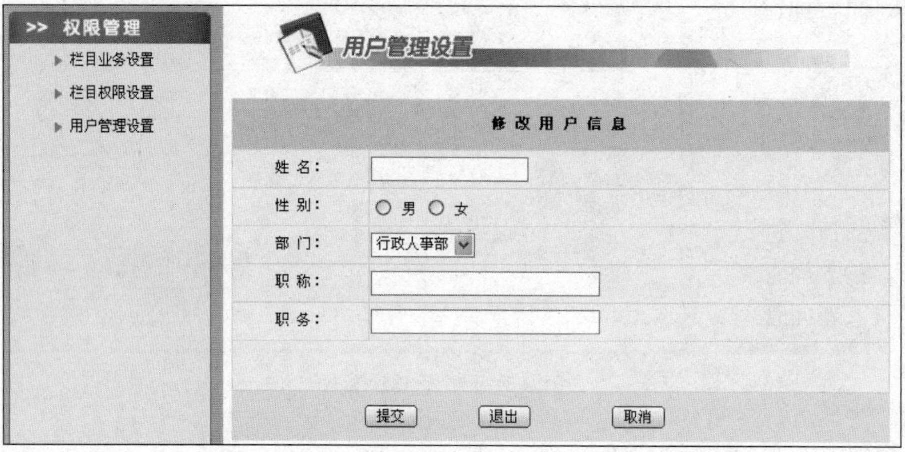

图 2-10　修改用户

(2) 前提条件：无。

(3) 主事件流：

① 用户登录该网站的登录页面(E1)。

② 显示登录页面信息，如用户名、密码。

③ 输入用户名和密码，单击“登录”按钮(E2)。

④ 验证登录信息。

⑤ 加载用户所拥有的权限信息，并显示在页面中。

(4) 异常事件流。

E1：输入非法的标识符，指明错误。

E2：用户账号被管理员屏蔽，无法登录。

2）用户界面图

输入用户名和密码，进入系统，如图 2-11 所示。

当用户进入系统时，应该看到自己的权限范围，不同的用户有不同的权限，如图 2-12 所示。

这个用户具有的权限是对 1 个栏目的内容管理权限，如果用另外一个用户登录，结果就不同了，如图 2-13 所示。

图 2-11　用户登录

图 2-12　用户权限页面 1

图 2-13　用户权限页面 2

这个用户所有的权限是对 1 个栏目的内容审核权限。

2. 新闻管理（新闻的编辑、修改、屏蔽、删除）

1）用例描述

(1) 角色：管理员和高级管理员。

(2) 前提条件：用户必须完成登录的用例。

(3) 主事件流：

① 用户通知进入系统。

② 单击新闻管理。

③ 进入新闻管理页面（新闻列表）。

④ 单击“新增”按钮（S1）、“修改”按钮（S2）和“删除”按钮（S3）。

(4) 分支事件流。

S1：单击“新增”按钮。

① 单击“新增”按钮。

② 进入新闻添加页面。

③ 填写通知资料（E1）。

④ 单击“保存”按钮。

⑤ 验证信息，保存数据。

⑥ 返回通知新闻页面（新闻列表）。

S2：单击“修改”按钮。

① 单击“修改”按钮。

② 进入新闻修改页面。

③ 更改新闻数据，单击“修改”按钮。

④ 验证信息，保存数据。

⑤ 返回新闻管理页面。

S3：单击“删除”按钮。

① 在要删除的记录前打勾，单击“删除”按钮。

② 删除信息。

③ 返回新闻管理页面。

(5) 异常事件流。

E1：输入非法的标识符或者格式不对，指明错误。

2) 用户界面图

新闻管理—新闻编辑：单击内容管理中的“综合新闻管理”进入新闻编辑，如图 2-14 所示。

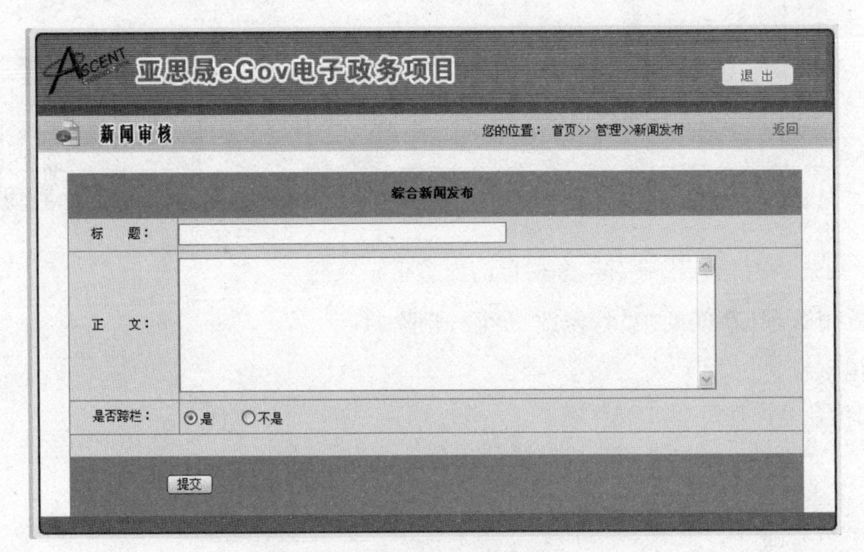

图 2-14 新闻发布编辑

大家不要忽略新闻发布的预览功能，如图 2-15 所示。

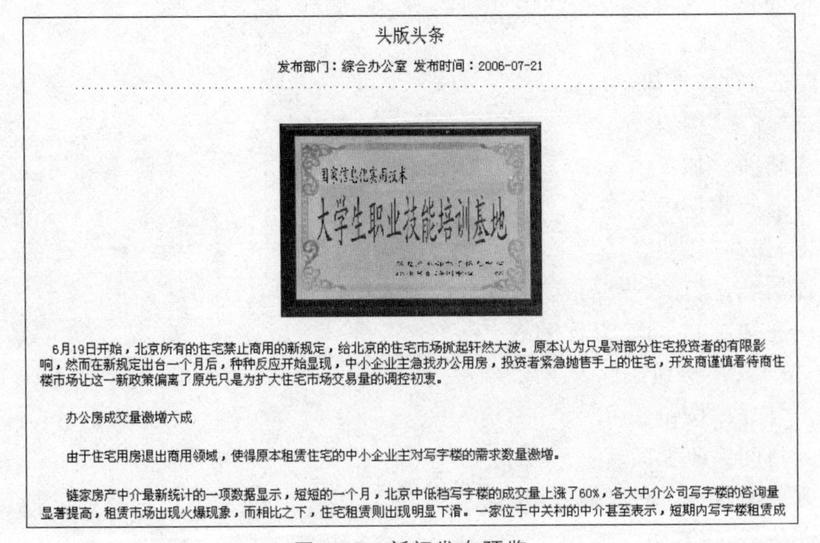

头版头条

发布部门：综合办公室 发布时间：2006-07-21

大学生职业技能培训基地

6月19日开始，北京所有的住宅禁止商用的新规定，给北京的住宅市场掀起轩然大波。原本认为只是对部分住宅投资者的有限影响，然而在新规定出台一个月后，种种反应开始显现，中小企业主急找办公用房，投资者紧急抛售手上的住宅，开发商谨慎看待商住楼市场让这一新政策偏离了原先只是为扩大住宅市场交易量的调控初衷。

办公房成交量激增六成

由于住宅用房退出商用领域，使得原本租赁住宅的中小企业主对写字楼的需求数量激增。

链家房产中介最新统计的一项数据显示，短短的一个月，北京中低档写字楼的成交量上涨了60%，各大中介公司写字楼的咨询量显著提高，租赁市场出现火爆现象，而相比之下，住宅租赁则出现明显下滑。一家位于中关村的中介甚至表示，短期内写字楼租赁成

图 2-15 新闻发布预览

预览效果和发布后的最终效果一样，这里如果符合标准，就可以提交了。

提交后的浏览页应该按时间倒序排列，以保证最后发布的新闻在第一条上。刚刚发布的新闻的发布状态是待审（已经提交了，但是要等待审核）。这时就要等待有审核权限的人审核这条新闻，通过后才能发布上去。

新闻管理—新闻修改：对于任何一个必须通过审核的新闻，都必须符合这里修改的规则，也就是当新闻处于发布状态的时候，任何人都不得修改新闻，只有新闻处于屏蔽状态，或者为待审时才可以修改。对于发布、待审、屏蔽等注释的数字，在数据字典中都有，大家可以查询。如果修改已经发布的新闻，应该给用户返回一个友好界面，如图 2-16 所示。

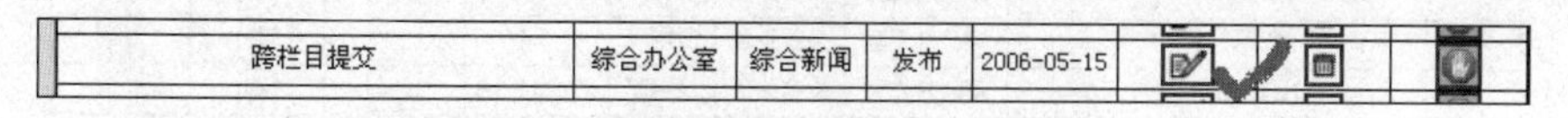

图 2-16　新闻修改

单击发布的新闻，如图 2-17 所示。

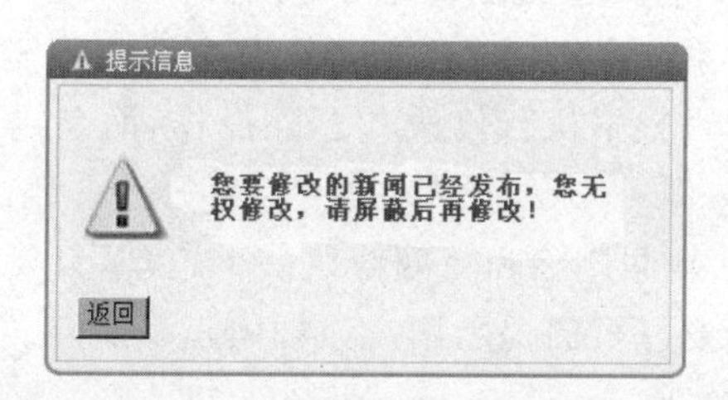

图 2-17　单击发布的新闻

如果新闻没有发布则可以修改，如图 2-18 所示。

标　题：Ajax

正　文：Ajax (Asynchronous JavaScript + XML) 的定义
基于web标准 (standards-based presentation) XHTML+CSS的表示；
使用 DOM (Document Object Model) 进行动态显示及交互；
使用 XML 和 XSLT 进行数据交换及相关操作；
使用 XMLHttpRequest 进行异步数据查询、检索；
使用 JavaScript 将所有的东西绑定在一起。英文参见Ajax的提出者Jesse James Garrett的原文，原文题目(Ajax: A New Approach to Web Applications)。
类似于DHTML或LAMP，AJAX不是指一种单一的技术，而是有机地利用了一系列相关的技

一审意见：没有问题

二核意见：内容不够新颖

是否跨栏：◉是　○不是

图 2-18　如果新闻没有发布则可以修改

新闻管理—新闻屏蔽：新闻屏蔽功能是当一个新闻要在首页新闻栏目中被撤下时所具有的功能，如图 2-19 所示。

标题	发布部门	栏目来源	发布状态	发布时间	修改	删除	屏蔽
头版头条	综合办公室	头版头条	待审				
文曲 郭子	综合办公室	头版头条	屏蔽	2006-07-03			
testgoogle	综合办公室	头版头条	屏蔽	2006-06-08			
需要审核	综合办公室	头版头条	屏蔽	2006-06-08			
ajax事例	综合办公室	头版头条	屏蔽	2006-05-27			
AJAX开发简略	综合办公室	头版头条	屏蔽	2006-05-15			
open-open介绍	综合办公室	头版头条	屏蔽	2006-05-15			
跨栏目提交	综合办公室	综合新闻	发布	2006-05-15			
新闻关于电子政务格式	综合办公室	头版头条	屏蔽				
Ajax简介	综合办公室	头版头条	屏蔽	2006-05-06			

图 2-19　新闻屏蔽

在浏览页上可以看到发布状态就是对新闻存在状态(statu)的标注，这时如果删除或者修改一个已经发布的新闻，系统会有一个友好界面提醒我们不能随便删除或者修改一个发布的新闻。如果状态为发布，就不执行修改和删除操作，而是跳转到一个友好界面上提示用户。

新闻管理—新闻删除：新闻删除和新闻修改一个道理，只要新闻不处于发布状态的时候就可以删除，否则跳转到友好页面提示用户该如何正确删除。

3. 通知管理(通知的编辑、修改、屏蔽、删除)

单击内容管理中的通知栏目，进入该模块，显示已发布的通知。

1) 用例描述

(1) 角色：管理员和高级管理员。

(2) 前提条件：用户必须完成登录的用例。

(3) 主事件流：

① 用户通知进入系统。

② 单击通知管理。

③ 进入通知管理页面(通知列表)。

④ 单击“新增”按钮(S1)、“修改”按钮(S2)和“删除”按钮(S3)。

(4) 分支事件流

S1：单击“新增”按钮。

① 单击“新增”按钮。

② 进入通知添加页面。

③ 填写通知资料(E1)。

④ 单击“保存”按钮。

⑤ 验证信息，保存数据。

⑥ 返回通知管理页面(通知列表)。

S2：单击“修改”按钮。

① 单击“修改”按钮。

② 进入通知修改页面。

③ 更改通知数据，单击“修改”按钮。

④ 验证信息，保存数据。

⑤ 返回通知管理页面。

S3：单击“删除”按钮。

① 在要删除的记录前打勾，单击“删除”按钮。

② 删除信息。

③ 返回通知管理页面。

(5) 异常事件流

E1：输入非法的标识符或者格式不对，指明错误。

2) 用户界面图

单击“新增”按钮，进入通知编辑页面，如图 2-20 所示。

通知管理　　您的位置：首页>>通知管理

序号	标题	发布部门	发布日期	修改	删除
1	公告标题公告标题公告标题公告标题公告标题	综合办公室	2005-1-10 12:00		
2	公告标题公告标题公告标题公告标题公告标题	综合办公室	2005-1-10 12:00		
3	公告标题公告标题公告标题公告标题公告标题	综合办公室	2005-1-10 12:00		
4	公告标题公告标题公告标题公告标题公告标题	综合办公室	2005-1-10 12:00		
5	公告标题公告标题公告标题公告标题公告标题	综合办公室	2005-1-10 12:00		
6	公告标题公告标题公告标题公告标题公告标题	综合办公室	2005-1-10 12:00		
7	公告标题公告标题公告标题公告标题公告标题	综合办公室	2005-1-10 12:00		
8	公告标题公告标题公告标题公告标题公告标题	综合办公室	2005-1-10 12:00		
9	公告标题公告标题公告标题公告标题公告标题	综合办公室	2005-1-10 12:00		
10	公告标题公告标题公告标题公告标题公告标题	综合办公室	2005-1-10 12:00		
11	公告标题公告标题公告标题公告标题公告标题	综合办公室	2005-1-10 12:00		
12	公告标题公告标题公告标题公告标题公告标题	综合办公室	2005-1-10 12:00		
13	公告标题公告标题公告标题公告标题公告标题	综合办公室	2005-1-10 12:00		
14	公告标题公告标题公告标题公告标题公告标题	综合办公室	2005-1-10 12:00		
15	公告标题公告标题公告标题公告标题公告标题	综合办公室	2005-1-10 12:00		

新增　共 2 页　第 1 页　上一页　下一页　转到　页

图 2-20　通知编辑页面 1

通知管理—通知编辑：通知业务虽然没有审核功能，但是必须上传附件，如图 2-21 所示。

这个模块在首页上位于右栏职能部门通知。通知添加页如上所示，附件 1、附件 2、附件 3 后面的框为附件名称，每个附件名称后面的 3 个框为要上传的 3 种文件，这里要说明的是每个附件只代表一个文件，也就是说，后面这 3 种文件(本地文件、政策法规、文件表格)只能选择一种上传。

通知管理—通知删除：因为通知不需要审核，所以通知删除业务不会有很多的判断，只要判断不是发布状态就可以删除，如图 2-22 所示。

通知管理—通知修改：本业务在任何时候都可以修改，可以修改所有项。

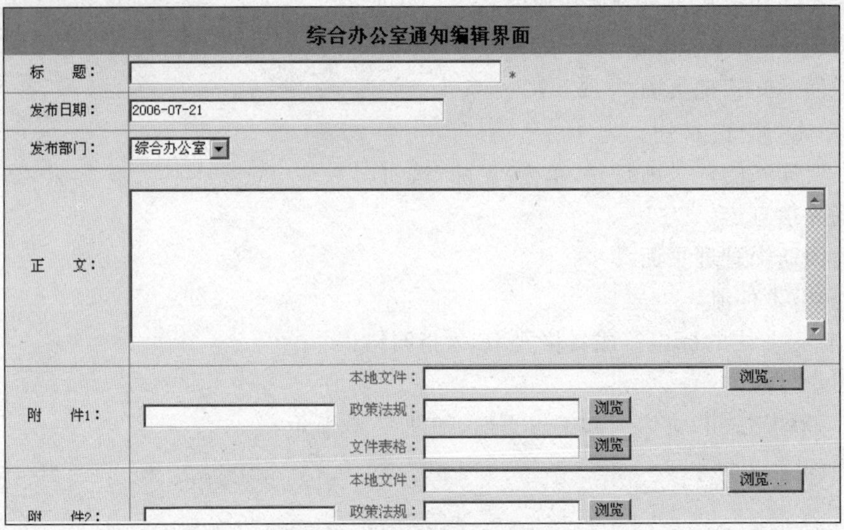

综合办公室通知编辑界面

标　题：*

发布日期：2006-07-21

发布部门：综合办公室

正　文：

附　件1：本地文件：浏览...　政策法规：浏览　文件表格：浏览

附　件2：本地文件：浏览...　政策法规：浏览

图 2-21　通知编辑页面 2

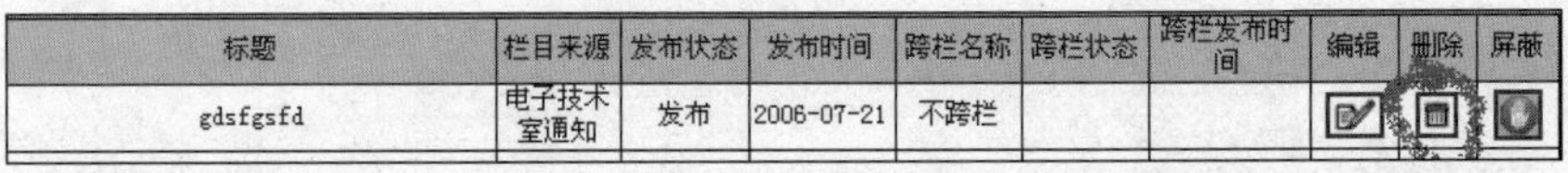

标题	栏目来源	发布状态	发布时间	跨栏名称	跨栏状态	跨栏发布时间	编辑	删除	屏蔽
gdsfgsfd	电子技术室通知	发布	2006-07-21	不跨栏					

图 2-22　通知删除

4. 新闻内容的审核

1）用例描述

（1）角色：高级管理员。

（2）前提条件：用户必须完成登录的用例。

（3）主事件流：

① 管理员通知进入系统。

② 单击内容审核列表里的新闻栏目。

③ 进入内容审核管理页面。

④ 单击"审核"按钮。

⑤ 进入审核页面。

⑥ 填写审批意见，单击"已阅"按钮(S1)、"同意"按钮(S2)或"退出"按钮(S3)。

（4）分支事件流

S1：单击"已阅"按钮。

① 单击"已阅"按钮。

② 返回内容审核管理页面，发布状态改变为"已审"。

③ 发布用户可以看到发布状态，单击"已审"按钮。

④ 查看管理员审批意见。

⑤ 单击"返回"按钮。

⑥ 返回内容审核管理页面。

⑦ 用户单击“修改”按钮，根据审批意见修改新闻。

⑧ 返回内容审核管理页面，发布状态改变为“待审”。

⑨ 管理员或审批人员再次审批，审批流程同步骤⑥。

S2：单击“同意”按钮。

① 单击“同意”按钮。

② 返回内容审核管理页面，发布状态改变为“发布”。

S3：单击“退出”按钮。

① 单击“退出”按钮。

② 返回内容审核管理页面。

(5) 异常事件流

E1：输入非法的标识符或者格式不对，指明错误。

E2：如果待审批的数据超过有效期，则指明不能审批，数据无效。

2) 用户界面图

新闻审核：单击内容审核列表里的新闻栏目，进入该模块，如图 2-23 所示。

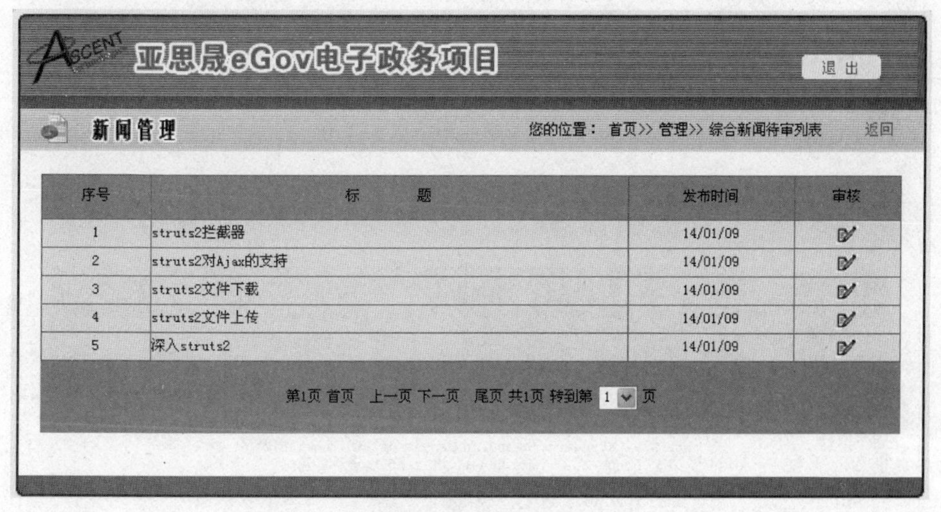

图 2-23 新闻审核页面 1

在审核的任务浏览页单击“审核”，如图 2-24 所示。

审核页面和正式的发布是一样的，审核者可以根据新闻是否可以发布选择按钮，这里的同意表示此新闻可以发布，已阅则是此新闻有问题不可以发布，并且可以在审核意见中输入文字说明。如果新闻为已阅，在发布者那里就可以看到没有通过的原因，如图 2-25 所示。

在新闻发布者那里能看到发布状态，单击发布状态栏目的一审，如图 2-26 所示。

新闻发布者可以看到审核后的意见，这时用户就可以修改这条新闻。修改后这条新闻状态发生了改变，变成了待审，这时需要等待审核者再次审核，审核者发现新闻没有问题，单击“同意”，这时新闻的状态变为“已发布”，如图 2-27 所示。

再看一下首页，如图 2-28 所示。

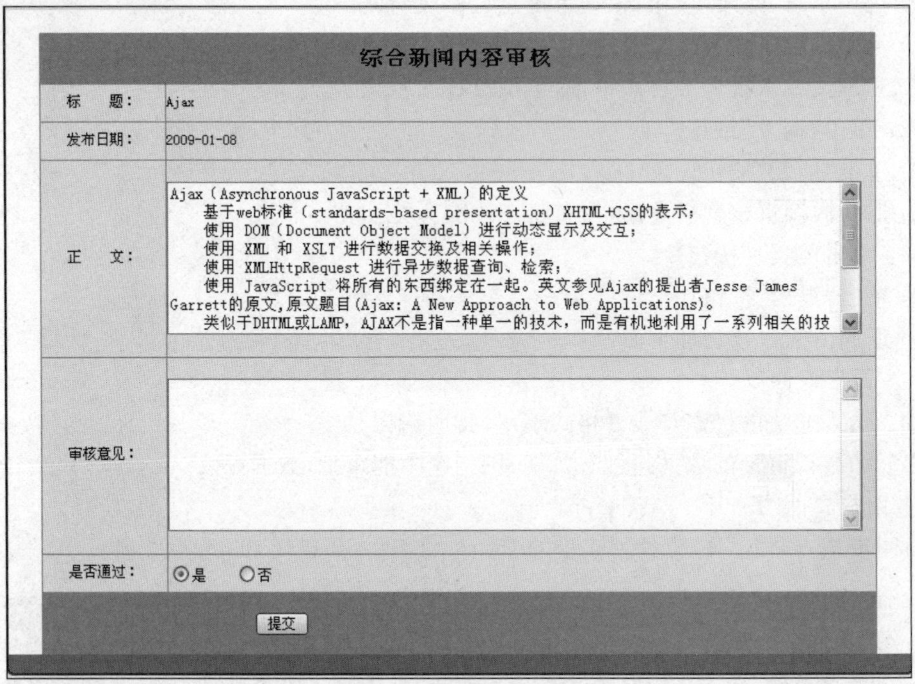

综合新闻内容审核

标　题：Ajax

发布日期：2009-01-08

正　文：

Ajax（Asynchronous JavaScript + XML）的定义
基于web标准（standards-based presentation）XHTML+CSS的表示；
使用 DOM（Document Object Model）进行动态显示及交互；
使用 XML 和 XSLT 进行数据交换及相关操作；
使用 XMLHttpRequest 进行异步数据查询、检索；
使用 JavaScript 将所有的东西绑定在一起。英文参见Ajax的提出者Jesse James Garrett的原文,原文题目(Ajax: A New Approach to Web Applications)。
类似于DHTML或LAMP，AJAX不是指一种单一的技术，而是有机地利用了一系列相关的技

审核意见：

是否通过：是　否

提交

图 2-24　新闻审核页面 2

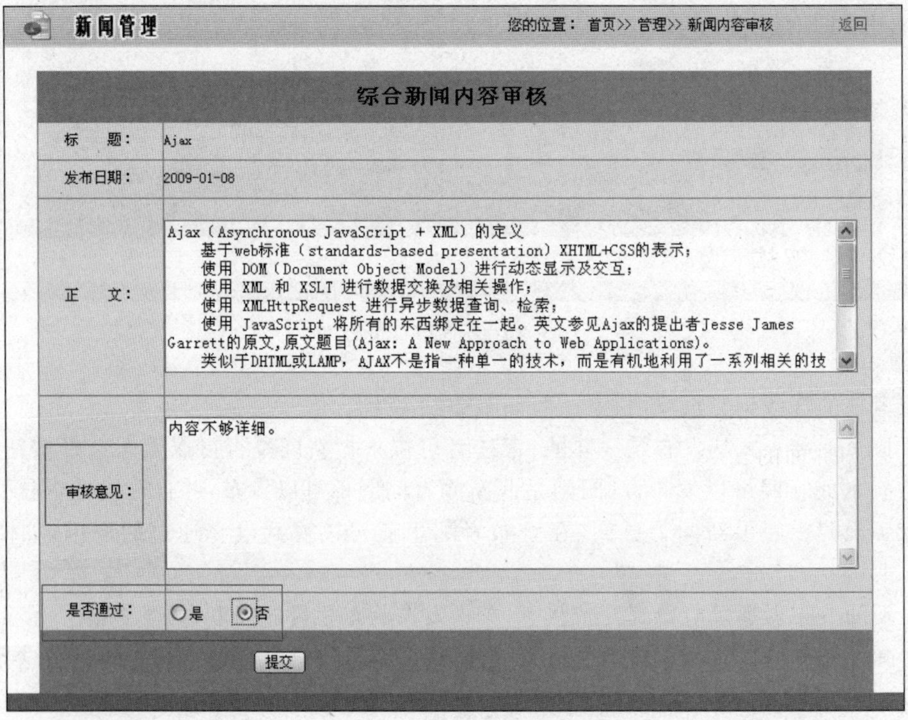

新闻管理　　您的位置：首页>> 管理>> 新闻内容审核　　返回

综合新闻内容审核

标　题：Ajax

发布日期：2009-01-08

正　文：

Ajax（Asynchronous JavaScript + XML）的定义
基于web标准（standards-based presentation）XHTML+CSS的表示；
使用 DOM（Document Object Model）进行动态显示及交互；
使用 XML 和 XSLT 进行数据交换及相关操作；
使用 XMLHttpRequest 进行异步数据查询、检索；
使用 JavaScript 将所有的东西绑定在一起。英文参见Ajax的提出者Jesse James Garrett的原文,原文题目(Ajax: A New Approach to Web Applications)。
类似于DHTML或LAMP，AJAX不是指一种单一的技术，而是有机地利用了一系列相关的技

审核意见：内容不够详细。

是否通过：是　否

提交

图 2-25　新闻审核页面 3

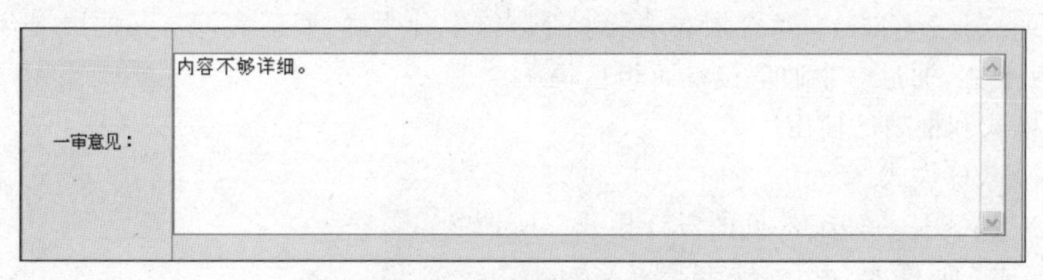

图 2-26 新闻审核意见

图 2-27 新闻的发布状态

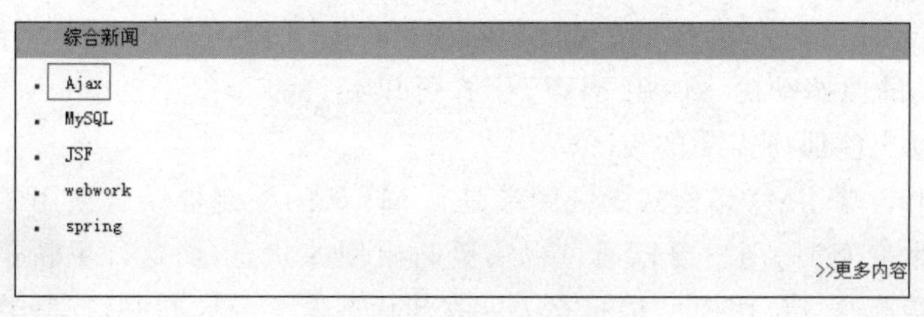

图 2-28 新闻发布结果

2.2 项目系统分析和设计

eGov 电子政务系统是由 Web 服务器、数据服务器和浏览器客户端组成的多层 Web 计算机服务系统，采用 Spring＋Spring MVC- MyBatis 架构，具有先进性、灵活性、可扩展性等特点。

2.2.1 架构设计

1. 系统整体方案

1）eGov 电子政务系统的主要特性

可以从以下 5 个方面确定目标系统的特性。

- 用户界面的复杂度：数据的静态显示/可定制视图。
- 用户界面的部署约束：基于独立的桌面计算机或专用工作站的浏览器。
- 用户的数量和类型：组织内的日常使用者，总共几百人。
- 系统接口类型：通过 HTTP 提供服务，未来可以使用 SOAP 的 SOA 技术。
- 性能：主要是独立的数据更新，有少量并发处理。

从上述特性可以判断 eGov 电子政务系统属于中大型项目，因此使用基于 Spring＋Spring MVC＋MyBatis 框架的分层架构设计方案。

2）架构分层

在 eGov 电子政务项目架构设计中使用分层模式。具体地，我们将 eGov 电子政务系统应用在职责上分成 3 层：表示层（Presentation Layer）、持久层（Persistence Layer）和业务层(Business Layser)。每个层在功能上都应该十分明确,而不应该与其他层混合。每个层要相互独立,通过一个通信接口而相互联系。

3) 模式和框架的使用

(1) MVC 模式。

MVC 模式是一种很常见的设计模式。所谓的 MVC 模式,即模型-视图-控制器(Model-View-Controller)模式。其结构如图 2-29 所示。

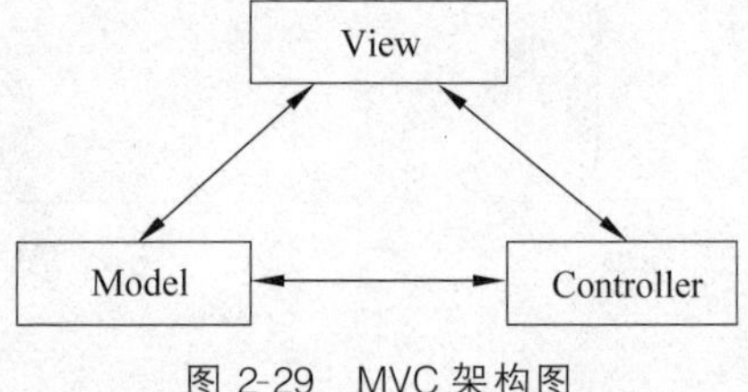

图 2-29 MVC 架构图

①Model 端。在 MVC 中,模型是执行某些任务的代码,而这部分代码并没有任何逻辑决定用户端的表示方法。Model 只有纯粹的功能性接口,也就是一系列的公共方法,通过这些公共方法,便可以取得模型端的所有功能。

② View 端。在 MVC 模式里,一个 Model 可以有几个 View 端,而实际上多个 View 端是使用 MVC 的原始动机。使用 MVC 模式可以允许多于一个的 View 端存在,并可以在需要的时候动态注册所需要的 View。

③ Controller 端。MVC 模式的视图端是与 MVC 的控制器结合使用的。当用户端与相应的视图发生交互时,用户可以通过视窗更新模型的状态,而这种更新是通过控制器端进行的。控制器端通过调用模型端的方法更改其状态值。与此同时,控制器端会通知所有注册了的视图刷新用户界面。

(2) 框架。

根据项目特点,使用 3 种开源框架:表示层用 Spring MVC;业务层用 Spring;持久层用 MyBatis,如图 2-30 所示。

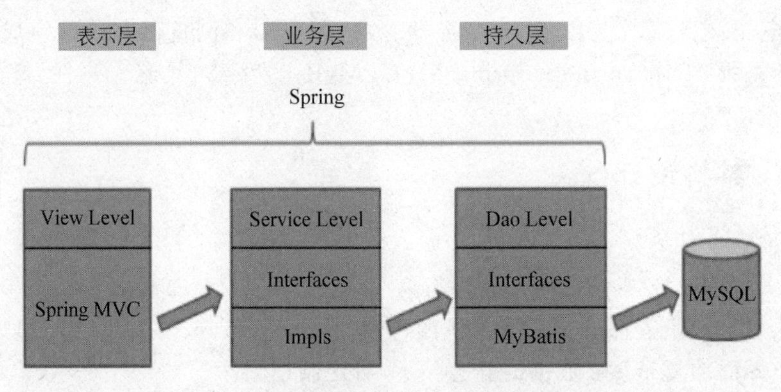

图 2-30 Spring MVC+Spring+MyBatis 架构

① 表示层。一般来讲,一个典型的 Web 应用的前端应该是表示层。这里可以使用 Spring MVC 框架。

下面是 Spring MVC 所负责的。

- 管理用户的请求,做出相应的响应。
- 提供一个流程控制器,委派调用业务逻辑和其他上层处理。
- 处理异常。
- 为显示提供一个数据模型。

- 用户界面的验证。

以下内容不该在 Spring MVC 表示层的编码中经常出现,与表示层无关。

- 与数据库直接通信。
- 与应用程序相关联的业务逻辑及校验。
- 事务处理。

在表示层引入这些代码,会带来高耦合和难以维护的后果。

② 持久层。典型的 Web 应用的后端是持久层。开发者总是低估构建他们自己的持久层框架的挑战性。系统内部的持久层不但需要大量调试时间,而且还经常因为缺少功能使之变得难以控制。这是持久层的通病。幸运的是,有几个对象/关系映射(Object/Relation Mapping,ORM)开源框架很好地解决了这类问题,尤其是 MyBatis——一个半自动化处理的 ORM 框架。MyBatis 是支持 SQL(用户自定义 SQL)、存储过程和高级映射的优秀持久层框架。MyBatis 消除了几乎所有的 JDBC 代码和参数的手工设置以及结果集的检索。MyBatis 使用简单的 XML 或注解用于配置和原始映射,将接口和 Java 的 POJOs 映射成数据库中的记录。

下面是 MyBatis 所负责的:

- 封装 SQL JDBC 操作。

MyBatis 是支持定制化 SQL、存储过程以及高级映射的优秀的持久层框架,通过 Java 接口成为 mapper 定义查询的方法,定义普通的 Java 的 POJO 封装查询结果,使用 XML 文件为每个 mapper 接口中的查询方法定义具体的 SQL 语句,使得开发人员可以更简便地开发 RDBMS 的 SQL 交互过程。

- 支持大部分主流关系型数据库,并且支持父表/子表(Parent/Child)关系、实体关系映射、事务处理等支撑功能。

③ 业务层。一个典型 Web 应用的中间部分是业务层或者服务层。从编码的视角看,这是最容易被忽视的一层。我们往往在用户界面层或持久层周围看到这些业务处理的代码,这其实是不正确的。因为它会造成程序代码的高耦合,这样,随着时间的推移,这些代码将很难维护。幸好,针对这一问题存在好几种框架。最受欢迎的两个框架是 Spring 和 PicoContainer。这些也被称为轻量级容器,它们能让你很好地把对象搭配起来。这两个框架都着手于"依赖注入"(还有我们知道的 IoC)这样的简单概念。这里将关注 Spring 的依赖注入和面向方面编程。另外,Spring 把程序中涉及的包含业务逻辑和数据存取对象(DataAccess Object)的 Objects(如 transaction management handler(事务管理控制)、Object Factories(对象工厂)、service objects(服务组件))都通过 XML 以及 XML 中的 Schema 功能命名空间、JDK 中的注解功能配置联系起来。

下面是 Spring 所负责的:

- 处理应用程序的业务逻辑和业务校验。
- 管理事务。
- 提供与其他层相互作用的接口。
- 管理业务层级别的对象的依赖。
- 在表示层和持久层之间增加一个灵活的机制,使得它们不直接联系在一起。
- 通过揭示从表示层到业务层之间的上下文(Context)得到业务逻辑。

● 管理程序的执行(从业务层到持久层)。

2. UML 视图

(1) 用例图如图 2-31 所示。

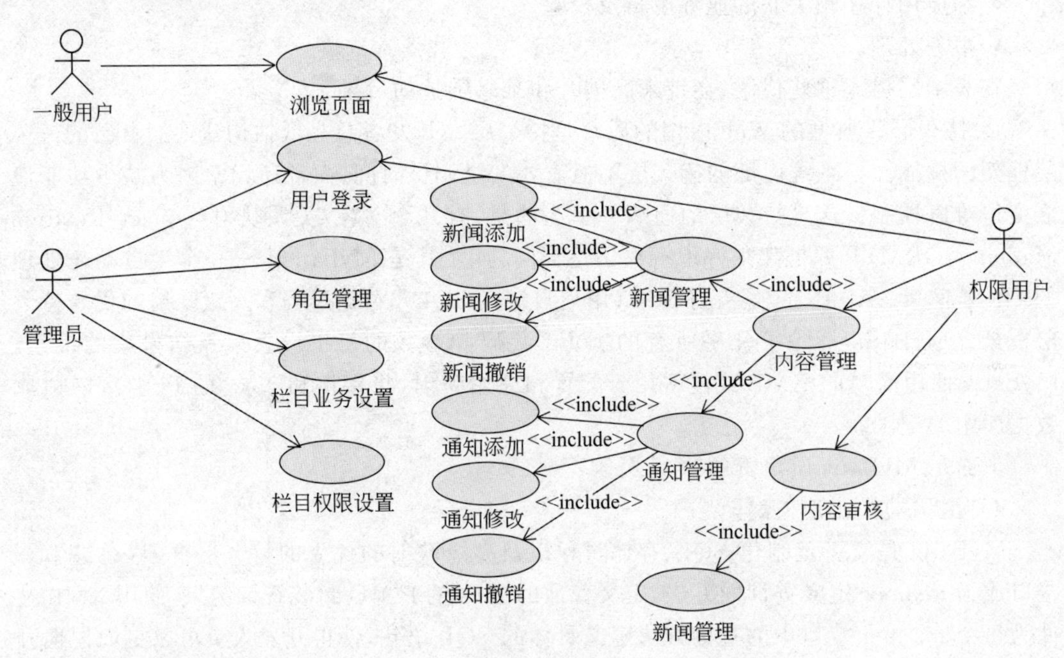

图 2-31 用例图

(2) 类图如图 2-32 所示。

2.2.2 数据库设计

1. 逻辑结构设计

逻辑结构设计图如图 2-33 所示。

2. 物理表的设计

用户表结构(usr)见表 2-1。

表 2-1 用户表结构(usr)

序号	列名	PK	FK	属性	长度	备注
1	id	Y		Integer	20	该表的主键,唯一标识,自动增长
2	name			Varchar	16	用户名
3	password			Varchar	16	用户密码
4	phone			Varchar	16	电话号
5	deptid		Y	Integer	20	部门号
6	address			Varchar	64	用户地址
7	title			Varchar	32	称呼

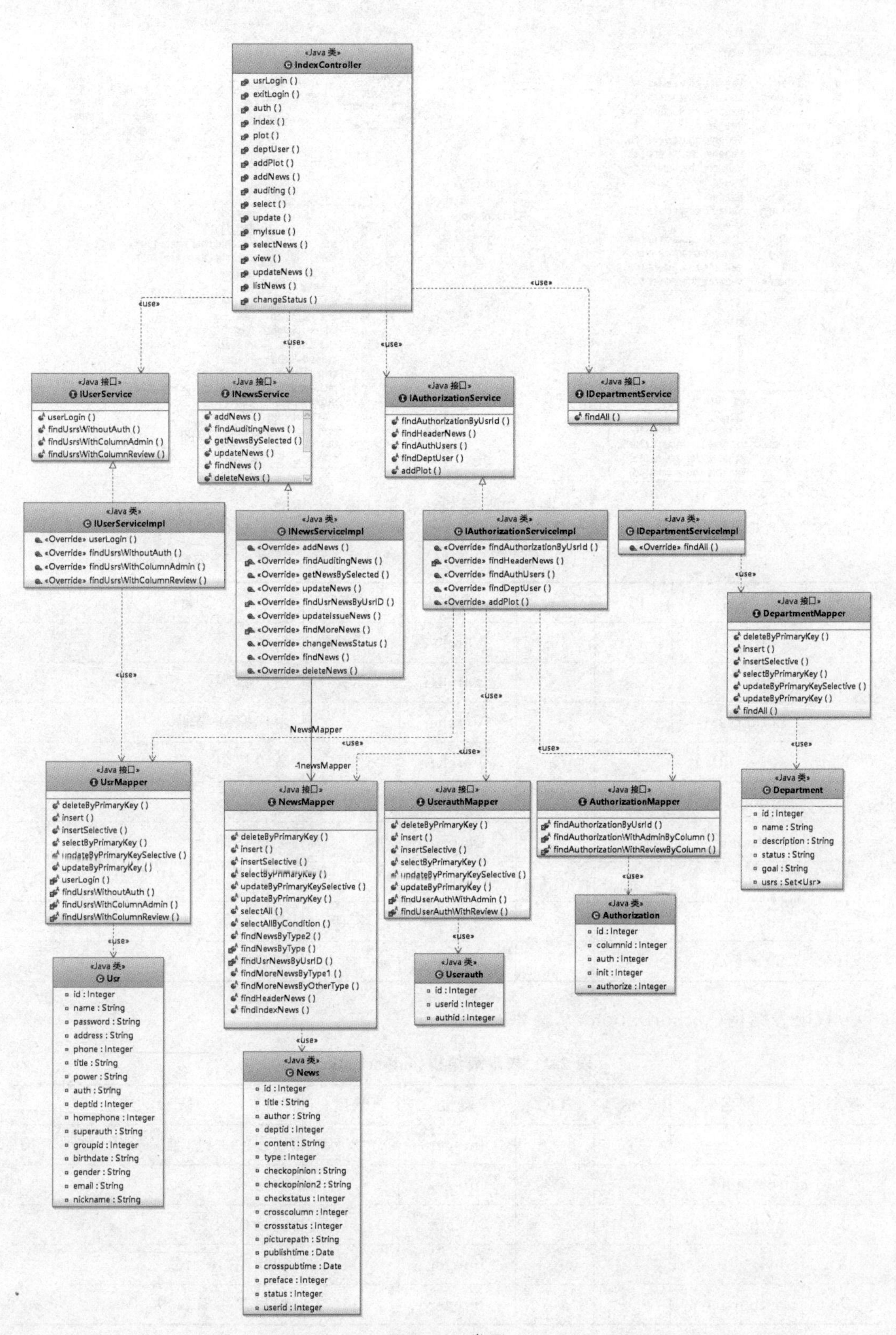

«Java 类»
IndexController
usrLogin ()
exitLogin ()
auth ()
index ()
plot ()
deptUser ()
addPlot ()
addNews ()
auditing ()
select ()
update ()
myIssue ()
selectNews ()
view ()
updateNews ()
listNews ()
changeStatus ()
«use»
«Java 接口»
IUserService
userLogin ()
findUsrsWithoutAuth ()
findUsrsWithColumnAdmin ()
findUsrsWithColumnReview ()
«Java 接口»
INewsService
addNews ()
findAuditingNews ()
getNewsBySelected ()
updateNews ()
findNews ()
deleteNews ()
«Java 接口»
IAuthorizationService
findAuthorizationByUsrId ()
findHeaderNews ()
findAuthUsers ()
findDeptUser ()
addPlot ()
«Java 接口»
IDepartmentService
findAll ()
«Java 类»
IUserServiceImpl
«Override» userLogin ()
«Override» findUsrsWithoutAuth ()
«Override» findUsrsWithColumnAdmin ()
«Override» findUsrsWithColumnReview ()
«Java 类»
INewsServiceImpl
«Override» addNews ()
«Override» findAuditingNews ()
«Override» getNewsBySelected ()
«Override» updateNews ()
«Override» findUsrNewsByUsrID ()
«Override» updateIssueNews ()
«Override» findMoreNews ()
«Override» changeNewsStatus ()
«Override» findNews ()
«Override» deleteNews ()
«Java 类»
IAuthorizationServiceImpl
«Override» findAuthorizationByUsrId ()
«Override» findHeaderNews ()
«Override» findAuthUsers ()
«Override» findDeptUser ()
«Override» addPlot ()
«Java 类»
IDepartmentServiceImpl
«Override» findAll ()
«Java 接口»
DepartmentMapper
deleteByPrimaryKey ()
insert ()
insertSelective ()
selectByPrimaryKey ()
updateByPrimaryKeySelective ()
updateByPrimaryKey ()
findAll ()
NewsMapper
-1newsMapper
«Java 接口»
UsrMapper
deleteByPrimaryKey ()
insert ()
insertSelective ()
selectByPrimaryKey ()
updateByPrimaryKeySelective ()
updateByPrimaryKey ()
userLogin ()
findUsrsWithoutAuth ()
findUsrsWithColumnAdmin ()
findUsrsWithColumnReview ()
«Java 接口»
NewsMapper
deleteByPrimaryKey ()
insert ()
insertSelective ()
selectByPrimaryKey ()
updateByPrimaryKeySelective ()
updateByPrimaryKey ()
selectAll ()
selectAllByCondition ()
findNewsByType2 ()
findNewsByType ()
findUsrNewsByUsrID ()
findMoreNewsByType1 ()
findMoreNewsByOtherType ()
findHeaderNews ()
findIndexNews ()
«Java 接口»
UserauthMapper
deleteByPrimaryKey ()
insert ()
insertSelective ()
selectByPrimaryKey ()
updateByPrimaryKeySelective ()
updateByPrimaryKey ()
findUserAuthWithAdmin ()
findUserAuthWithReview ()
«Java 接口»
AuthorizationMapper
findAuthorizationByUsrId ()
findAuthorizationWithAdminByColumn ()
findAuthorizationWithReviewByColumn ()
«Java 类»
Department
id : Integer
name : String
description : String
status : String
goal : String
usrs : Set<Usr>
«Java 类»
Authorization
id : Integer
columnid : Integer
auth : Integer
init : Integer
authorize : Integer
«Java 类»
Userauth
id : Integer
userid : Integer
authid : Integer
«Java 类»
Usr
id : Integer
name : String
password : String
address : String
phone : Integer
title : String
power : String
auth : String
deptid : Integer
homephone : Integer
superauth : String
groupid : Integer
birthdate : String
gender : String
email : String
nickname : String
«Java 类»
News
id : Integer
title : String
author : String
deptid : Integer
content : String
type : Integer
checkopinion : String
checkopinion2 : String
checkstatus : Integer
crosscolumn : Integer
crossstatus : Integer
picturepath : String
publishtime : Date
crosspubtime : Date
preface : Integer
status : Integer
userid : Integer

图 2-32　类图

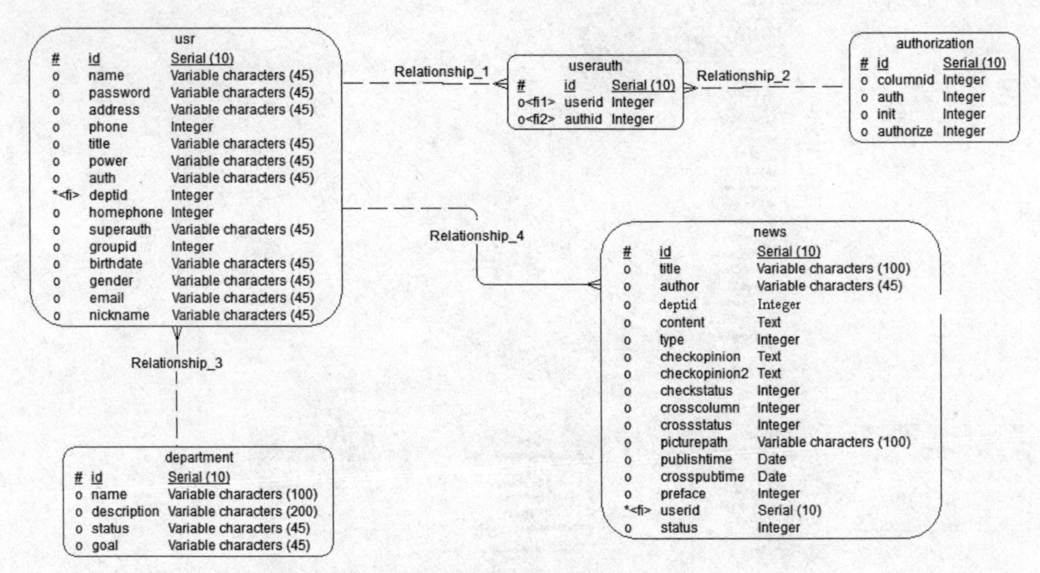

图 2-33 逻辑结构设计图

续表

序号	列名	PK	FK	属性	长度	备 注
8	power			Varchar	32	权力
9	auth			Varchar	32	用户权限
10	homephone			Varchar	16	用户家庭号码
11	superauth			Varchar	8	高级权限
12	groupid			Integer	20	组号
13	birthdate			Date		生日日期
14	gender			Varchar	8	用户性别
15	email			Varchar	255	电子信箱
16	nickname			Varchar	45	用户昵称

权限表结构(authorization)见表 2-2。

表 2-2 权限表结构(authorization)

序号	列名	PK	FK	属性	长度	备 注
1	id	Y		Integer	20	该表的主键,唯一标识,自动增长
2	columnid		Y	Integer	20	栏目编号
3	auth			Integer	11	栏目权限
4	init			Integer	11	初始值
5	authorize			Integer	11	栏目是否有权限

用户权限表结构(userauth)见表 2-3。

表 2-3 用户权限表结构(userauth)

序号	列名	PK	FK	属性	长度	备　注
1	id	Y		Integer	20	该表的主键,唯一标识,自动增长
2	userid		Y	Integer	20	用户 ID(外键)
3	authid		Y	Integer	20	栏目权限 ID(外键)

部门表结构(department)见表 2-4。

表 2-4 部门表结构(department)

序号	列名	PK	FK	属性	长度	备　注
1	id	Y		Integer	20	该表的主键,唯一标识,自动增长
2	name			Varchar	32	部门名
3	status			Varchar	255	部门状态
4	description			Varchar	255	部门描述
5	goal			Varchar	255	目标

新闻表结构(news)见表 2-5。

表 2-5 新闻表结构(news)

序号	列名	PK	FK	属性	长度	备　注
1	id	Y		Integer	20	该表的主键,唯一标识,自动增长
2	title			Varchar	255	新闻标题
3	author			Varchar	32	作者
4	deptid			Integer	20	部门号
5	content			Longtext		新闻内容
6	type			Integer	11	新闻类型
7	checkopinion			Varchar	255	一审意见
8	checkopinion2			Varchar	255	二审意见
9	checkstatus			Integer	11	当前新闻审核状态
10	crosscolumn		Y	Integer	11	跨栏栏目(值来自数据字典)
11	crossstatus			Integer	11	跨栏状态(值来自数据字典)
12	picturepath			Varchar	128	图片路径
13	publishtime			Date		发布时间
14	crosspubtime			Date		跨栏日期(值来自数据字典)
15	preface			Integer	11	前言
16	userid		Y	Integer	20	用户 ID(外键)
17	status			Integer	11	发布状态

2.3 项目运行指南

(1) 需要的环境：

① MySQL 8.0.19；

② Tomcat 8.5.34；

③ 集成开发环境(IDE)：MyEclipse 2017 CI 7。

注意：这些软件的版本很重要，版本太高或太低都可能带来部署和运行问题。请读者特别留意，需要和以上软件的版本保持一致！

(2) 创建数据库。

首先需要建立数据库并导入数据。具体步骤如下：

① 启动 MySQL 命令行，输入正确的数据库密码，按回车键进入 MySQL，如图 2-34 所示。

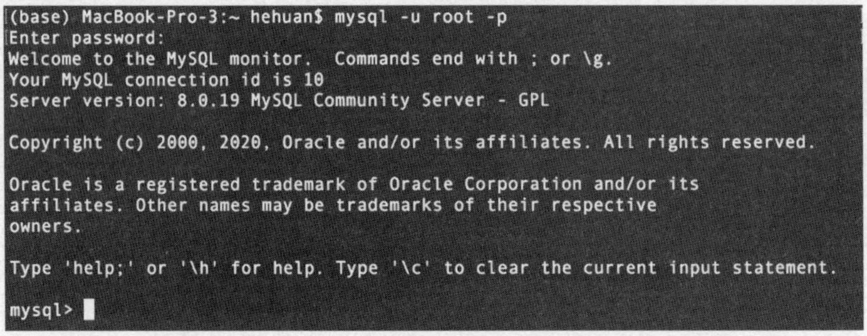

```
(base) MacBook-Pro-3:~ hehuan$ mysql -u root -p
Enter password:
Welcome to the MySQL monitor.  Commands end with ; or \g.
Your MySQL connection id is 10
Server version: 8.0.19 MySQL Community Server - GPL

Copyright (c) 2000, 2020, Oracle and/or its affiliates. All rights reserved.

Oracle is a registered trademark of Oracle Corporation and/or its
affiliates. Other names may be trademarks of their respective
owners.

Type 'help;' or '\h' for help. Type '\c' to clear the current input statement.

mysql>
```

图 2-34 启动 MySQL

② 创建 electrones 数据库，并使用 electrones 数据库，具体如图 2-35 所示。

③ 执行导入命令 `mysql> source /Users/hehuan/Desktop/electrones.sql;` ，其中/Users/hehuan/Desktop/electrones.sql 是 SQL 脚本，可以把它放在任意目录下，本例放在/Users/hehuan/Desktop 下，按回车键执行导入命令，具体如图 2-36 所示。

```
mysql> create database electrones;
Query OK, 1 row affected (0.00 sec)

mysql> use electrones;
Database changed
```

图 2-35 创建并使用数据库

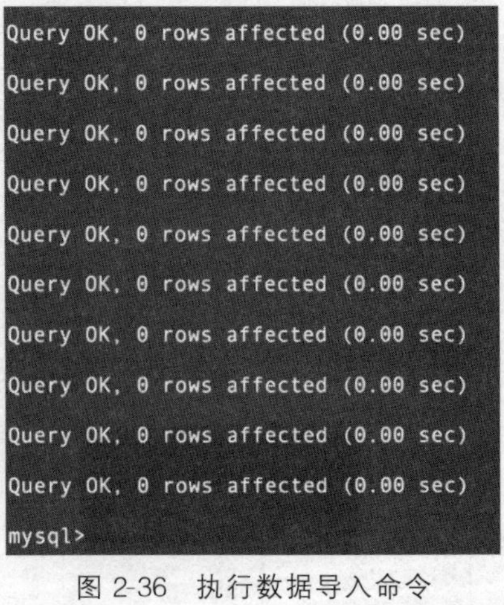

```
Query OK, 0 rows affected (0.00 sec)

Query OK, 0 rows affected (0.00 sec)

Query OK, 0 rows affected (0.00 sec)

Query OK, 0 rows affected (0.00 sec)

Query OK, 0 rows affected (0.00 sec)

Query OK, 0 rows affected (0.00 sec)

Query OK, 0 rows affected (0.00 sec)

Query OK, 0 rows affected (0.00 sec)

Query OK, 0 rows affected (0.00 sec)

Query OK, 0 rows affected (0.00 sec)

mysql>
```

图 2-36 执行数据导入命令

成功导入后，数据库建立成功。读者也可以使用 MySQL GUI 客户端，在其中进行类似操作。

（3）将 electrone.war 解压后的 electrone 文件夹复制到 tomcat\webapps 下，找到 tomcat\webapps\electrone\WEB-INF\classes\config\db.conf 文件，打开并修改下面代码中的数据库驱动信息，将数据库的用户名和密码改成自己的用户名和密码。

```
jdbc.url=jdbc:mysql://localhost:3306/electrones?useUnicode=true&characterEncoding=
utf-8&autoReconnect=true&serverTimezone=UTC
jdbc.driver=com.mysql.jdbc.Driver
jdbc.username=数据库的用户名
jdbc.password=数据库的密码
```

至此工程就可以启动运行了。

注意：*在修改过程中不要破坏 db.conf 文件格式，否则项目无法正常启动。*

（4）启动 Tomcat，输入 http://localhost:8080/electrone，项目就正确启动并运行了。

（5）管理员的用户名为 admin，密码为 123，登录试运行。

（6）用户还可以作为普通人员登录网站试运行。常见的用户（登录名和密码）信息见表 2-6。

表 2-6 用户信息

登录名	密码
lixing	lixing
ascent	ascent
q1	1
q2	1
q3	1
q4	1
q5	1
q6	1

具体信息可查询数据库中的 usr 表。

习题

1. eGov 电子政务系统主要包括哪些模块？
2. eGov 电子政务系统的主要特性是什么？
3. eGov 电子政务系统的数据库主要包括哪些表？

第3章 Spring技术和Spring IoC

学习目的与学习要求

学习目的：深入了解 Spring 框架及模块，掌握基本配置及 IoC 机制。

学习要求：扎实掌握 Spring 的核心配置及 IoC 机制，熟练搭建 Spring 环境的流程，学会配置和管理 bean。

本章主要内容

本章介绍 Spring 框架的结构概述，重点讲解 Spring IoC 原理、bean 的各种配置方式及作用域、使用 BeanFactory 和 ApplicationContext 加载 Spring 配置和管理 bean。

接下来讨论 Spring 框架，它是连接 Spring MVC 与 MyBatis 的桥梁，同时很好地处理了业务逻辑层。

3.1 Spring 简介

Spring 框架是一个分层架构，由 7 个定义好的模块组成。Spring 模块构建在核心容器之上，核心容器定义了创建、配置和管理 bean 的方式，如图 3-1 所示。

组成 Spring 框架的每个模块（或组件）都可以单独存在，或者与其他一个或多个模块联合实现。每个模块的功能如下。

核心容器：提供 Spring 框架的基本功能。核心容器的主要组件是 BeanFactory，它是工厂模式的实现。BeanFactory 使用 IoC 模式将应用程序的配置和依赖性规范与实际的应用程序代码分开。

Spring 上下文：是一个配置文件，向 Spring 框架提供上下文信息。Spring 上下文包括企业服务，如 JNDI、EJB、电子邮件、国际化、校验和调度功能。

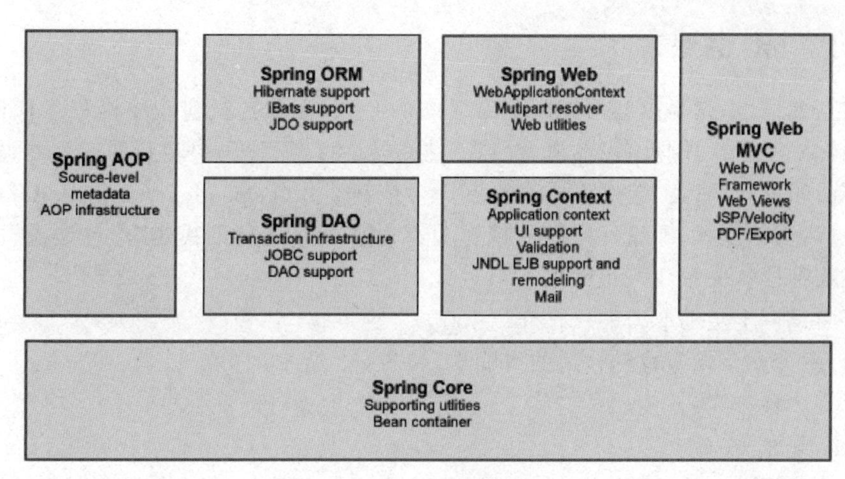

图 3-1　Spring 框架的模块

Spring AOP：通过配置管理特性，Spring AOP 模块直接将面向方面的编程功能集成到 Spring 框架中，所以可以很容易地使 Spring 框架管理的任何对象支持 AOP。Spring AOP 模块为基于 Spring 的应用程序中的对象提供了事务管理服务。使用 Spring AOP，不依赖 EJB 组件，就可以将声明性事务管理集成到应用程序中。

Spring DAO：JDBC DAO（Data Access Object）抽象层提供了有意义的异常层次结构，可用该结构管理异常处理和不同数据库供应商抛出的错误消息。异常层次结构简化了错误处理，并且极大地降低了需要编写的异常代码数量。Spring DAO 的面向 JDBC 的异常遵从通用的 DAO 异常层次结构。

Spring ORM：Spring 框架插入了若干个 Object/Relation Mapping 框架，从而提供了 ORM 的对象关系映射工具，其中包括 JDO、MyBatis 和 iBatis SQL Map。所有这些都遵从 Spring 的通用事务和 DAO 异常层次结构。

Spring Web 模块：Web 上下文模块建立在应用程序上下文模块之上，为基于 Web 的应用程序提供了上下文。所以，Spring 框架支持与 Jakarta Spring MVC 的集成。Web 模块还简化了处理大部分请求以及将请求参数绑定到域对象的工作。

Spring MVC 框架：MVC 框架是一个全功能的构建 Web 应用程序的 MVC 实现。通过策略接口，MVC 框架变成高度可配置的，MVC 容纳了大量视图技术，其中包括 JSP、Velocity、Tiles、iText 和 POI。

Spring 框架的功能可以用在任何的 J2EE 服务器中，大多数功能也适用于不受管理的环境。Spring 的核心要点是：支持不绑定到特定 J2EE 服务的可重用业务和数据访问对象。毫无疑问，这样的对象可以在不同 J2EE 环境（Web 或 EJB）、独立应用程序、测试环境之间重用。

3.2　Spring IoC

首先介绍 Spring IoC 这个最核心、最重要的概念。

3.2.1 IoC 的原理

IoC,直观地讲,就是由容器控制程序之间的关系,而非传统实现中由程序代码直接操控。这也就是所谓"控制反转"的概念所在:控制权由应用代码中转到外部容器,控制权的转移是所谓的反转。IoC 还有另外一个名字:"依赖注入(Dependency Injection)"。从名字上理解,所谓依赖注入,即组件之间的依赖关系由容器在运行期决定,形象地说,即由容器动态地将某种依赖关系注入组件中。

下面通过一个生动形象的例子介绍 IoC。

例如,一个女孩希望找到合适的男朋友,如图 3-2 所示。

可以有 3 种方式:

(1) 青梅竹马;

(2) 亲友介绍;

(3) 父母包办。

第一种方式是青梅竹马,如图 3-3 所示。

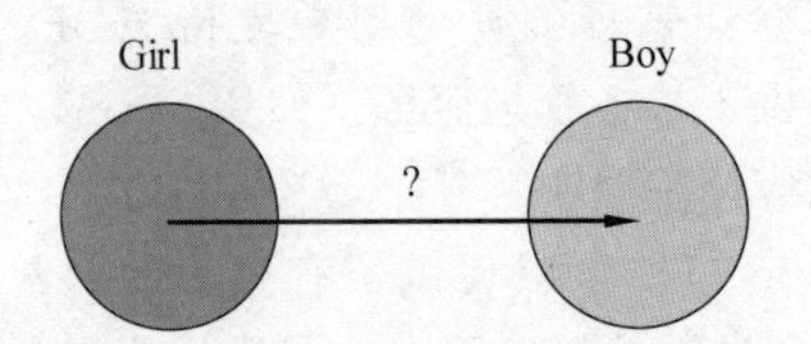

图 3-2 一个女孩希望找到合适的男朋友

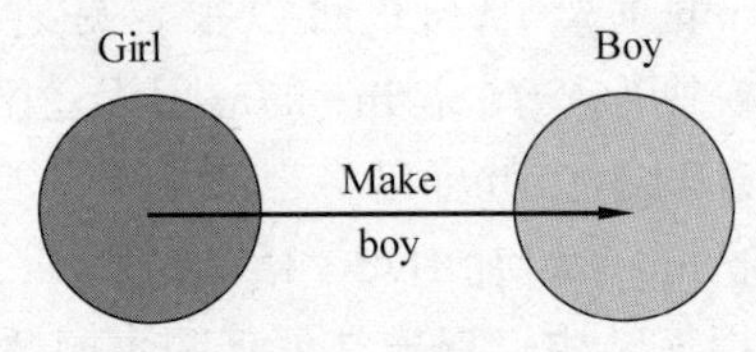

图 3-3 青梅竹马

通过代码表示如下:

```
public class Girl {
  void kiss(){
    Boy boy=new Boy();
  }
}
```

第二种方式是亲友介绍,如图 3-4 所示。

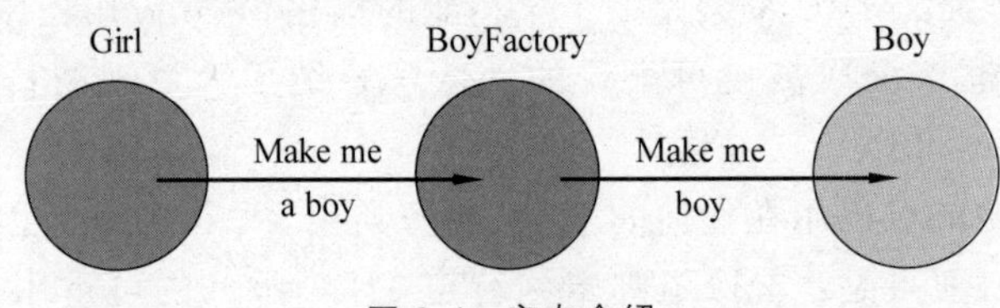

图 3-4 亲友介绍

通过代码表示如下:

```
public class Girl {
  void kiss(){
    Boy boy=BoyFactory.createBoy();
  }
}
```

第三种方式是父母包办，如图 3-5 所示。

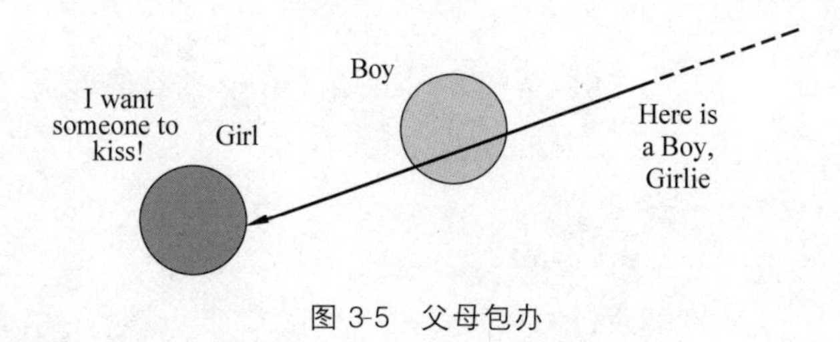

图 3-5 父母包办

通过代码表示如下：

```
public class Girl {
  void kiss(Boy boy) {
    //kiss boy
    boy.kiss();
  }
}
```

哪种方式为 IoC 呢？虽然在现实生活中我们都希望青梅竹马，但在 Spring 世界里选择的却是父母包办，它就是 IoC，而这里具有控制力的父母就是 Spring 的所谓的容器概念。

典型的 IoC 如图 3-6 所示。

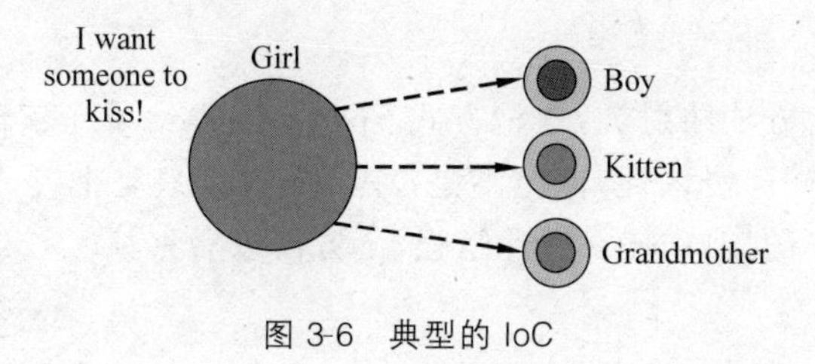

图 3-6 典型的 IoC

IoC 的 3 种依赖注入类型：

第一种是通过接口注射，这种方式要求类必须实现容器给定的一个接口，然后容器会利用这个接口给这个类注射它所依赖的类。

```
public class Girl implements Servicable{
  Kissable kissable;
  public void service(ServiceManager mgr) {
    kissable=(Kissable) mgr.lookup("kissable");
  }
  public void kissYourKissable() {
    kissable.kiss();
  }
}

<container>
  <component name="kissable" class="Boy">
    <configuration>…</configuration>
```

```
  </component><component name="girl" class="Girl" />
</container>
```

第二种是通过 setter()方法注射,这种方式也是 Spring 推荐的方式。

```
public class Girl {
  private Kissable kissable;
  public void setKissable(Kissable kissable) {
    this.kissable=kissable;
  }
  public void kissYourKissable() {
    kissable.kiss();
  }
}
```

```
<beans>
  <bean id="boy" class="Boy"/>
  <bean id="girl" class="Girl">
    <property name="kissable">
      <ref bean="boy"/>
    </property>
  </bean>
</beans>
```

第三种是通过构造方法注射类,这种方式 Spring 同样给予了实现,它和通过 setter()方法一样,都在类里无任何侵入性,但是不是没有侵入性,只是把侵入性转移了。显然,第一种方式要求实现特定的接口,侵入性非常强,不方便以后移植。

```
public class Girl {
  private Kissable kissable;
  public Girl(Kissable kissable) {
    this.kissable=kissable;
  }
  public void kissYourKissable() {
    kissable.kiss();
  }
}
```

```
PicoContainer container=new DefaultPicoContainer();
container.registerComponentImplementation(Boy.class);
container.registerComponentImplementation(Girl.class);
Girl girl=(Girl) container.getComponentInstance(Girl.class);
girl.kissYourKissable();
```

3.2.2 Bean Factory

Spring IoC 设计的核心是 org.springframework.beans 包,它的设计目标是与 JavaBean

组件一起使用。这个包通常不是由用户直接使用，而是由服务器将其用作其他多数功能的底层中介。下一个最高级抽象是 BeanFactory 接口，它是工厂设计模式的实现，允许通过名称创建和检索对象。BeanFactory 也可以管理对象之间的关系。

BeanFactory 支持两个对象模型。

- 单态模型：它提供了具有特定名称的对象的共享实例，可以在查询时对其进行检索。Singleton 是默认的，也是最常用的对象模型，对于无状态服务对象很理想。
- 原型模型：它确保每次检索都会创建单独的对象。在每个用户都需要自己的对象时，原型模型最适合。

bean 工厂的概念是 Spring 作为 IoC 容器的基础，IoC 将处理事情的责任从应用程序代码转移到框架。Spring 框架使用 JavaBean 属性和配置数据指出必须设置的依赖关系。

1. BeanFactory

BeanFactory 实际上是实例化、配置和管理众多 bean 的容器。这些 bean 通常彼此合作，因而它们之间会产生依赖。BeanFactory 使用的配置数据可以反映这些依赖关系（一些依赖可能不像配置数据一样可见，而是在运行期作为 bean 之间程序交互的函数）。

一个 BeanFactory 可以用接口 org.springframework.beans.factory.BeanFactory 表示，这个接口有多个实现。最常使用的简单的 BeanFactory 实现是 org.springframework.beans.factory.xml.XmlBeanFactory。（这里提醒一下：ApplicationContext 是 BeanFactory 的子类，所以大多数用户更喜欢使用 ApplicationContext 的 XML 形式。）

虽然大多数情况下，几乎所有被 BeanFactory 管理的用户代码都不需要知道 BeanFactory，但是 BeanFactory 还是以某种方式实例化。可以使用下面的代码实例化 BeanFactory：

```
InputStream is=new FileInputStream("beans.xml");
XmlBeanFactory factory=new XmlBeanFactory(is);
```

或者

```
ClassPathResource res=new ClassPathResource("beans.xml");
XmlBeanFactory factory=new XmlBeanFactory(res);
```

或者

```
ClassPathXmlApplicationContext appContext=new ClassPathXmlApplicationContext(
    new String[] {"applicationContext.xml", "applicationContext-part2.xml"});
//of course, an ApplicationContext is just a BeanFactory
BeanFactory factory=(BeanFactory) appContext;
```

很多情况下，用户代码不需要实例化 BeanFactory，因为 Spring 框架代码会做这件事。例如，Web 层提供支持代码，在 J2EE Web 应用启动过程中自动载入一个 Spring ApplicationContext。这个声明过程在这里描述：

编程操作 BeanFactory 将会在后面提到，下面集中描述 BeanFactory 的配置。

一个最基本的 BeanFactory 配置由一个或多个它所管理的 Bean 定义组成。在一个 XmlBeanFactory 中，根节点 Beans 中包含一个或多个 Bean 元素。

```
<?xml version="1.0" encoding="UTF-8"?>
<!DOCTYPE beans PUBLIC "-//SPRING//DTD BEAN//EN" "http://www.springframework.
org/dtd/spring-beans.dtd">

<beans>

  <bean id="..." class="...">
    ...
  </bean>
  <bean id="..." class="...">
    ...
  </bean>

  ...

</beans>
```

2. Bean Definition

一个 XmlBeanFactory 中的 Bean 定义包括的内容有：

(1) classname，这通常是 Bean 的真正的实现类。但是，如果一个 Bean 使用一个静态工厂方法创建，而不是被普通的构造函数创建，那么这实际上就是工厂类的 classname。

(2) Bean 行为配置元素，它声明这个 Bean 在容器的行为方式(如 prototype 或 singleton，自动装配模式，依赖检查模式，初始化和析构方法)。

构造函数的参数和新创建 Bean 需要的属性：举一个例子，一个管理连接池的 Bean 使用的连接数目(既可以指定为一个属性，也可以作为一个构造函数参数)，或者池的大小限制和这个 Bean 工作相关的其他 Bean：比如它的合作者(同样可以作为属性或者构造函数的参数)。这也被叫作依赖。

上面列出的概念直接转化为组成 Bean 定义的一组元素。这些元素(见表 3-1)都有更详细的说明的链接。

表 3-1　Bean 定义的解释

特　性	说　明	特　性	说　明
class	Bean 的类	自动装配模式	自动装配协作对象
id 和 name	Bean 的标识符(id 与 name)	依赖检查模式	依赖检查
singleton 或 prototype	Singleton 的使用与否	初始化模式	生命周期接口
构造函数参数	设置 Bean 的属性和合作者	析构方法	生命周期接口
Bean 的属性	设置 Bean 的属性和合作者		

注意：Bean 定义可以表示为真正的接口 org.springframework.beans.factory.config.BeanDefinition 以及它的各种子接口和实现。然而，绝大多数的用户代码不需要与 BeanDefinition 直接接触。

3. Bean 类

class 属性通常是强制性的，有两种用法。在绝大多数情况下，BeanFactory 直接调用 Bean 的构造函数创建一个 Bean(相当于调用 new 的 Java 代码)。class 属性指定了需要创建的 Bean 的类。在比较少的情况下，BeanFactory 调用某个类的静态的工厂方法创建 Bean。class 属性指定了实际包含静态工厂方法的那个类。(至于静态工厂方法返回的 Bean 的类型是同一个类，还是完全不同的另一个类，这并不重要。)

1) 通过构造函数创建 Bean

当使用构造函数创建 Bean 时，所有普通的类都可以被 Spring 使用并且和 Spring 兼容。这就是说，被创建的类不需要实现任何特定的接口或者按照特定的样式进行编写，仅指定 Bean 的类就足够了。然而，根据 Bean 使用的 IoC 类型，你可能需要一个默认的(空的)构造函数。

另外，BeanFactory 并不局限于管理真正的 JavaBean，它也能管理任何你想让它管理的类。虽然很多使用 Spring 的人喜欢在 BeanFactory 中用真正的 JavaBean (仅包含一个默认的(无参数的)构造函数，在属性后面定义相对应的 setter()和 getter()方法)，但是在 BeanFactory 中也可以使用特殊的非 Bean 样式的类。例如，如果需要使用一个遗留下来的完全没有遵守 JavaBean 规范的连接池，不要担心，Spring 同样能够管理它。

使用 XmlBeanFactory 可以像下面这样定义 Bean class：

```
<bean id="exampleBean"
      class="examples.ExampleBean"/>
<bean name="anotherExample"
      class="examples.ExampleBeanTwo"/>
```

至于为构造函数提供(可选的)参数，以及在对象实例创建后设置实例属性，将会在后面叙述。

2) 通过静态工厂方法创建 Bean

当定义一个使用静态工厂方法创建的 Bean，同时使用 class 属性指定包含静态工厂方法的类，这时需要 factory-method 属性指定工厂方法名。Spring 调用这个方法(包含一组可选的参数)并返回一个有效的对象，之后这个对象就完全和构造方法创建的对象一样。用户可以使用这样的 Bean 定义在遗留代码中调用静态工厂。

下面是一个 Bean 定义的例子，声明这个 Bean 要通过 factory-method 指定的方法创建。注意，这个 Bean 定义并没有指定返回对象的类型，只指定包含工厂方法的类。在这个例子中，createInstance 必须是 static()方法。

```
<bean id="exampleBean"
      class="examples.ExampleBean2"
      factory-method="createInstance"/>
```

至于为工厂方法提供(可选的)参数，以及在对象实例被工厂方法创建后设置实例属性，将会在后面叙述。

3) 通过实例工厂方法创建 Bean

使用一个实例工厂方法(非静态的)创建 Bean 和使用静态工厂方法非常类似，调用一个已存在的 Bean(这个 Bean 应该是工厂类型)的工厂方法创建新的 Bean。

使用这种机制,class 属性必须为空,而且 factory-bean 属性必须指定一个 Bean 的名字,这个 Bean 一定要在当前的 Bean 工厂或者父 Bean 工厂中,并包含工厂方法。而工厂方法本身仍然要通过 factory-method 属性设置。

下面是一个例子：

```
<!--The factory bean, which contains a method called
    createInstance -->
<bean id="myFactoryBean"
     class="...">
...
</bean>
<!--The bean to be created via the factory bean -->
<bean id="exampleBean"
     factory-bean="myFactoryBean"
     factory-method="createInstance"/>
```

虽然我们要在后面讨论设置 Bean 的属性,但是这个方法意味着工厂 Bean 本身能够被容器通过依赖注射管理和配置。

4. Bean 的标识符(id 与 name)

每个 bean 都有一个或多个 id(也叫作标识符,或名字;这些名词说的是一回事)。这些 id 在管理 Bean 的 BeanFactory 或 ApplicationContext 中必须是唯一的。一个 Bean 差不多总是只有一个 id,但是如果一个 Bean 有超过一个的 id,那么另外的那些本质上可以认为是别名。

在一个 XmlBeanFactory 中(包括 ApplicationContext 的形式),可以用 id 或者 name 属性指定 Bean 的 id(s),并且在这两个或其中一个属性中至少指定一个 id。id 属性允许指定一个 id,并且它在 XML DTD(定义文档)中作为一个真正的 XML 元素的 ID 属性被标记,所以 XML 解析器能够在其他元素指向它的时候做一些额外的校验。正因如此,用 id 属性指定 Bean 的 id 是一个比较好的方式。然而,XML 规范严格限定了在 XML ID 中合法的字符。通常这并不是真正限制你,但是如果有必要使用这些字符(在 ID 中的非法字符),或者想给 Bean 增加其他的别名,可以通过 name 属性指定一个或多个 id(用逗号,或者分号;分隔)。

5. Singleton 的使用与否

Beans 被定义为两种部署模式中的一种：singleton 或 non-singleton(后一种也叫作 prototype,尽管这个名词用得不精确)。如果一个 Bean 是 singleton 形态的,就只有一个共享的实例存在,所有和这个 Bean 定义的 id 符合的 Bean 请求都会返回这个唯一的、特定的实例。

如果 Bean 以 non-singleton、prototype 模式部署,对这个 Bean 的每次请求都会创建一个新的 Bean 实例。这对于每个 user 需要一个独立的 user 对象这样的情况是非常理想的。

Beans 默认被部署为 singleton 模式,除非有指定。把部署模式变为 non-singletion (prototype)后,每次对这个 Bean 的请求都会导致一个新创建的 Bean,而这可能并不是你真正想要的。所以,仅在绝对需要的时候才把模式改成 prototype。

在下面这个例子中,两个 Bean 中的一个被定义为 singleton,另一个被定义为 non-singleton(prototype)。客户端每次向 BeanFactory 请求都会创建新的 exampleBean,而 AnotherExample 仅被创建一次;每次对它请求都会返回这个实例的引用。

```
<bean id="exampleBean"
      class="examples.ExampleBean" singleton="false"/>
<bean name="yetAnotherExample"
      class="examples.ExampleBeanTwo" singleton="true"/>
```

注意:当部署一个 Bean 为 prototype 模式,这个 Bean 的生命周期就会有稍许改变。通过定义,Spring 无法管理一个 non-singleton/prototype Bean 的整个生命周期,因为当它创建之后,它被交给客户端,而且容器根本不再跟踪它了。当说起 non-singleton/prototype Bean 的时候,可以把 Spring 的角色想象成 new 操作符的替代品。之后的任何生命周期方面的事情都由客户端处理。

3.2.3 ApplicationContext

Beans 包提供了以编程的方式管理和操控 Bean 的基本功能,而 context 包增加了 ApplicationContext,它以一种更加面向框架的方式增强了 BeanFactory 的功能。多数用户可以以一种完全的声明式方式使用 ApplicationContext,甚至不用手工创建它,但是却依赖像 ContextLoader 的支持类,在 J2EE 的 Web 应用的启动进程中用它启动 ApplicationContext。当然,这种情况下还是可以以编程的方式创建一个 ApplicationContext。

context 包的基础是位于 org.springframework.context 包中的 ApplicationContext 接口。它由 BeanFactory 接口集成而来,提供 BeanFactory 所有的功能。为了以一种更像面向框架的方式工作,context 包使用分层和有继承关系的上下文类,包括:

- MessageSource,提供对 i18n 消息的访问。
- 资源访问,如 URL 和文件。
- 事件传递给实现了 ApplicationListener 接口的 Bean。
- 载入多个(有继承关系)上下文类,使得每个上下文类都专注于一个特定的层次,如应用的 Web 层。

因为 ApplicationContext 包括 BeanFactory 所有的功能,所以通常建议先于 BeanFactory 使用,除有限的一些场合(如在一个 Applet 中,内存的消耗是关键的),每千字节都很重要。接下来介绍 ApplicationContext 在 BeanFactory 的基本能力上增加的功能。

1. 使用 MessageSource

ApplicationContext 接口继承 MessageSource 接口,所以提供了 messaging 功能(i18n 或者国际化)。它同 NestingMessageSource 一起使用,就能处理分级的信息,这些是 Spring 提供的处理信息的基本接口。这里定义的方法有如下 3 个。

StringgetMessage (String code,Object[] args,String default,Locale loc):这个方法是从 MessageSource 取得信息的基本方法。如果对于指定的 locale 没有找到信息,则使用

默认的信息。传入的参数 args 用来代替信息中的占位符，这是通过 Java 标准类库的 MessageFormat 实现的。

StringgetMessage (String code，Object[] args，Locale loc)：本质上和上一个方法一样，区别只是没有默认值可以指定；如果找不到信息，就会抛出一个 NoSuchMessage Exception。

StringgetMessage(MessageSourceResolvable resolvable，Locale locale)：上面两个方法使用的所有属性都封装到一个叫作 MessageSourceResolvable 的类中，可以通过这个方法直接使用它。

当 ApplicationContext 被加载的时候，它会自动查找在 context 中定义的 MessageSource Bean。这个 Bean 必须叫作 messageSource。如果找到了这样一个 Bean，所有对上述方法的调用将会被委托给找到的 message source。如果没有找到 message source，ApplicationContext 将会尝试查它的父亲是否包含这个名字的 Bean。如果有，它将会把找到的 Bean 作为 Message Source。如果它最终没有找到任何信息源，一个空的 StaticMessageSource 将会被实例化，使它能够接受上述方法的调用。

Spring 目前提供了两个 MessageSource 的实现，分别是 ResourceBundleMessageSource 和 StaticMessageSource。两个都实现了 NestingMessageSource，以便能够嵌套地解析信息。StaticMessageSource 很少被使用，但是它提供以编程的方式向 source 增加信息。Resource BundleMessageSource 用得更多一些，下面是它的一个例子：

```
<beans>
    <bean id="messageSource"
    class="org.springframework.context.support.ResourceBundleMessageSource">
        <property name="basenames">
            <list>
                <value>format</value>
                <value>exceptions</value>
                <value>windows</value>
            </list>
        </property>
    </bean>
</beans>
```

这段配置假定你在 classpath 有 3 个 resource bundle，分别为 format、exceptions 和 windows。使用 JDK 通过 ResourceBundle 解析信息的标准方式，任何解析信息的请求都会被处理。

2. 事件传递

ApplicationContext 中的事件处理是通过 ApplicationEvent 类和 ApplicationListener 接口提供的。如果上下文中部署了一个实现了 ApplicationListener 接口的 Bean，每次一个 ApplicationEvent 发布到 ApplicationContext 时，那个 Bean 就会被通知。实质上，这是标准的 Observer 设计模式。Spring 提供了 3 个标准事件，见表 3-2。

表 3-2　内置事件

事　　件	解　　释
ContextRefreshedEvent	当 ApplicationContext 已经初始化或刷新后发送的事件。这里的初始化意味着所有的 Bean 被装载，singleton 被预实例化，以及 ApplicationContext 已准备好
ContextClosedEvent	当使用 ApplicationContext 的 close()方法结束上下文的时候发送的事件。这里的结束意味着 singleton 被销毁
RequestHandledEvent	一个与 Web 相关的事件，告诉所有的 Bean 一个 HTTP 请求已经被响应了(这个事件将会在一个请求结束后被发送)。注意，这个事件只能应用于使用了 Spring 的 DispatcherServlet 的 Web 应用

同样，也可以实现自定义的事件。通过调用 ApplicationContext 的 publishEvent()方法，并且指定一个参数，这个参数是你自定义的事件类的一个实例。下面看一个例子。

首先是 ApplicationContext:

```
<bean id="emailer" class="example.EmailBean">
    <property name="blackList">
        <list>
            <value>black@list.org</value>
            <value>white@list.org</value>
            <value>john@doe.org</value>
        </list>
    </property>
</bean>

<bean id="blackListListener" class="example.BlackListNotifier">
    <property name="notificationAddress">
        <value>spam@list.org</value>
    </property>
  </bean>
```

然后是实际的 Bean:

```
public class EmailBean implements ApplicationContextAware {

    /** the blacklist */
    private List blackList;

    public void setBlackList(List blackList) {
        this.blackList=blackList;
}

    public void setApplicationContext(ApplicationContext ctx) {
        this.ctx=ctx;
}

    public void sendEmail(String address, String text) {
        if (blackList.contains(address)) {
```

```
            BlackListEventevt=new BlackListEvent(address, text);
            ctx.publishEvent(evt);
            return;
        }
        //send email
    }
}

public class BlackListNotifier implement ApplicationListener{

    /** notification address * /
    private String notificationAddress;

    public void setNotificationAddress(String notificationAddress) {
        this.notificationAddress=notificationAddress;
    }

    public void onApplicationEvent(ApplicationEvent evt) {
        if (evtinstanceofBlackListEvent) {
            //notify appropriate person
        }
    }
}
```

3. 在 Spring 中使用资源

很多应用程序都需要访问资源。Spring 提供了一个清晰透明的方案,以一种协议无关的方式访问资源。ApplicationContext 接口包含一个方法(getResource(String))负责这项工作。

Resource 类定义了几个方法,见表 3-3。这几个方法被所有的 Resource 实现所共享。

表 3-3 资源功能

方　法	解　释
getInputStream()	用 InputStream 打开资源,并返回这个 InputStream
exists()	检查资源是否存在,如果不存在,则返回 false
isOpen()	如果这个资源不能打开多个流将会返回 true。常见的资源实现一般返回 false
getDescription()	返回资源的描述,通常是全限定文件名或者实际的 URL

Spring 提供了几个 Resource 的实现。它们都需要一个 String 表示的资源的实际位置。依据这个 String,Spring 将会自动为你选择正确的 Resource 实现。当向 ApplicationContext 请求一个资源时,Spring 首先检查你指定的资源位置,之后寻找任何前缀。根据不同的 Application Context 的实现,不同的 Resource 实现可被使用的 Resource 最好使用 ResourceEditor 配置,如 XmlBeanFactory。

接下来看一个 Spring IoC 实例。

(1) 新建 Java 项目,如图 3-7 所示。

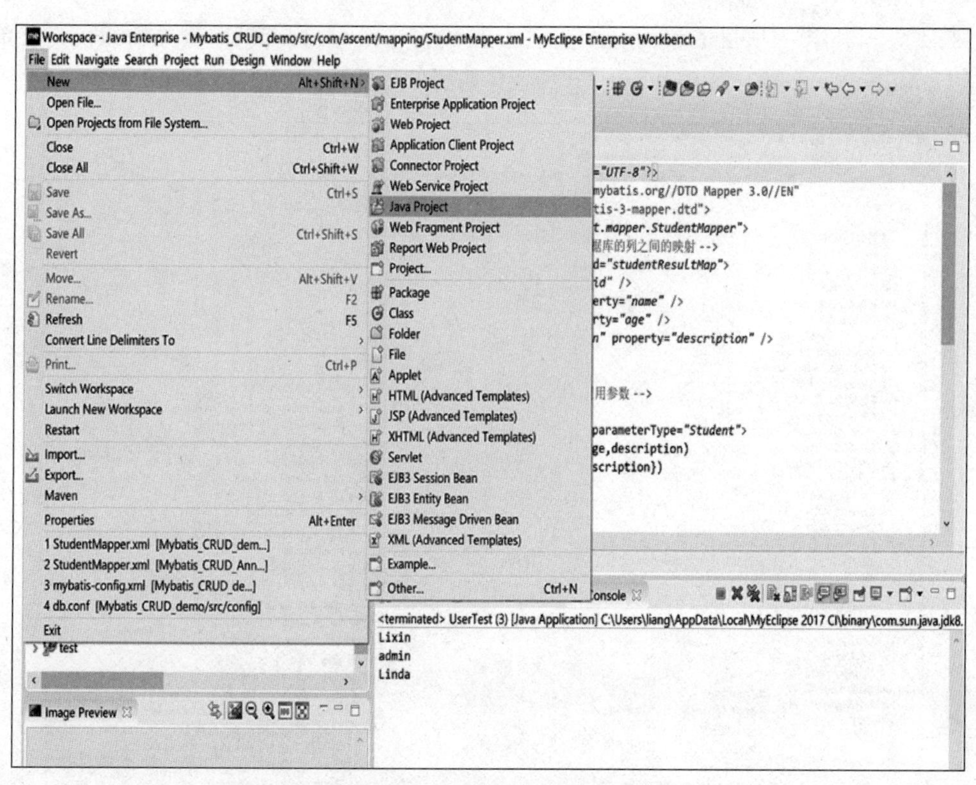

图 3-7 新建 Java 项目

在 Project name 中输入 Spring_IoC_demo，如图 3-8 所示。

New Java Project
Create a Java Project
Create a Java project in the workspace or in an external location.
Project name: Spring_IoC_demo
Use default location
Location: D:\Workspace\Spring_IoC_demo Browse...
JRE
Use an execution environment JRE: JavaSE-1.8
Use a project specific JRE: Java JDK VM 1.8.0_v112 Windows
Use default JRE (currently 'Java JDK VM 1.8.0_v112 Windows') Configure JREs...
Project layout
Use project folder as root for sources and class files
Create separate folders for sources and class files Configure default...
Working sets
Add project to working sets New...
Working sets: Select...
< Back Next > Finish Cancel

图 3-8 命名项目

(2) 导入 Spring 项目依赖的 jar 包，右击 Spring_IoC_demo 项目，从弹出的快捷菜单中选择 Configure Facets→Install Spring Facet，如图 3-9 所示。

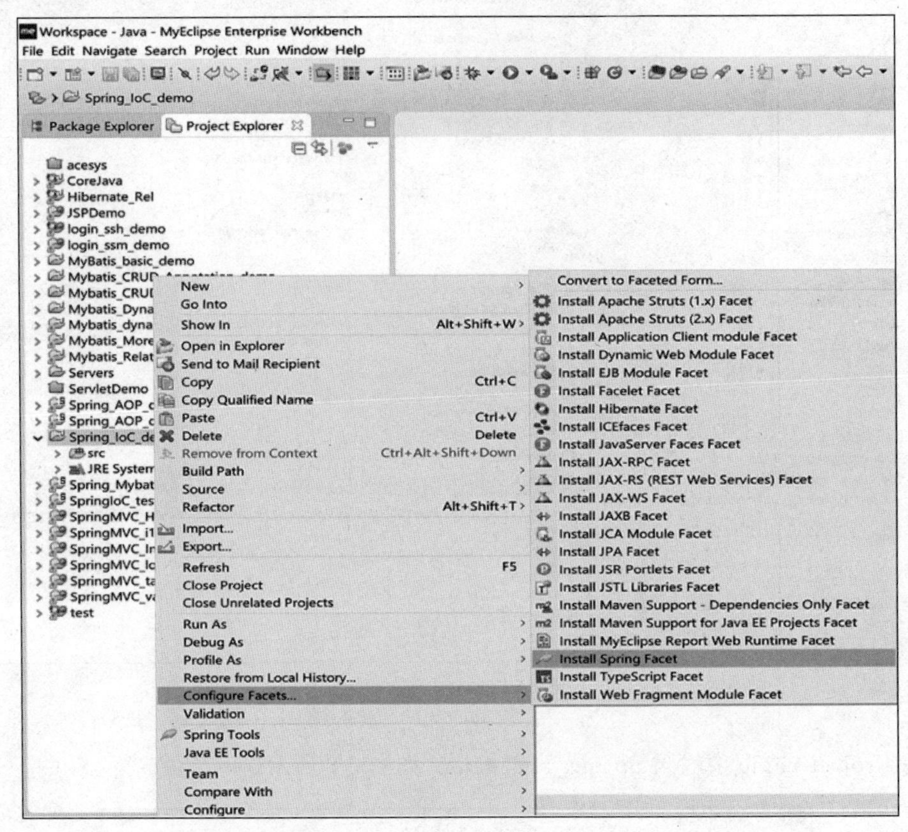

图 3-9　Install Spring Facet 界面 1

单击 Next 按钮，不需修改，直到单击 Finish 按钮，如图 3-10 所示。

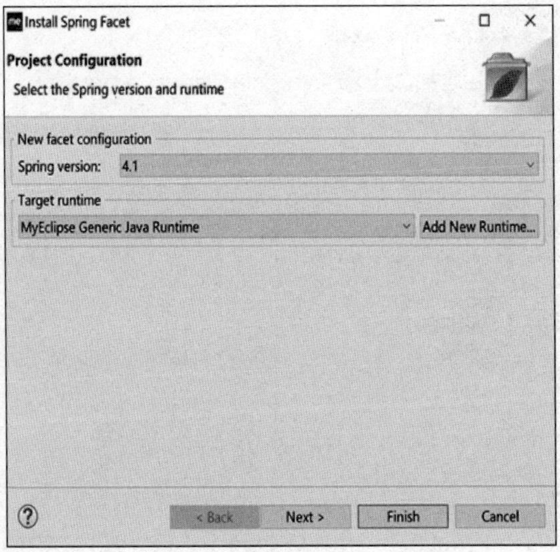

图 3-10　Install Spring Facet 界面 2

（3）建立包结构。右击 src，从弹出的快捷菜单中选择 New→Package，如图 3-11 所示。

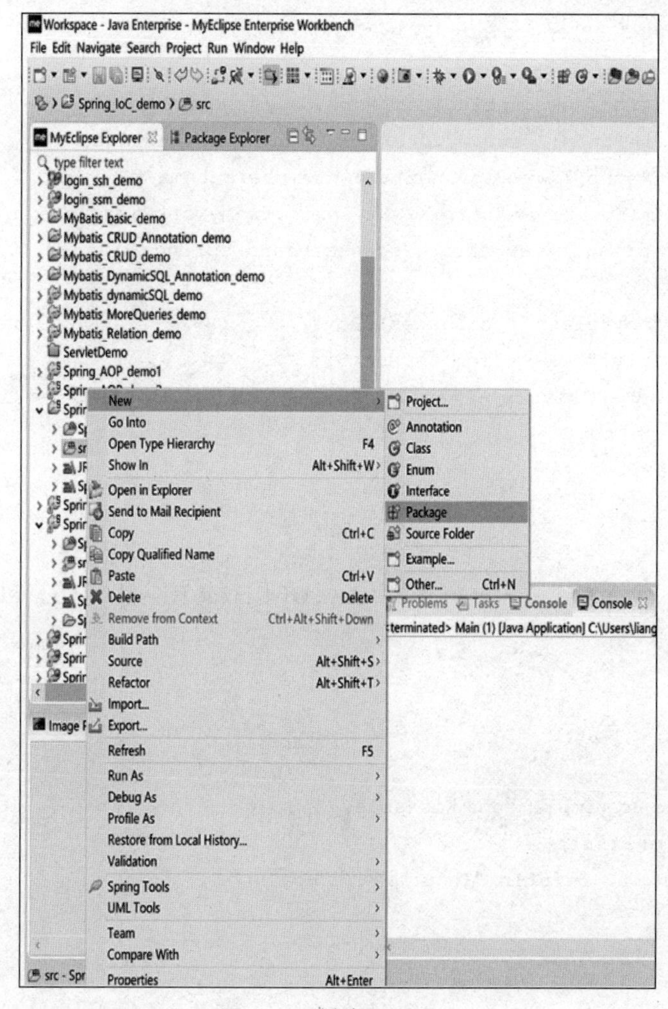

图 3-11　新建 Package

在 Name 处输入 com.ascent，之后单击 Finish 按钮，如图 3-12 所示。

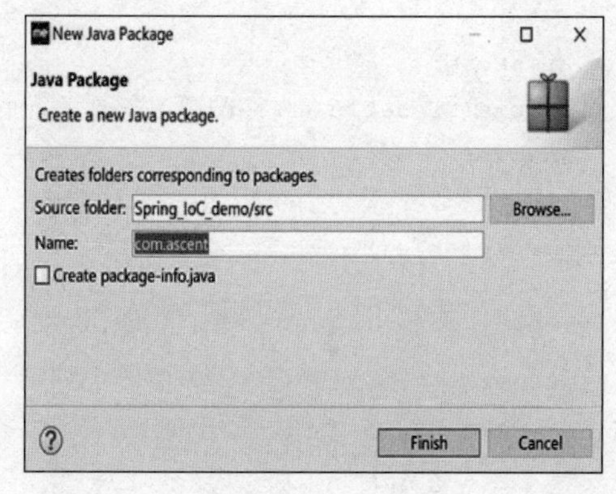

图 3-12　命名 Package

（4）在 src 目录下编写配置文件 applicationContext.xml，内容如下：

```
<?xmlversion="1.0"encoding="UTF-8"?>
<beans
    xmlns="http://www.springframework.org/schema/beans"
    xmlns:xsi="http://www.w3.org/2001/XMLSchema-instance"
    xmlns:p="http://www.springframework.org/schema/p"
    xsi:schemaLocation="http://www.springframework.org/schema/beans
http://www.springframework.org/schema/beans/spring-beans-4.1.xsd">

  <bean id="boy"class="com.ascent.Boy"/>
  <bean id="girl"class="com.ascent.Girl">
  <propertyname="kissable">
    <ref bean="boy"/>
      </property>
    </bean>
  </beans>
```

（5）在 com.ascent 目录下分别编写 Boy、Girl.java、Kissable.java 和 Test.java，代码如下：

```
//Boy.java
package com.ascent;

public class Boy implements Kissable{
  public void kiss(){
      System.out.println("This is Kiss Boy");
  }
}

//Girl.java
package com.ascent;

public class Girl {
    private Kissable kissable;
    public void setKissable(Kissable kissable) {
    this.kissable=kissable;
    }
    public void kissYourKissable() {
    kissable.kiss();
    }
}

//Kissable.java
package com.ascent;
```

```
public interface Kissable {
  public void kiss();
}

  //Test.java
  package com.ascent;

import org.springframework.context.ApplicationContext;
import org.springframework.context.support.ClassPathXmlApplicationContext;

public class Test {

    public static void main(String[] args) {
        ApplicationContext apc=new ClassPathXmlApplicationContext(
            "applicationContext.xml");
        Girl g=(Girl) apc.getBean("girl");
        g.kissYourKissable();
    }

}
```

(6) 右击 Test 类，从弹出的快捷菜单中选择 Run As→Java Application，如图 3-13 所示。

得到如下结果：

```
This is Kiss Boy
```

3.3 项目案例

3.3.1 学习目标

使用 Spring 框架，为 eGov 项目提供功能组件对象，并装配功能组件为应用上层提供封装对象，在实践中学习如何使用 ApplicationContext 对象，获取 Spring 配置文件中声明的功能组件。如果是在 Web 环境中，请使用 WebApplicationContext 对象替代 ApplicatiionContext 对象。

3.3.2 案例描述

本章案例为 eGov 中一般用户浏览新闻信息中的两部分内容：头版头条新闻和综合新闻。其中，头版头条需要取一条新闻数据，而综合新闻则需要按一定条件获取 6 条新闻数据。

获取头版头条的一条新闻数据和获取综合新闻的 6 条新闻数据为一个业务功能，用 Java 中的一个类的一个方法实现，将两种结果合并到 HashMap 中。

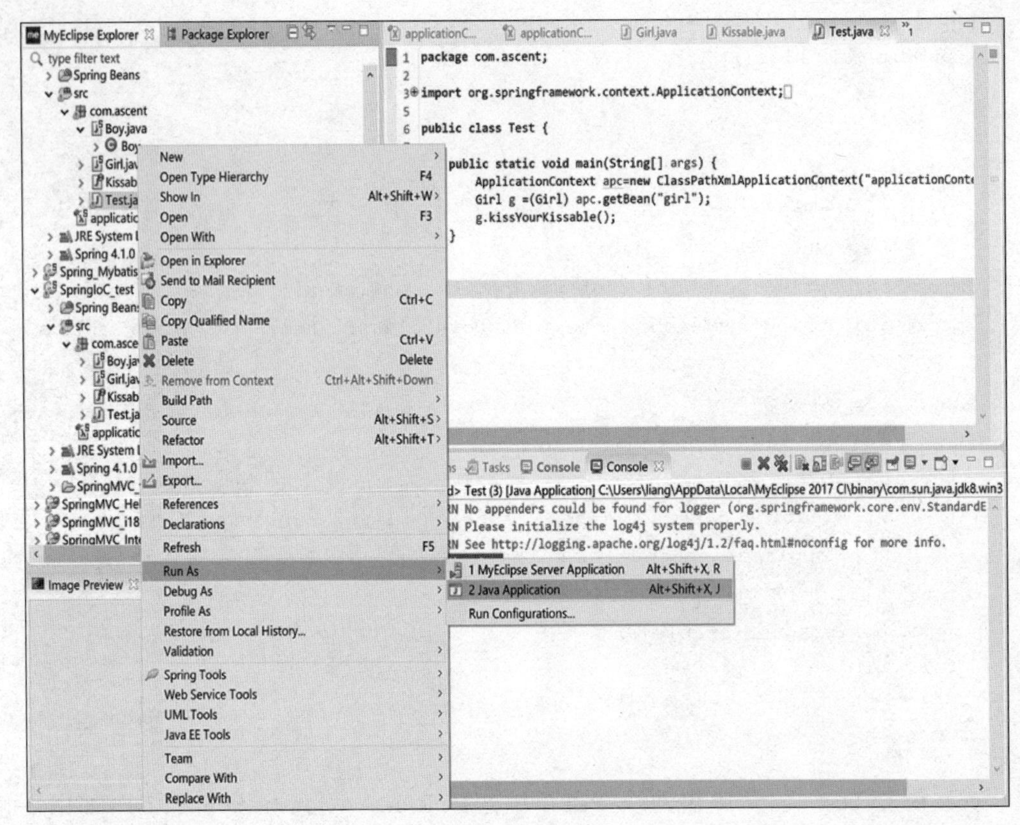

图 3-13　运行 class

3.3.3　案例要点

用户输入浏览器的 Web 项目地址，请求到达 Spring MVC 控制器，控制器再转向 index.jsp 页面。在 index.jsp 中像调用普通 Java 功能类一样，使用 JSTL 标签调用控制器中的相应业务控制方法，获取头条新闻和综合新闻的数据集合，然后再传给 index.jsp，在 index.jsp 中完成头版头条新闻和综合新闻数据的处理。由于没有使用到 Spring MVC 和 MyBatis，所以在这个案例中重点是 Spring 框架的搭建，以及如何使用 Spring DI/IoC 的依赖注入和控制反转配置业务类，使用 main( ) 方法模拟 Spring MVC 控制器，借助 ApplicationContext 对象获取业务对象，执行业务方法，将执行结果输出到控制台终端。

3.3.4　案例实施

(1) 在 MyEclipse 中新建 electrone Java Web 项目。

(2) 导入 mysql-connector-java-5.1.47-bin.jar 到项目 buildpath 中，如图 3-14 所示。

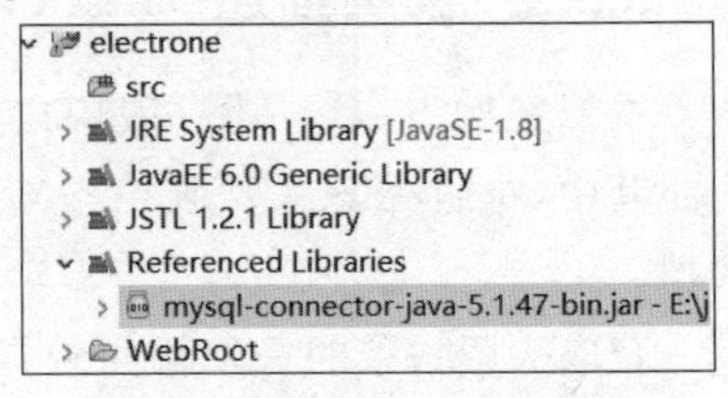

图 3-14　导入 mysql 驱动包

(3) 为项目添加 Spring 的支持。

右击工程，从弹出的快捷菜单中选择 Spring Tools→Add Spring Runtime Dependencies，如图 3-15 所示。

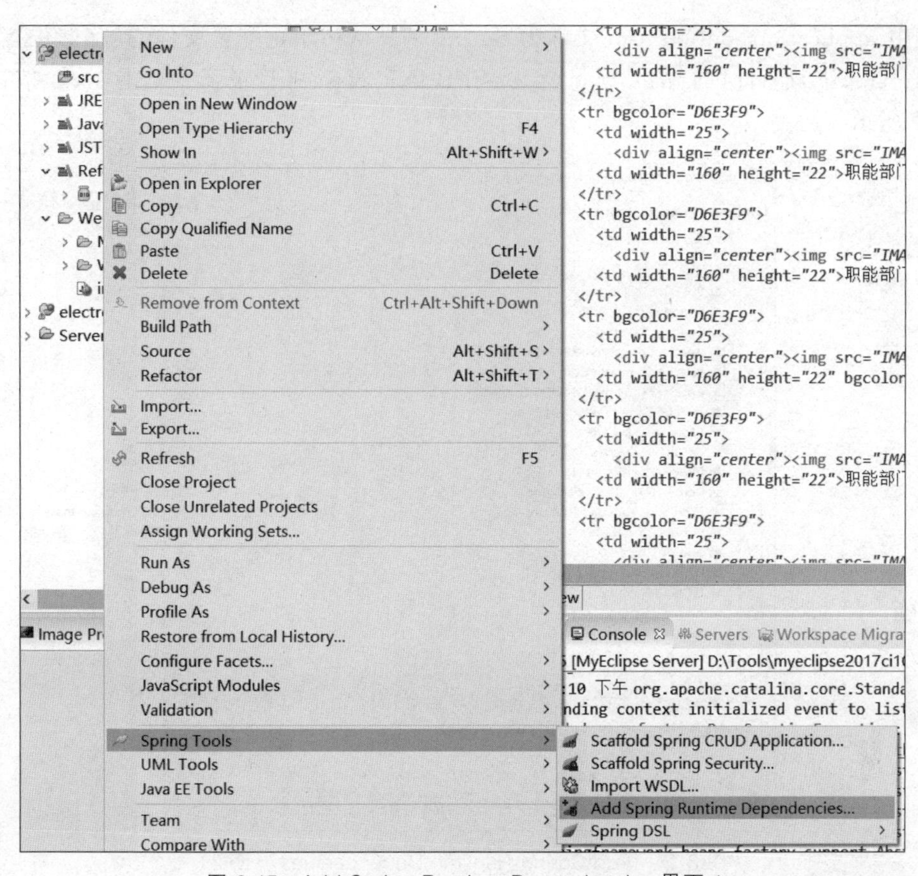

图 3-15　Add Spring Runtime Dependencies 界面 1

单击 Next 按钮，无须修改，Spring Version 为 4.1，再单击 Next 按钮，如图 3-16 所示。

Add Spring Runtime Dependencies

Add Spring Capabilities

Configure Spring project

Configure Spring libraries

Type: MyEclipse Library

Libraries will be provided through MyEclipse Library System

Spring Framework version: 4.1.0

Spring Security version: 3.2.5

Spring Webflow version: 2.4.0

☑ Add Spring-Web support (TLDs copied to project, ContextLoaderListener added to web.xml)

☑ Enable AOP Builder

i Larger projects that do not use Spring AOP may see a performance gain by leaving this disabled.

☑ Specify new or existing Spring bean configuration file?

Bean configuration type: ◉ New ○ Existing

Folder: src　Browse

File: applicationContext.xml

< Back　Next >　Finish　Cancel

图 3-16　Add Spring Runtime Dependencies 界面 2

其中 applicationContext.xml 文件为 Spring 的默认初始化配置文件，其存储位置在项目的 src 目录下，运行时的 Web 项目就在 WEB-INF/classes 目录下。单击 Next 按钮，如图 3-17 所示。

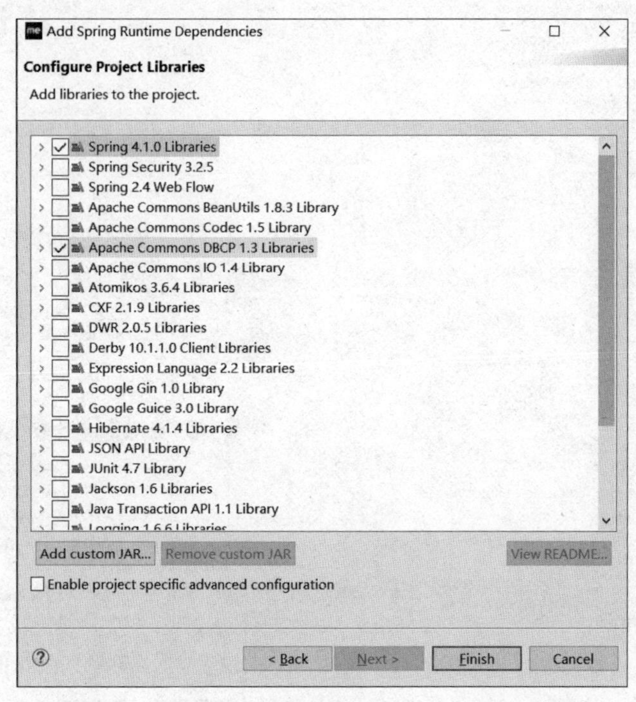

图 3-17　Add Spring Runtime Dependencies 界面 3

勾选 Spring 4.1.0 Libraries 和 Apache Commons DBCP 1.3 Libraries，其他不必勾选。Spring 库添加成功，如图 3-18 所示。

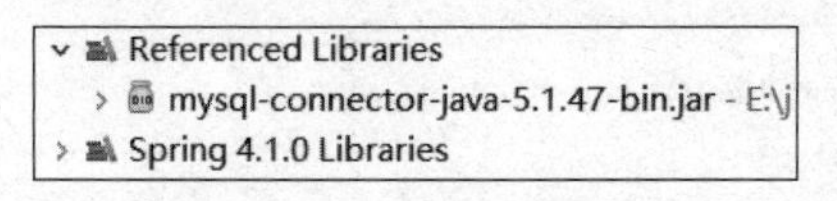

图 3-18　Spring 库添加成功

其中，Spring 的核心配置文件 applicationContext.xml 如图 3-19 所示。

```
applicationContext.xml
1 <?xml version="1.0" encoding="UTF-8"?>
2 <beans
3     xmlns="http://www.springframework.org/schema/beans"
4     xmlns:xsi="http://www.w3.org/2001/XMLSchema-instance"
5     xmlns:p="http://www.springframework.org/schema/p"
6     xsi:schemaLocation="http://www.springframework.org/schema/beans http://www.springframework.org/schema/beans/spring
7
8
9 </beans>
```

图 3-19　Spring 的核心配置文件 applicationContext.xml

MyEclipse 默认创建的文件中仅包含两个 XML 标签：beans 和 p。为了以后配置更多的功能，需要重新创建该文件，该文件的 Schema 标签定义不能更改，所以需要删除并新建

applicationContext.xml。

右击 electrone 项目，从弹出的快捷菜单中选择 New→Other，如图 3-20 所示。

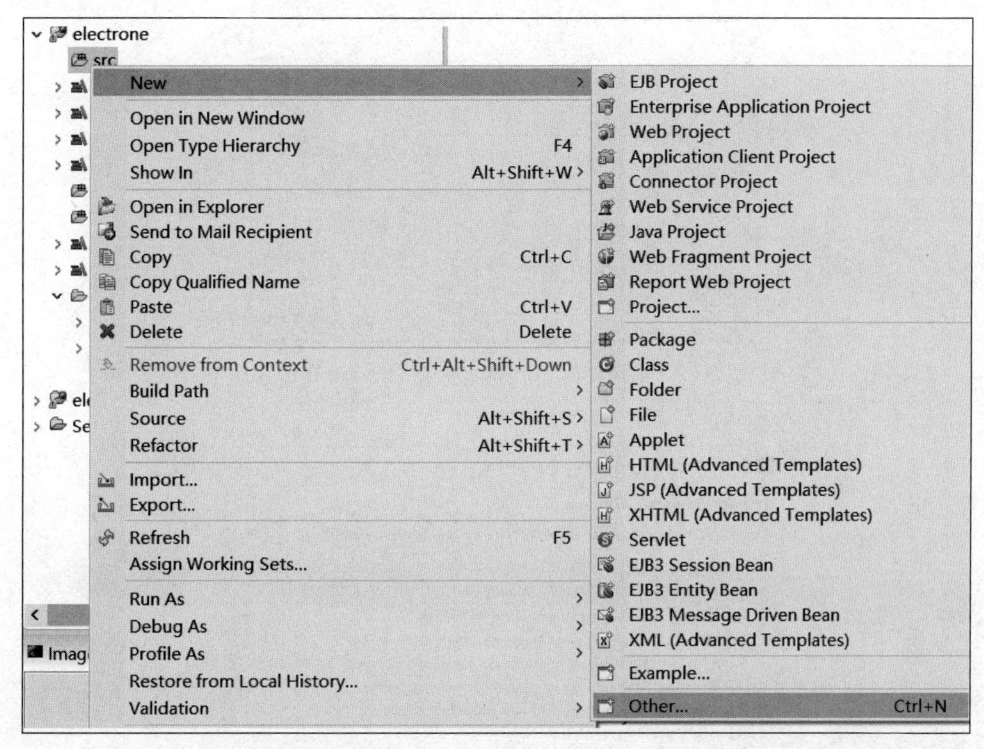

图 3-20 创建新的 applicationContext.xml 界面 1

找到 Spring 选项，展开并选择 Spring Bean Definition，如图 3-21 所示。

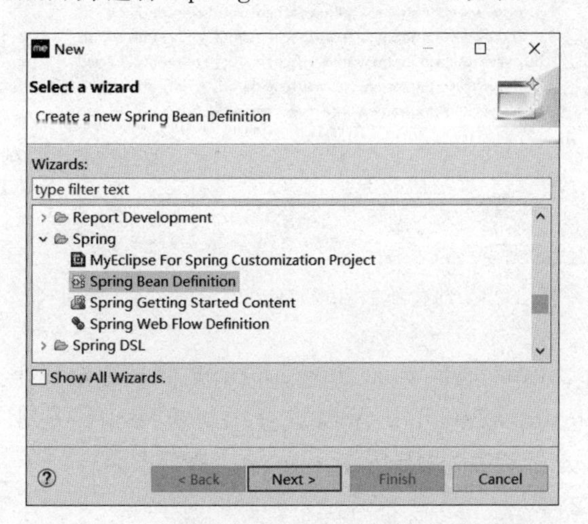

图 3-21 创建新的 applicationContext.xml 界面 2

单击 Next 按钮，如图 3-22 所示。

保持 applicationContext.xml 默认在项目源代码 src 目录下，单击 Next 按钮之后选择 xml 中使用的标签以及对应的版本文件，如图 3-23 所示。

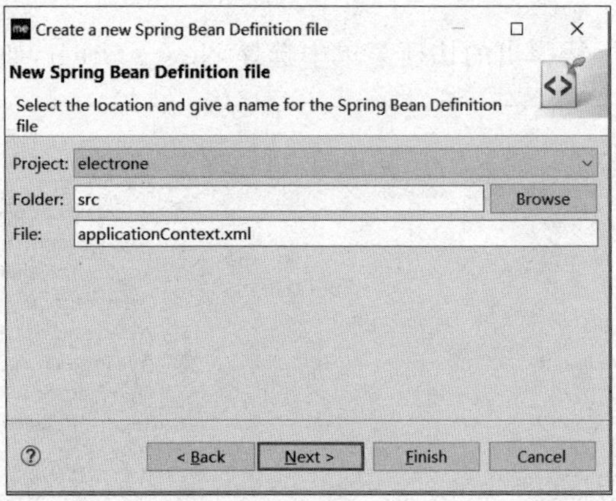

图 3-22　创建新的 applicationContext.xml 界面 3

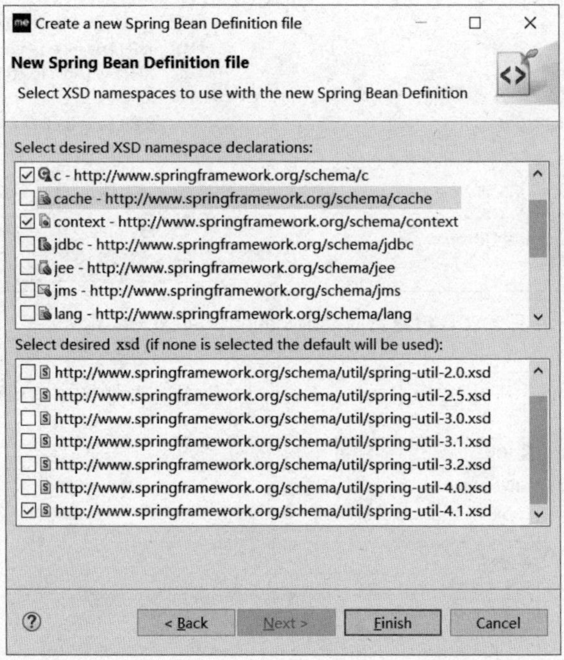

图 3-23　创建新的 applicationContext.xml 界面 4

这里需要选择 aop、beans、c、context、mvc、p、tx 和 util，并选择下方 xsd 对应的 Spring 版本号，本例中使用 Spring MVC 4.1，所以选择的对应 xsd 都是 4.1 版本。C 标签没有 xsd。其中 mvc 可选，为了以后便于使用 Spring MVC 框架，加上 MVC 也可以。单击 Finish 按钮，出现图 3-24。

至此，完成了 electrones 项目在 MyEclipse 中的环境支持。

(4) 编写 Java 类，具体包括：

- News.java 新闻类，其属性对应 mysql 中的 news 表中的字段，用于封装查询后的新闻数据。
- IAuthorizationService.java 接口，定义用于执行头版头条新闻的方法 findHeaderNews()。

```
*applicationContext.xml
1 <?xml version="1.0" encoding="UTF-8"?>
2 <beans xmlns="http://www.springframework.org/schema/beans"
3     xmlns:xsi="http://www.w3.org/2001/XMLSchema-instance"
4     xmlns:aop="http://www.springframework.org/schema/aop"
5     xmlns:c="http://www.springframework.org/schema/c"
6     xmlns:context="http://www.springframework.org/schema/context"
7     xmlns:mvc="http://www.springframework.org/schema/mvc"
8     xmlns:p="http://www.springframework.org/schema/p"
9     xmlns:tx="http://www.springframework.org/schema/tx"
0     xmlns:util="http://www.springframework.org/schema/util"
1     xsi:schemaLocation="http://www.springframework.org/schema/beans http://www.springframework.org/schema/beans/spring
2         http://www.springframework.org/schema/aop http://www.springframework.org/schema/aop/spring-aop-4.1.xsd
3         http://www.springframework.org/schema/context http://www.springframework.org/schema/context/spring-context-4.1
4         http://www.springframework.org/schema/mvc http://www.springframework.org/schema/mvc/spring-mvc-4.1.xsd
5         http://www.springframework.org/schema/tx http://www.springframework.org/schema/tx/spring-tx-4.1.xsd
6         http://www.springframework.org/schema/util http://www.springframework.org/schema/util/spring-util-4.1.xsd">
7
8
9 </beans>
0
```

图 3-24　创建新的 applicationContext.xml 界面 5

- IAuthorizationServiceImpl.java 是上述接口的实现类，使用 NewsDAO 完成数据库操作。
- NewsDAO 数据库对象访问类，该类使用 spring jdbc 模块提供的 JdbcTemplate 类执行传统的 SQL 语句，简化传统使用 JDBC 的烦琐代码。
- NewsRowMapper.java 类，使用 JdbcTemplate 过程中需要使用 RowMapper＜T＞泛型类的实现，以自定义需要封装的新闻数据，并将封装过程(即数据库表)中的字段对应 News.java 中的字段。
- IndexTestCase.java 测试类模仿控制器的功能，用于测试配置在 Spring 环境中的 IAuthorizationService 的实现类是否能正常工作，取得头版头条新闻和综合新闻。

(5) News.java。

```
package com.ascent.po;
import java.util.Date;
public class News {
    private Integer id;
    private String title;
    private String author;
    private Integer deptid;
    private String content;
    private Integer type
    private String checkopinion;
    private String checkopinion2;
    private Integer checkstatus;
    private Integer crosscolumn;
    private Integer crossstatus;
    private String picturepath;
    private Date publishtime;
    private Date crosspubtime;
    private Integer preface;
    private Integer status;
```

```
    private Integer userid;

    public Integer getId() {
        return id;
    }
    public void setId(Integer id) {
        this.id=id;
    }
    public String getTitle() {
        return title;
    }
    public void setTitle(String title) {
        this.title=title==null ? null : title.trim();
    }
    public String getAuthor() {
        return author;
    }
    public void setAuthor(String author) {
        this.author=author==null ? null : author.trim();
    }
    public Integer getDeptid() {
        return deptid;
    }
    public void setDeptid(Integer deptid) {
        this.deptid=deptid;
    }
    public String getContent() {
        return content;
    }
    public void setContent(String content) {
        this.content=content==null ? null : content.trim();
    }
    public Integer getType() {
        return type;
    }
    public void setType(Integer type) {
        this.type=type;
    }
    public String getCheckopinion() {
        return checkopinion;
    }
    public void setCheckopinion(String checkopinion) {
        this.checkopinion=checkopinion==null ? null: checkopinion.trim();
    }
    public String getCheckopinion2() {
```

```
        return checkopinion2;
    }
    public void setCheckopinion2(String checkopinion2) {
        this.checkopinion2=checkopinion2==null ? null : checkopinion2.trim();
    }
    public Integer getCheckstatus() {
        return checkstatus;
    }
    public void setCheckstatus(Integer checkstatus) {
        this.checkstatus=checkstatus;
    }
    public Integer getCrosscolumn() {
        return crosscolumn;
    }
    public void setCrosscolumn(Integer crosscolumn) {
        this.crosscolumn=crosscolumn;
    }
    public Integer getCrossstatus() {
        return crossstatus;
    }
    public void setCrossstatus(Integer crossstatus) {
        this.crossstatus=crossstatus;
    }
    public String getPicturepath() {
        return picturepath;
    }
    public void setPicturepath(String picturepath) {
        this.picturepath=picturepath==null ? null: picturepath.trim();
    }
    public Date getPublishtime() {
        return publishtime;
    }
    public void setPublishtime(Date publishtime) {
        this.publishtime=publishtime;
    }
    public Date getCrosspubtime() {
        return crosspubtime;
    }
    public void setCrosspubtime(Date crosspubtime) {
        this.crosspubtime=crosspubtime;
    }
    public Integer getPreface() {
        return preface;
    }
    public void setPreface(Integer preface) {
```

```
        this.preface=preface;
    }
    public Integer getStatus() {
        return status;
    }
    public void setStatus(Integer status) {
        this.status=status;
    }
    public Integer getUserid() {
        return userid;
    }
    public void setUserid(Integer userid) {
        this.userid=userid;
    }
}
```

(6) NewsRowMapper.java。

```
package com.ascent.jdbc.rowmapper;
import java.sql.ResultSet;
import java.sql.SQLException;
import org.springframework.jdbc.core.RowMapper;
import com.ascent.po.News;

public class NewsRowMapper implements RowMapper<News>{
/*
 *
 * rs为执行某SQL查询后的查询结果集,rowNum为查询的行号
 * 每行为单位,每行查询结果用该方法进行封装
 */
@Override
public News mapRow(ResultSetrs, introwNum) throws SQLException{
    News news=new News();              //创建空的新闻对象
    //从查询结果行数据中一一取出新闻数据,封装到News对象中
    news.setId(rs.getInt("id"));
    news.setAuthor(rs.getString("author"));
    news.setTitle(rs.getString("title"));
    news.setContent(rs.getString("content"));
    news.setDeptid(rs.getInt("deptid"));
    news.setPicturepath(rs.getString("picturepath"));
    news.setPreface(rs.getInt("preface"));
    news.setPublishtime(rs.getDate("publishtime"));
    news.setStatus(rs.getInt("status"));
    news.setTitle(rs.getString("title"));
    news.setType(rs.getInt("type"));
    news.setUserid(rs.getInt("userid"));
```

```
    news.setCheckopinion(rs.getString("checkopinion"));
    news.setCheckopinion2(rs.getString("checkopinion2"));
    news.setCheckstatus(rs.getInt("checkstatus"));
    news.setCrosscolumn(rs.getInt("crosscolumn"));
    news.setCrosspubtime(rs.getDate("crosspubtime"));
    news.setCrossstatus(rs.getInt("crossstatus"));
    return news;
    }

}
```

该类用以封装 news 新闻表中查询出的每行数据。

(7) 编写用于执行 SQL 操作的 NewsDAO 类,该类需要继承 JdbcDaoSupport 类,并使用 JdbcTemplate 完成数据库的增、删、改、查、操作,本案例使用的是查询方法。该类包含 findHeaderNews()方法,该方法返回一条头版头条新闻,以及 findIndexNews()方法,该方法返回 6 条综合新闻。

```
NewsDAO.java
package com.ascent.jdbc.dao;

import java.sql.SQLException;
import java.util.List;

import org.springframework.jdbc.core.support.JdbcDaoSupport;
import org.springframework.jdbc.datasource.DataSourceTransactionManager;
import org.springframework.transaction.TransactionStatus;
import org.springframework.transaction.support.DefaultTransactionDefinition;

import com.ascent.jdbc.rowmapper.NewsRowMapper;
import com.ascent.po.News;
/**
  * 该类使用 Spring 提供的 jdbcTempate 完成 SQL 操作
  * @author gary
  *
  */
public class NewsDAO extends JdbcDaoSupport{

/*
  * @paramsql 执行查询头条新闻的 SQL 语句
  * @paramrowMapper 告知 jdbcTemplate 用什么 RowMapper 封装查询结果
  * @return News 返回 News 新闻对象
  */
public News findHeaderNews(String sql, NewsRowMapperrowMapper){

    //配置数据库操作所需要的事务控制
    DefaultTransactionDefinition td=new DefaultTransactionDefinition();
```

```
        td.setIsolationLevel(DefaultTransactionDefinition.ISOLATION_READ_COMMITTED);
            DataSourceTransactionManager dm=new
            DataSourceTransactionManager(this.getJdbcTemplate().getDataSource());
            TransactionStatus ts=dm.getTransaction(td);

            News news=null;

            try{

            news=this.getJdbcTemplate().queryForObject(sql, rowMapper);
            dm.commit(ts);

              }catch(Exception e) {

            dm.rollback(ts);
            throw new RuntimeException(e.getMessage());
            }

        return news;

        }
    /**
     *
     * @paramsql 执行查询综合新闻的 SQL 语句
     * @return 返回所有符合查询条件的 News 新闻集合 List 对象
     */
    public List<News> findIndexNews(String sql) {

        //配置数据库操作所需要的事务控制
        DefaultTransactionDefinition td=new DefaultTransactionDefinition();
        td.setIsolationLevel(DefaultTransactionDefinition.ISOLATION_READ_COMMITTED);
        DataSourceTransactionManager dm=new
        DataSourceTransactionManager(this.getJdbcTemplate().getDataSource());
        TransactionStatus ts=dm.getTransaction(td);

        List<News>list=null;

        try{
            list=this.getJdbcTemplate().query(sql,new NewsRowMapper());
            dm.commit(ts);
        }catch(Exception e) {

            dm.rollback(ts);
            throw new RuntimeException(e.getMessage());
            }
```

```
        return list;
    }

}
```

注意：该类不需要初始化 JdbcTemplate，需要使用 Spring 框架配置和注入。

(8) 编写 IAuthorizationService 接口及其 IAuthorizationServiceImpl 实现类。

IAuthorizationService 接口定义了查询头版头条新闻和综合新闻的抽象方法 findHeaderNews()。该方法由 IAuthorizationServiceImpl 类具体实现，在具体实现类中依赖 NewsDAO 完成数据库操作，获取 6 条数据以及合并数据在具体实现类的 findHeaderNews() 方法中完成。

```
IAuthorizationService.java
package com.ascent.service;
import java.util.HashMap;
import java.util.List;

import com.ascent.po.News;
/**
 * Java 接口，定义查询头版头条新闻和综合新闻的方法 findHeaderNews()
 * @author Administrator
 *
 * /
public interface IAuthorizationService {

    /**
     * @return 直接返回包含头版头条新闻和综合新闻的 HashMap
     * 由于需要在前端分别处理，所以使用 HashMap 保存不同业务属性的新闻数据
     * 每个 key 对应不同业务属性的新闻数据
     * /
    public HashMap<String, List<News>> findHeaderNews();
}

IAuthorizationServiceImpl.java
package com.ascent.service.impl;
import java.util.ArrayList;
import java.util.HashMap;
import java.util.List;
import com.ascent.jdbc.dao.NewsDAO;
import com.ascent.jdbc.rowmapper.NewsRowMapper;
import com.ascent.po.News;
import com.ascent.service.IAuthorizationService;

public class IAuthorizationServiceImpl implements IAuthorizationService {

    private NewsDAOnewsDao;

    public NewsDAOgetNewsDao() {
```

```
        return newsDao;
    }
    public void setNewsDao(NewsDAOnewsDao) {
        this.newsDao=newsDao;
    }
    //声明使用的 NewsDAO
    /**
     * 使用 NewsDAO 完成获取头版头条新闻和综合新闻前 6 条的业务过程
     *
     * @return 返回包含头版头条新闻和综合新闻的 HashMap
     * 头版头条新闻对应一个 key:headerNews,综合新闻对应一个 Key:indexNews
     */
    @Override
    public HashMap<String, List<News>>findHeaderNews() {

        //查询新闻头版头条仅取一条数据
        String headerSql="select "+"id, title, author, deptid, content, type," +"
        checkopinion, checkopinion2, checkstatus, "+" crosscolumn, crossstatus,
        picturepath, publishtime," +" crosspubtime, preface, status, userid "+"
        from News n"+" where (n.status=1 and n.type=1)"+"or (n.type=2 and n.
        checkstatus=2 and n.status=1)"+" order by n.id desc limit 1";
        //status=1 表示已经发布
        //type=1 表示头版头条新闻,2 表示综合,3 表示通知,以此表示不同的新闻业务属性
        //checkstatus=2 表示申请头版头条新闻审核状态通过

        News headerNews=newsDao.findHeaderNews(headerSql, new NewsRowMapper());

        List<News>headerNewsList=new ArrayList<News>(1);
                              //构造空的 List<News>集合对象,集合成员必须为 News 对象
        headerNewsList.add(headerNews);    //将查询得到的头版头条新闻保存到 List 中

        //PageHelper.startPage(1, 6);
        String indexSql="select "+"id, title, author, deptid, content, type," +"
        checkopinion, checkopinion2, checkstatus, "+" crosscolumn, crossstatus,
        picturepath, publishtime," +" crosspubtime, preface, status, userid "+"
        from News n "+" where n.status=1 and n.type=2 " +"and n.checkstatus!=2
        order by n.id desc";

        List<News>indexNews=newsDao.findIndexNews(indexSql);
        //取前 6 条新闻
        //当 List 中的成员数量大于 6 时,截取前 6 条数据,即下标为 0~5 的元素
        if(indexNews !=null &&indexNews.size() >6) {
            indexNews=indexNews.subList(0, 5);
        }

        //定义并初始化 HashMap<String,List<News>>对象
        HashMap< String, List< News> > headerAndIndexNews = new HashMap< String,
        List<News>>(2);
```

```
        //将头版头条新闻保存在 Map 中
        headerAndIndexNews.put("headerNews", headerNewsList);
        //将综合新闻保存在 Map 中
        headerAndIndexNews.put("indexNews", indexNews);

        return headerAndIndexNews;
    }

}
```

（9）在 Spring 的核心配置文件 applicationContext.xml 中配置需要的数据源、IAuthorizationService、NewsDAO、JdbcTemplate。

省去 xml 头信息和 beans 根标签信息，主要内容如下：

```
<!--配置 DataSource 数据源 -->
<bean
id="dataSource" class="org.springframework.jdbc.datasource.
                                    DriverManagerDataSource">
        <property name="driverClassName" value="com.mysql.jdbc.Driver">
        </property>
        <property name="url" value="jdbc:mysql://localhost:3306/electrones">
        </property>
        <property name="username" value="root"></property><!--数据库用户名 -->
        <property name="password" value="root"></property><!--输入设置的密码 -->
    </bean>
    <!--配置 JdbcTemplate -->
    <bean id="jdbcTemplate" class="org.springframework.jdbc.core.JdbcTemplate">
        <property name="dataSource" ref="dataSource"></property>
    </bean>
    <!--配置 NewsDAO -->
    <bean id="newsDao" class="com.ascent.jdbc.dao.NewsDAO">
        <property name="jdbcTemplate" ref="jdbcTemplate"></property>
    </bean>
    <!--配置 IAuthorizationService -->
    <bean id="iAuthorizationService" class="com.ascent.service.impl.
                                    IAuthorizationServiceImpl" >
        <property name="newsDao" ref="newsDao"></property>
    </bean>
```

（10）编写 IndexTestCase.java。

```
package com.ascent.test;
import java.util.HashMap;
import java.util.List;
import org.springframework.context.ApplicationContext;
import org.springframework.context.support.ClassPathXmlApplicationContext;
import com.ascent.po.News;
import com.ascent.service.IAuthorizationService;
```

```
public class IndexTestCase{

    public static void main(String[] args) {
        //读取 class 路径下保存的 applicationContext.xml,构造对象
        ApplicationContextapplicationContext=
                new ClassPathXmlApplicationContext("applicationContext.xml");

        //从 applicationContext 对象中获取 Spring 创建的 IAuthorizationService 对
          象,bean 的 id iAuthoriaztionService
        IAuthorizationService authorizationService=(IAuthorizationService)
           applicationContext.getBean("iAuthorizationService");

        HashMap<String,List<News>>newsMap=authorizationService.findHeaderNews();

        //输出到控制台
        //头版头条新闻 headerNews,综合新闻 indexNews
        List<News>header=newsMap.get("headerNews");
        System.out.println(header);
        List<News>indexNews=newsMap.get("indexNews");
        System.out.println(indexNews);

    }
}
```

运行 IndexTestCase 类,结果如图 3-25 所示。

```
Problems  Tasks  Console ☒  Servers  Workspace Migration
<terminated> IndexTestCase [Java Application] D:\Tools\myeclipse2017ci10\binary\com.sun.java.jdk8.win32.x86_64_1.8.0.v112\bin\javaw.exe (2020年4月8
log4j:WARN No appenders could be found for logger (org.springframework.core.env.StandardEnvironment).
log4j:WARN Please initialize the log4j system properly.
log4j:WARN See http://logging.apache.org/log4j/1.2/faq.html#noconfig for more info.
Wed Apr 08 12:54:04 CST 2020 WARN: Establishing SSL connection without server's identity verification is not recommended. Ac
Wed Apr 08 12:54:04 CST 2020 WARN: Establishing SSL connection without server's identity verification is not recommended. Ac
[News [id=21, title=JavaScript, author=测试1, deptid=0, content=Javascript是一种由Netscape的LiveScript发展而来的脚本语言，主要
    JavaScript 的正式名称是 "ECMAScript"。这个标准由 ECMA 组织发展和维护。ECMA-262 是正式的 JavaScript 标准。这个标准基于 JavaScr
    在 1998 年，该标准成为了国际 ISO 标准 (ISO/IEC 16262)。这个标准仍然处于发展之中。, type=2, checkopinion=, checkopinion2=, ch
[News [id=22, title=Ajax, author=测试1, deptid=0, content=Ajax（Asynchronous JavaScript + XML）的定义
    基于web标准（standards-based presentation）XHTML+CSS的表示；
    使用 DOM（Document Object Model）进行动态显示及交互；
    使用 XML 和 XSLT 进行数据交换及相关操作；
    使用 XMLHttpRequest 进行异步数据查询、检索；
    使用 JavaScript 将所有的东西绑定在一起。英文参见Ajax的提出者Jesse James Garrett的原文,原文题目(Ajax: A New Approach to Web A
    类似于DHTML或LAMP，AJAX不是指一种单一的技术，而是有机地利用了一系列相关的技术。事实上，一些基于AJAX的“派生/合成”式（derivative
    AJAX的应用使用支持以上技术的web浏览器作为运行平台。这些浏览器目前包括：Mozilla、Firefox、Internet Explorer、Opera、Konqueror
    2、Java Server Faces (JSF)
    JSF全称Java Server Faces ，是一种用于构建Java Web 应用程序的标准框架(是Java Community Process 规定的JSR-127标准)。它提供了一
    JSF(Java Server Faces)技术为开发基于网络用户界面的Java开发者提供了标准的编程接口API以及标签库。就像Struts框架一样，JSF定义了一
    WebWork2的特性包括：
    灵活的Validation框架：允许你在XML文件中定义验证内容，在运行时通过Interceptor自动应用，因此完全脱离Action类。新版支持客户端验证
    Type conversion：允许你在类之间很容易转换对象。
    OGNL(Object Graphical Navigation Language)表达式语言：允许动态对象图表遍历和方法执行，使用ValueStack透明访问多Beans的属性。W
    IoC(Inversion of Control)容器：管理组件的生命周期，使客户获得组件实例不需要创建注册类（与容器环境无关）。
    FreeMarker Templates：使UI组件可重用，从而允许开发者容易定制Web页面的look & feel。
    Interceptors：在Actions处理的前后动态拦截，以简单化Action代码，增加减少代码的机会。
    支持I18N。
    容易和第三方软件集成，包括Hibernate, Spring, Pico, Sitemesh等。
    支持多种视图技术，如JSP，Velocity，FreeMarker，JasperReports，XML等。
    支持Packages和Namespaces，来管理Actions。, type=2, checkopinion=, checkopinion2=null, checkstatus=1, crosscolumn=0, cross
```

图 3-25　运行结果

至此,Spring 配置组件以及如何从 ApplicationContext 中获取对象已经完成。

3.3.5 特别提示

获取新闻信息(头版头条新闻和综合新闻)功能在真实项目中使用 Spring+Spring MVC-MyBatis 整合完成,在此我们模拟案例,使用 JdbcTemplate 技术替代 MyBatis,前端使用 IndexTestCase 测试类实现,提供加载 Spring 配置文件,获取 Spring IoC 配置的 IAuthorizationService 的业务类。

注意: 查询头版头条新闻和综合新闻的 SQL 语句的条件和语义。

3.3.6 拓展与提高

(1) 模拟完成查询更多头版头条新闻和更多综合新闻功能。

(2) 使用 BeanFactory 实现加载 Spring。

习题

1. Spring 分层结构中的主要模块分别是什么?
2. 如何理解 Spring IoC? 请举例说明。
3. Spring 配置 Bean 的作用域有哪些?
4. Spring 创建 Bean 实例的方式有哪几种?
5. Spring 属性注入有哪几种方式?
6. BeanFactory 和 ApplicationContext 的关系是什么?

第4章 Spring面向方面编程和事务处理

学习目的与学习要求

学习目的：深入了解 Spring 的面向方面编程（AOP）原理，掌握基本概念及配置，学习声明式事务处理和编程式事务处理。

学习要求：扎实掌握 AOP 的基本概念，重点学会使用 Spring 的声明式事务处理。

本章主要内容

本章详细介绍 Spring AOP 概念，其中重点讲解 Spring 的切入点、通知类型和 Advisor，还介绍使用 ProxyFactoryBean 创建 AOP。该章还重点讲解事务处理，其中包括声明式事务处理和编程式事务处理。

介绍完 IoC 之后，会介绍另外一个重要的概念：AOP（Aspect Oriented Programming），也就是面向方面编程的技术。AOP 基于 IoC 基础之上，是对 OOP 的有益补充。

AOP 将应用系统分为两部分：核心业务逻辑（Core Business Concerns）及横向的通用逻辑（Crosscutting Enterprise Concerns），也就是所谓的方面，例如所有大中型应用都要涉及的持久化（Persistentce）管理、事务（Transaction）管理、安全（Security）管理、日志（Logging）管理和调试（Debugging）管理等。

4.1 AOP 概念

下面是一些重要的 AOP 概念。

- 方面（Aspect）：一个关注点的模块化，这个关注点实现可能另外横切多个对象。事务管理是 J2EE 应用中一个很好的横切关注点例子。方面用 Spring 的 Advisor 或拦截器实现。

- 连接点(Joinpoint)：程序执行过程中明确的点，如方法的调用或特定的异常被抛出。
- 通知(Advice)：在特定的连接点，AOP框架执行的动作。各种类型的通知包括Around、Before和Throws通知。通知类型将在下面讨论。许多AOP框架(包括Spring)都是以拦截器做通知模型，维护一个"围绕"连接点的拦截器链。
- 切入点(Pointcut)：指定一个通知将被引发的一系列连接点的集合。AOP框架必须允许开发者指定切入点，例如使用正则表达式。
- 引入(Introduction)：添加方法或字段到被通知的类。Spring允许引入新的接口到任何被通知的对象。例如，可以使用一个引入使任何对象实现IsModified接口，简化缓存。
- 目标对象(Target Object)：包含连接点的对象，也被称作被通知或被代理对象。
- AOP代理(AOP Proxy)：AOP框架创建的对象，包含通知。在Spring中，AOP代理可以是JDK动态代理或者CGLIB代理。
- 编织(Weaving)：从组装方面创建一个被通知对象。这可以在编译时完成(例如，使用AspectJ编译器)，也可以在运行时完成。Spring和其他纯Java AOP框架一样，在运行时完成织入。

通知类型包括

- Around通知：包围一个连接点的通知，如方法调用。这是最强大的通知。Aroud通知在方法调用前后完成自定义的行为。它们负责选择继续执行连接点或通过返回它们自己的返回值或抛出异常短路执行。
- Before通知：在一个连接点之前执行的通知，但这个通知不能阻止连接点前的执行(除非它抛出一个异常)。
- Throws通知：在方法抛出异常时执行的通知。Spring提供强制类型的Throws通知，因此可以书写代码捕获感兴趣的异常(和它的子类)，不需要从Throwable或Exception强制类型转换。
- After Returning通知：在连接点正常完成后执行的通知。例如，一个方法正常返回，没有抛出异常。

其中Around通知是最通用的通知类型。大部分基于拦截的AOP框架(如Nanning和JBoss4)，只提供Around通知。

如同AspectJ，Spring提供所有类型的通知，推荐使用最合适的通知类型实现需要的行为。例如，如果只是需要用一个方法的返回值更新缓存，最好实现一个After Returning通知，而不是Around通知，虽然Around通知也能完成同样的事情。使用最合适的通知类型使编程模型变得简单，并能减少潜在错误。例如，由于不需要调用在Around通知中使用MethodInvocation的proceed()方法，因此就调用失败。

切入点的概念是AOP的关键，使AOP区别于其他使用拦截的技术。切入点使通知独立于OO的层次选定目标。例如，提供声明式事务管理的Around通知可以被应用到跨越多个对象的一组方法上。因此，切入点构成了AOP的结构要素。

下面为实现一个Spring AOP的例子。在这个例子中，我们将实现一个before advice，这意味着advice的代码在被调用的public()方法开始前被执行。以下是这个before advice

的实现代码。

```
package com.ascenttech.springaop.test;
import java.lang.reflect.Method;
import org.springframework.aop.MethodBeforeAdvice;
public class TestBeforeAdvice implements MethodBeforeAdvice{
public void before(Method m, Object[] args, Object target)
  throws Throwable{
    System.out.println("Hello world! (by "
      +this.getClass().getName()
      +")");
  }
}
```

接口 MethodBeforeAdvice 只有一个方法 before()需要实现，它定义了 advice 的实现。before()方法共有 3 个参数，它们提供了相当丰富的信息。参数 Method m 是 advice 开始后执行的方法。方法名称可以用作判断是否执行代码的条件。Object[] args 是传给被调用的 public()方法的参数数组。当需要记日志时，参数 args 和被执行方法的名称都是非常有用的信息。也可以改变传给 m 的参数，但要小心使用这个功能；编写最初主程序的程序员并不知道主程序可能和传入的参数发生冲突。Object target 是执行方法 m 对象的引用。

在下面的 BeanImpl 类中，每个 public()方法调用前都会执行 advice：

```
package com.ascenttech.springaop.test;
public class BeanImpl implements Bean {
  public void theMethod() {
    System.out.println(this.getClass().getName()
      +"."+new Exception().getStackTrace()[0].getMethodName()
      +"()"
      +" says HELLO!");
  }
}
```

类 BeanImpl 实现了下面的接口 Bean：

```
package com.ascenttech.springaop.test;
  public interface Bean {
  public void theMethod();
}
```

虽然不是必须使用接口，但面向接口而不是面向实现编程是良好的编程实践，Spring 也鼓励这样做。

pointcut 和 advice 通过配置文件实现，因此，接下来只需编写主方法的 Java 代码：

```
package com.ascenttech.springaop.test;
import org.springframework.context.ApplicationContext;
import org.springframework.context.support.FileSystemXmlApplicationContext;
public class Main {
```

```
    public static void main(String[] args) {
      //Read the configuration file
      ApplicationContextctx= new FileSystemXmlApplicationContext("springconfig.
      xml");
      //Instantiate an object
      Bean x= (Bean) ctx.getBean("bean");
      //Execute the public method of the bean (the test)
      x.theMethod();
    }
  }
```

从读入和处理配置文件开始，接下来马上创建它。这个配置文件将作为黏合程序不同部分的“胶水”。读入和处理配置文件后，会得到一个创建工厂 ctx。任何一个 Spring 管理的对象都必须通过这个工厂创建。对象通过工厂创建后便可正常使用。

仅用配置文件便可把程序的每一部分组装起来。

代码如下：

```
<?xml version="1.0" encoding="UTF-8"?>
<!DOCTYPE beans PUBLIC "-//SPRING//DTD BEAN//EN" "http://www.springframework.
org/dtd/spring-beans.dtd">
<beans>
  <!--CONFIG-->
  <bean id="bean" class="org.springframework.aop.framework.ProxyFactoryBean">
  <property name="proxyInterfaces">
    <value>com.ascenttech.springaop.test.Bean</value>
  </property>
  <property name="target">
    <ref local="beanTarget"/>
  </property>
    <property name="interceptorNames">
      <list>
        <value>theAdvisor</value>
      </list>
    </property>
  </bean>
<!--CLASS-->
  <bean id="beanTarget" class="com.ascenttech.springaop.test.BeanImpl"/>
  <!--ADVISOR-->
  <!--Note: An advisor assembles pointcut and advice-->
  <bean id="theAdvisor" class="org.springframework.aop.support.
                                                RegexpMethodPointcutAdvisor">
    <property name="advice">
      <ref local="theBeforeAdvice"/>
    </property>
    <property name="pattern">
      <value>com\.ascenttech\.springaop\.test\.Bean\.theMethod</value>
    </property>
```

```
    </bean>
    <!--ADVICE-->
    <bean id="theBeforeAdvice" class="com.ascenttech.springaop.test.
                                                    TestBeforeAdvice"/>
</beans>
```

4 个 bean 定义的次序并不重要。现在有了一个 advice,一个包含了正则表达式 pointcut 的 advisor,一个主程序类和一个配置好的接口,通过工厂 ctx,这个接口返回自己本身实现的一个引用。

BeanImpl 和 TestBeforeAdvice 都是直接配置。我们用一个唯一的 ID 创建一个 Bean 元素,并指定一个实现类,这就是全部的工作。

advisor 通过 Spring Framework 提供的一个 RegexpMethodPointcutAdvisor 类实现。我们用 advisor 的一个属性指定它所需的 advice-bean。第二个属性则用正则表达式定义了 pointcut,确保良好的性能和易读性。

最后配置的是 Bean,它可以通过一个工厂创建。Bean 的定义看起来比实际要复杂。Bean 是 ProxyFactoryBean 的一个实现,它是 Spring Framework 的一部分。这个 Bean 的行为通过以下 3 个属性定义。

(1) 属性 proxyInterfaces 定义了接口类。

(2) 属性 target 指向本地配置的一个 Bean,这个 Bean 返回一个接口的实现。

(3) 属性 interceptorNames 是唯一允许定义一个值列表的属性。这个列表包含所有需要在 beanTarget 上执行的 advisor。注意,advisor 列表的次序非常重要。

4.2 Spring 的切入点

让我们看看 Spring 如何处理切入点这个重要的概念。

1. 概念

Spring 的切入点模型能够使切入点独立于通知类型被重用。同样的切入点有可能接收不同的通知。

org.springframework.aop.Pointcut 接口是重要的接口,用来指定通知到特定的类和方法目标。完整的接口定义如下:

```
public interface Pointcut{

ClassFilter getClassFilter();

MethodMatcher getMethodMatcher();

}
```

将 Pointcut 接口分成两部分有利于重用类和方法的匹配部分,并且组合细粒度的操作(如和另一个方法匹配器执行一个"并"的操作)。

ClassFilter 接口用来将切入点限制到一个给定的目标类的集合。如果 matches()永远返回 true,所有的目标类都将被匹配。

```
public interface ClassFilter{

    boolean matches(Class clazz);
}
```

MethodMatcher 接口通常更加重要。完整的接口如下:

```
public interface MethodMatcher{

    boolean matches(Method m, Class targetClass);

    boolean isRuntime();

    boolean matches(Method m, Class targetClass, Object[] args);
}
```

matches(Method,Class)方法用来测试这个切入点是否匹配目标类的给定方法。这个测试可以在 AOP 代理创建的时候执行,避免在所有方法调用时都需要进行测试。如果 2 个参数的匹配方法对某个方法返回 true,并且 MethodMatcher 的 isRuntime()也返回 true,那么 3 个参数的匹配方法将在每次方法调用的时候被调用。这使切入点能够在目标通知被执行之前立即查看传递给方法调用的参数。

大部分 MethodMatcher 都是静态的,意味着 isRuntime()方法返回 false。这种情况下 3 个参数的匹配方法永远不会被调用。

2. 切入点的运算

Spring 支持的切入点的运算有并和交。

并表示只要任何一个切入点匹配的方法。

交表示两个切入点都要匹配的方法。

并通常比较有用。

切入点可以用 org.springframework.aop.support.Pointcuts 类的静态方法组合,或者使用同一个包中的 ComposablePointcut 类。

3. 实用切入点实现

Spring 提供了几个实用的切入点实现。一些可以直接使用,另一些需要子类化来实现应用相关的切入点。

1) 静态切入点

静态切入点只基于方法和目标类,而不考虑方法的参数。静态切入点足够满足大多数情况的使用。Spring 可以只在方法第一次被调用的时候计算静态切入点,不需要在每次方法调用的时候计算。

让我们看一下 Spring 提供的一些静态切入点的实现。

(1) 正则表达式切入点。

一个很显然的指定静态切入点的方法是正则表达式。除了 Spring 以外,其他的 AOP

框架也实现了这一点。org.springframework.aop.support.RegexpMethodPointcut 是一个通用的正则表达式切入点,它使用的是 Perl 5 的正则表达式的语法。

使用这个类可以定义一个模式的列表。如果任何一个匹配,那个切入点将被计算成 true。(所以结果相当于是这些切入点的并集。)

用法如下：

```
<bean id="settersAndAbsquatulatePointcut"
    class="org.springframework.aop.support.RegexpMethodPointcut">
    <property name="patterns">
        <list>
            <value>.*get.*</value>
            <value>.*absquatulate</value>
        </list>
    </property>
</bean>
```

RegexpMethodPointcut 的一个实用子类 RegexpMethodPointcutAdvisor 允许同时引用一个通知。(记住：通知可以是拦截器、Before 通知、Throws 通知等)这简化了 bean 的装配,因为一个 bean 可以同时当作切入点和通知,如下所示。

```
<bean id="settersAndAbsquatulateAdvisor"
    class="org.springframework.aop.support.RegexpMethodPointcutAdvisor">
    <property name="interceptor">
        <ref local="beanNameOfAopAllianceInterceptor"/>
    </property>
    <property name="patterns">
        <list>
            <value>.*get.*</value>
            <value>.*absquatulate</value>
        </list>
    </property>
</bean>
```

RegexpMethodPointcutAdvisor 可用于任何通知类型。

RegexpMethodPointcut 类需要 Jakarta ORO 正则表达式包。

(2) 属性驱动的切入点。

一类重要的静态切入点是元数据驱动的切入点,它使用元数据属性的值,典型地,使用源代码级元数据。

2) 动态切入点

动态切入点的演算代价比静态切入点高得多。它们不仅考虑静态信息,还要考虑方法的参数。这意味着它们必须在每次方法调用的时候都被计算;并且不能缓存结果,因为参数是变化的。

Spring 的控制流切入点概念上和 AspectJ 的 cflow 切入点一致,虽然没有其那么强大(当前没有办法指定一个切入点在另一个切入点后执行)。一个控制流切入点匹配当前的调用栈。例如,连接点被 com.mycompany.web 包或者 SomeCaller 类中的一个方法调用的时候,触发该切入点。控制流切入点的实现类是 org.springframework.aop.support.ControlFlowPointcut。

4. 切入点超类

Spring 提供了非常实用的切入点的超类,可帮助你实现自己的切入点。

因为静态切入点非常实用,你很可能子类化 StaticMethodMatcherPointcut,如下所示。这只需要实现一个抽象方法(虽然可以改写其他的方法来自定义行为)。

```
class TestStaticPointcut extends StaticMethodMatcherPointcut {

    public boolean matches(Method m, Class targetClass) {
        //return true if custom criteria match
    }
}
```

当然,也有动态切入点的超类。

5. 自定义切入点

因为 Spring 中的切入点是 Java 类,而不是语言特性(如 AspectJ),因此,无论静态,还是动态,可以定义自定义切入点。但是,没有直接支持用 AspectJ 语法书写的复杂的切入点表达式。不过,Spring 的自定义切入点也可以任意地复杂。

4.3 Spring 的通知类型

下面看 Spring AOP 是如何处理通知的。

1. 通知的生命周期

Spring 的通知可以跨越多个被通知对象共享,或者每个被通知对象有自己的通知。这分别对应 per-class 或 per-instance 通知。

per-class 通知使用最广泛。它适合于通用的通知,如事务 advisor。它们不依赖被代理的对象的状态,也不添加新的状态,仅作用于方法和方法的参数。

per-instance 通知适合于导入,以支持混入(mixin)。在这种情况下,通知添加状态到被代理的对象。

可以在同一个 AOP 代理中混合使用共享和 per-instance 通知。

2. Spring 中的通知类型

Spring 提供了几种现成的通知类型并可扩展提供任意的通知类型。下面看基本概念和标准的通知类型。

1) Interception Around 通知

Spring 中最基本的通知类型是 Interception Around Advice。

Spring 使用方法拦截器的 Around 通知是和 AOP 联盟接口兼容的。实现 Around 通知的类需要实现接口 MethodInterceptor:

```
public interface MethodInterceptor extends Interceptor {

    Object invoke(MethodInvocation invocation) throws Throwable;
}
```

invoke()方法的 MethodInvocation 参数暴露将被调用的方法、目标连接点、AOP 代理和传递给被调用方法的参数。invoke()方法应该返回调用的结果：连接点的返回值。

一个简单的 MethodInterceptor 实现看起来如下：

```
public class DebugInterceptor implements MethodInterceptor{

    public Object invoke(MethodInvocation invocation) throws Throwable{
        System.out.println("Before: invocation=["+invocation+"]");
        Object rval=invocation.proceed();
        System.out.println("Invocation returned");
        return rval;
    }
}
```

注意 MethodInvocation 的 proceed()方法的调用。这个调用会应用到目标连接点的拦截器链中的每一个拦截器。大部分拦截器会调用这个方法，并返回它的返回值。但是，一个 MethodInterceptor 和任何 Around 通知一样，可以返回不同的值或者抛出一个异常，而不调用 proceed()方法。

MethodInterceptor 提供了和其他 AOP 联盟的兼容实现的交互能力。这一节下面要讨论的其他的通知类型实现了 AOP 公共的概念，但是以 Spring 特定的方式。虽然使用特定通知类型有很多优点，但如果可能需要在其他的 AOP 框架中使用，请坚持使用 MethodInterceptor Around 通知类型。注意，目前切入点不能和其他框架交互操作，并且 AOP 联盟目前也没有定义切入点接口。

2) Before 通知

Before 通知是一种简单的通知类型。这个通知不需要一个 MethodInvocation 对象，因为它只在进入一个方法前被调用。

Before 通知的主要优点是不需要调用 proceed()方法，因此没有无意中忘掉继续执行拦截器链的可能性。

MethodBeforeAdvice 接口如下所示。(Spring 的 API 设计允许成员变量的 Before 通知，虽然一般的对象都可以应用成员变量拦截，但 Spring 有可能永远不会实现它。)

```
public interface MethodBeforeAdvice extends BeforeAdvice{

    void before(Method m, Object[] args, Object target) throws Throwable;
}
```

注意返回类型是 void。Before 通知可以在连接点执行前插入自定义的行为，但是不能改变返回值。如果一个 Before 通知抛出一个异常，这将中断拦截器链的进一步执行。这个异常将沿着拦截器链后退着向上传播。如果这个异常是 unchecked 的，或者出现在被调用的方法的签名中,它将被直接传递给客户代码;否则,它将被AOP代理包装到一个unchecked的异常里。

下面是Spring中一个Before通知的例子,这个例子计数所有正常返回的方法:

```
public class CountingBeforeAdvice implements MethodBeforeAdvice{
    private int count;
    public void before(Method m, Object[] args, Object target) throws Throwable{
        ++count;
    }

    public int getCount() {
        return count;
    }
}
```

Before通知可以用于任何类型的切入点。

3) Throws通知

如果连接点抛出异常,则Throws通知在连接点返回后被调用。Spring提供强类型的Throws通知。注意,这意味着org.springframework.aop.ThrowsAdvice接口不包含任何方法:它是一个标记接口,标识给定的对象实现了一个或多个强类型的Throws通知方法。这些方法的形式如下:

```
afterThrowing([Method], [args], [target], subclassOfThrowable)
```

只有最后一个参数是必需的。这样,从一个参数到四个参数,依赖于通知是否对方法和方法的参数感兴趣。下面是Throws通知的例子。

如果抛出RemoteException异常(包括子类),则这个通知会被调用。

```
public  class RemoteThrowsAdvice implements ThrowsAdvice{

    public void afterThrowing(RemoteException ex) throws Throwable{
        //Do something with remote exception
    }
}
```

如果抛出ServletException异常,下面的通知会被调用。和上面的通知不一样,它声明了四个参数,所以它可以访问被调用的方法,方法的参数和目标对象如下:

```
public static class ServletThrowsAdviceWithArguments implements ThrowsAdvice{

    public void afterThrowing(Method m, Object[] args, Object target,
        ServletException ex) {
        //Do something will all arguments
    }
}
```

最后一个例子演示了如何在一个类中使用两个方法同时处理RemoteException和

ServletException 异常。任意个数的 throws()方法可以被组合在一个类中。

```
public static class CombinedThrowsAdvice implements ThrowsAdvice{

    public void afterThrowing(RemoteException ex) throws Throwable{
        //Do something with remote exception
    }

    public void afterThrowing(Method m, Object[] args, Object target,
    ServletException ex) {
        //Do something will all arguments
    }
}
```

Throws 通知可用于任何类型的切入点。

4) After Returning 通知

Spring 中的 After Returning 通知必须实现 org.springframework.aop.AfterReturningAdvice 接口,代码如下所示。

```
public interface AfterReturningAdvice extends Advice {

    void afterReturning(Object returnValue, Method m, Object[] args, Object target)
            throws Throwable;
}
```

After Returning 通知可以访问返回值(不能改变)、被调用的方法、方法的参数和目标对象。

下面的 After Returning 通知统计所有成功的没有抛出异常的方法调用:

```
public class CountingAfterReturningAdvice implements AfterReturningAdvice {
    private int count;

    public void afterReturning (Object returnValue, Method m, Object[] args,
    Object target) throws Throwable{
        ++count;
    }

    public int getCount() {
        return count;
    }
}
```

这个方法不改变执行路径。如果它抛出一个异常,而这个异常不是返回值,则将被沿着拦截器链向上抛出。

After Returning 通知可用于任何类型的切入点。

5) Introduction 通知

Spring 将 Introduction 通知看作一种特殊类型的拦截通知。

Introduction 需要实现 IntroductionAdvisor 和 IntroductionInterceptor 接口。

```
public interface IntroductionInterceptor extends MethodInterceptor{

    boolean implementsInterface(Class intf);
}
```

继承自 AOP 联盟 MethodInterceptor 接口的 invoke()方法必须实现导入。也就是说，如果被调用的方法在导入的接口中，则导入拦截器负责处理这个方法调用，它不能调用 proceed()方法。

Introduction 通知不能用于任何切入点，因为它只能作用于类层次上，而不是方法。可以只用 InterceptionIntroductionAdvisor 实现导入通知，它有下面的方法：

```
public interface InterceptionIntroductionAdvisor extends InterceptionAdvisor{

    ClassFiltergetClassFilter();

    IntroductionInterceptor getIntroductionInterceptor();

    Class[] getInterfaces();
}
```

这里没有 MethodMatcher，因此也没有和导入通知关联的切入点，只有类过滤是合乎逻辑的。

getInterfaces()方法返回 advisor 导入的接口。

下面看一个来自 Spring 测试套件中的简单例子。假设想导入下面的接口到一个或者多个对象中：

```
public interface Lockable {
    void lock();
    void unlock();
    boolean locked();
}
```

在这个例子中，我们想将被通知对象的类型转换为 Lockable，不管它们的类型，并且调用 lock()和 unlock()方法。如果调用 lock()方法，我们希望所有的 setter()方法都抛出 LockedException 异常。这样就能添加一个方面，使得对象不可变，而它们不需要知道这一点：这是一个很好的 AOP 例子。

首先，需要一个做大量转化的 IntroductionInterceptor。这里，我们继承 org.springframework.aop.support.DelegatingIntroductionInterceptor 实用类。可以直接实现 IntroductionInterceptor 接口，但是大多数情况下 DelegatingIntroductionInterceptor 是最合适的。

DelegatingIntroductionInterceptor 的设计是将导入委托到真正实现导入接口的接口，隐藏完成这些工作的拦截器。委托可以使用构造方法参数设置到任何对象中；默认的委托就是自己(当无参数的构造方法被使用时)。这样，在下面的例子里，委托是 DelegatingIntroductionInterceptor 的子类 LockMixin。给定一个委托(默认是自身)的

DelegatingIntroductionInterceptor 实例，寻找被这个委托（而不是 IntroductionInterceptor）实现的所有接口，并支持它们中的任何一个导入。子类（如 LockMixin）也可能调用 suppressInterflace(Class intf)方法隐藏不应暴露的接口。然而，不管 IntroductionInterceptor 准备支持多少个接口，IntroductionAdvisor 都将控制哪个接口将被实际暴露。一个导入的接口将隐藏目标的同一个接口的所有实现。

这样，LockMixin 继承 DelegatingIntroductionInterceptor 并自己实现 Lockable。父类自动选择支持导入的 Lockable，所以我们不需要指定它。用这种方法可以导入任意数量的接口。

注意 locked 实例变量的使用。这有效地添加了额外的状态到目标对象。

```
public class LockMixin extends DelegatingIntroductionInterceptor
    implements Lockable {

    private boolean locked;

    public void lock() {
        this.locked=true;
    }

    public void unlock() {
        this.locked=false;
    }

    public boolean locked() {
        return this.locked;
    }

    public Object invoke(MethodInvocation invocation) throws Throwable{
        if (locked() && invocation.getMethod().getName().indexOf("set")==0)
            throw new LockedException();
        return super.invoke(invocation);
    }

}
```

通常不需要改写 invoke()方法：实现 DelegatingIntroductionInterceptor 就足够了，如果是导入的方法，DelegatingIntroductionInterceptor 实现会调用委托方法，否则继续沿着连接点处理。在现在的情况下，需要添加一个检查：在上锁状态下不能调用 setter()方法。

所需的导入 advisor 很简单，只需保存一个独立的 LockMixin 实例，并指定导入的接口，在这里就是 Lockable。一个稍微复杂的例子可能需要一个导入拦截器（可以定义成 prototype）的引用：在这种情况下，LockMixin 没有相关配置，所以我们简单地使用 new 创建它。

```
public class LockMixinAdvisor extends DefaultIntroductionAdvisor {
```

```
    public LockMixinAdvisor() {
        super(new LockMixin(), Lockable.class);
    }
}
```

可以非常简单地使用这个 advisor：它不需要任何配置(但是,有一点是必要的：不可能在没有 IntroductionAdvisor 的情况下使用 IntroductionInterceptor)。和导入一样,通常 advisor 必须是针对每个实例的,并且是有状态的。对于每个被通知的对象,就像 LockMixin 一样,我们都需要不同的 LockMixinAdvisor,advisor 就是被通知对象状态的一部分。

和其他 advisor 一样,可以使用 Advised.addAdvisor() 方法以编程的方式使用这种 advisor,或者在 XML 中配置(推荐这种方式)。下面讨论所有代理创建,包括"自动代理创建者",选择代理创建以正确地处理导入和有状态的混入。

4.4 Spring 中的 advisor

在 Spring 中,一个 advisor 就是一个 aspect 的完整的模块化表示。一般地,一个 advisor 包括通知和切入点。

撇开导入这种特殊情况,任何 advisor 可被用于任何通知。org.springframework.aop.support.DefaultPointcutAdvisor 是最通用的 advisor 类。例如,它可以和 MethodInterceptor、BeforeAdvice 或者 ThrowsAdvice 一起使用。

Spring 中可以将 advisor 和通知混合在一个 AOP 代理中。例如,可以在一个代理配置中使用一个对 Around 通知、Throws 通知和 Before 通知的拦截：Spring 将自动创建必要的拦截器链。

4.5 用 ProxyFactoryBean 创建 AOP 代理

如果你在为自己的业务对象使用 Spring 的 IoC 容器(如 ApplicationContext 或者 BeanFactory),应该会或者愿意会使用 Spring 的 AOP FactoryBean(记住,Factory Bean 引入了一个间接层,它能创建不同类型的对象)。

在 Spring 中创建 AOP Proxy 的基本途径是使用 org.springframework.aop.framework.ProxyFactoryBean,这样可以对 pointcut 和 advice 作精确控制。但是,如果不需要这种控制,那些简单的选择可能更适合你。

1. 基本概念

ProxyFactoryBean,和其他 Spring 的 FactoryBean 实现一样,引入一个间接的层次。如果定义一个名字为 foo 的 ProxyFactoryBean,引用 foo 的对象看到的不是 ProxyFactoryBean 实例本身,而是由实现 ProxyFactoryBean 的类的 getObject()方法所创建的对象。这个方法将创建一个包装了目标对象的 AOP 代理。

使用 ProxyFactoryBean 或者其他 IoC 可知的类创建 AOP 代理的最重要的优点之一是 IoC 可以管理通知和切入点。这是一个非常强大的功能,能够实现其他 AOP 框架很难实现的特定的方法。例如,一个通知本身可以引用应用对象(除了目标对象,它在任何AOP 框架中都可以引用应用对象),这完全得益于依赖注入提供的可插入性。

2. JavaBean 的属性

类似于 Spring 提供的绝大部分 FactoryBean 实现,ProxyFactoryBean 也是一个JavaBean,可以利用它的属性指定你将要代理的目标,或指定是否使用 CGLIB。

一些关键属性来自 org.springframework.aop.framework.ProxyConfig:它是所有AOP 代理工厂的父类。这些关键属性如下。

proxyTargetClass:如果应该代理目标类,而不是接口,则这个属性的值为 true。如果这个属性的值是 true,则需要使用 CGLIB。

optimize:是否使用强优化创建代理。不要使用这个设置,除非了解相关的 AOP 代理是如何处理优化的。目前这只对 CGLIB 代理有效,对 JDK 动态代理无效(默认)。

frozen:是否禁止通知的改变,一旦代理工厂已经配置完成,是否禁止通知的改变,默认是 false。

exposeProxy:当前代理是否要暴露在 ThreadLocal 中,以便它可以被目标对象访问。(它可以通过 MethodInvocation 得到,不需要 ThreadLocal。)如果一个目标需要获得它的代理并且 exposeProxy 的值是 ture,则可以使用 AopContext.currentProxy()方法。

aopProxyFactory:所使用的 AopProxyFactory 具体实现。这个参数提供了一条途径定义是使用动态代理、CGLIB,还是其他代理策略。默认实现将适当地选择动态代理或CGLIB。一般不需要使用这个属性;它的意图是允许 Spring 1.1 使用另外新的代理类型。

其他 ProxyFactoryBean 特定的属性如下。

proxyInterfaces:接口名称的字符串数组。如果没有提供,CGLIB 代理将被用于目标类。

interceptorNames:advisor、interceptor 或其他被应用的通知名称的字符串数组。顺序很重要的。这里的名称是当前工厂中 Bean 的名称,包括来自祖先工厂的 Bean 的名称。

singleton:工厂是否返回一个单独的对象,无论 getObject()被调用多少次。许多FactoryBean 的实现都提供这个方法,默认值是 true。如果想使用有状态的通知(例如,用于有状态的 mixin)将这个值设为 false,使用 prototype 通知。

3. 代理接口

下面看一个简单的 ProxyFactoryBean 的例子。这个例子涉及以下几个方面。

(1) 一个将被代理的目标 Bean,在这个例子里,这个 Bean 被定义为“personTarget”。

(2) 一个 advisor 和一个 interceptor 提供 advice。

(3) 一个 AOP 代理 Bean 定义,该 Bean 指定目标对象(这里是 personTarget bean)、代理接口和使用的 advice。

```
<bean id="personTarget" class="com.mycompany.PersonImpl">
    <property name="name"><value>Tony</value></property>
    <property name="age"><value>51</value></property>
</bean>
```

```
<bean id="myAdvisor" class="com.mycompany.MyAdvisor">
    <property name="someProperty"><value>Custom string property value</value>
   </property>
</bean>

<bean id=" debugInterceptor" class=" org. springframework. aop. interceptor.
NopInterceptor">
</bean>

<bean id="person"
    class="org.springframework.aop.framework.ProxyFactoryBean">
    <property name="proxyInterfaces"><value>com.mycompany. Person</value>
   </property>

    <property name="target"><ref local="personTarget"/></property>
    <property name="interceptorNames">
        <list>
            <value>myAdvisor</value>
            <value>debugInterceptor</value>
        </list>
    </property>
</bean>
```

注意：person bean 的 interceptorNames 属性提供一个 String 列表，列出的是该 ProxyFactoryBean 使用的，在当前 bean 工厂定义的 interceptor 或者 advisor 的名字(advisor、interceptor、before、after returning 和 throws advice 对象皆可)。advisor 在该列表中的次序很重要。

你也许会对该列表为什么不采用 bean 的引用存有疑问。原因在于如果 ProxyFactoryBean 的 singleton 属性被设置为 false，那么 bean 工厂必须能返回多个独立的代理实例。如果有任何一个 advisor 本身是 prototype 的，那么它就需要返回独立的实例，也就有必要从 bean 工厂获取 advisor 的不同实例，bean 的引用在这里显然是不够的。

上面定义的“person”bean 定义可以作为 Person 接口的实现使用，如下所示：

```
Person person=(Person) factory.getBean("person");
```

在同一个 IoC 的上下文中，其他的 bean 可以依赖于 Person 接口，就像依赖于一个普通的 Java 对象。

```
<bean id="personUser" class="com.mycompany.PersonUser">
    <property name="person"><ref local="person" /></property>
</bean>
```

在这个例子里，PersonUser 类暴露了一个类型为 Person 的属性。只要是在用到该属性的地方，AOP 代理都能透明地替代一个真实的 Person 实现。但是，这个类可能是一个动态代理类。也就是有可能把它的类型转换为一个 Advised 接口(该接口在下面的章节中论述)。

4. 代理类

如果需要代理的是类,而不是一个或多个接口,又该怎么办呢?

想象一下上面的例子,如果没有 Person 接口,我们需要通知一个叫 Person 的类,而且该类没有实现任何业务接口。在这种情况下,可以配置 Spring 使用 CGLIB 代理,而不是动态代理。只要在上面的 ProxyFactoryBean 定义中把它的 proxyTargetClass 属性改成 true 就行了。

只要愿意,即使在有接口的情况下,也可以强迫 Spring 使用 CGLIB 代理。

CGLIB 代理是通过在运行期产生目标类的子类进行工作的。Spring 可以配置这个生成的子类,代理原始目标类的方法调用。这个子类是用 Decorator 设计模式置入到 advice 中的。

CGLIB 代理对于用户来说应该是透明的。然而,还有以下一些因素需要考虑:

Final()方法不能被通知,因为不能被重写。

需要在 classpath 中包括 CGLIB 的二进制代码,而动态代理对任何 JDK 都是可用的。

CGLIB 和动态代理在性能上有微小的区别,对 Spring 1.0 来说,后者稍快。另外,以后可能会有变化。在这种情况下性能不是决定性因素。

接下来看一个实例。

(1) 新建 Java 项目 Spring_AOP_demo1。

(2) 导入 Spring 项目所依赖的 jar 包。

右击 Spring_ AOP_demo1 项目,从弹出的快捷菜单中选择 Configure Facets→Install Spring Facet。

(3) 建立 com.ascent.aop 目录。

(4) 在 src 下编写配置文件 applicationContext.xml,内容如下。

```
<?xml version="1.0"encoding="UTF-8"?>
<!DOCTYPE beans PUBLIC"-//SPRING//DTD BEAN//EN"
"http://www.springframework.org/dtd/spring-beans.dtd">
<beans>
  <!--CONFIG-->
  <bean id="bean" class="org.springframework.aop.framework.ProxyFactoryBean">
  <property name="proxyInterfaces">
    <value>com.ascent.aop.Bean</value>
  </property>
  <property name="target">
    <ref local="beanTarget"/>
  </property>
  <property name="interceptorNames">
    <list>
      <value>theAdvisor</value>
    </list>
  </property>
</bean>
<!--CLASS-->
```

```
<bean id="beanTarget"class="com.ascent.aop.BeanImpl"/>
<!--ADVISOR-->
<!--Note: An advisor assembles pointcut and advice-->
<bean id="theAdvisor"
class="org.springframework.aop.support.RegexpMethodPointcutAdvisor">
  <property name="advice">
    <ref local="theBeforeAdvice"/>
  </property>
  <property name="pattern">
    <value>com\.ascent\.aop\.Bean\.theMethod</value>
  </property>
  </bean>
  <!--ADVICE-->
  <bean id="theBeforeAdvice"class="com.ascent.aop.TestBeforeAdvice"/>
</beans>
```

(5) 在 com.ascent.aop 目录下分别编写 Bean.java、BeanImpl.java、TestBeforeAdvice.java 和 AOPTest.java,代码如下。

```
//Bean.java
package com.ascent.aop;
public interface Bean {
public void theMethod();
}

//BeanImpl.java
package com.ascent.aop;
public class BeanImpl implements Bean {
public void theMethod() {
  System.out.println(this.getClass().getName()
    +"."+new Exception().getStackTrace()[0].getMethodName()
    +"()"
    +" says this is the method!");
  }
}

//TestBeforeAdvice.java
package com.ascent.aop;
import java.lang.reflect.Method;
import org.springframework.aop.MethodBeforeAdvice;
public class TestBeforeAdvice implements MethodBeforeAdvice{
public void before(Method m, Object[] args, Object target)
throws Throwable{
  System.out.println("Hello world from Spring AOP! (by "
    +this.getClass().getName()
    +")");
```

```
    }
}

//AOPTest.java
package com.ascent.aop;
import org.springframework.context.ApplicationContext;
import org.springframework.context.support.FileSystemXmlApplicationContext;
public class AOPTest{
  public static void main(String[] args) {
    //Read the configuration file
    ApplicationContext ctx=new FileSystemXmlApplicationContext("/src/
        applicationContext.xml");
    //Instantiate bean object
    Bean x=(Bean) ctx.getBean("bean");
    //Execute the method in the bean
    x.theMethod();
  }
}
```

(6) 右击 AOPTest 类,从弹出的快捷菜单中选择 Run As→Java Application。

得到如下结果:

```
Hello world from Spring AOP! (by com.ascent.aop.TestBeforeAdvice)
com.ascent.aop.BeanImpl.theMethod() says this is the method!
```

下面再看一个实例。

(1) 新建 Java 项目 Spring_AOP_demo2。

(2) 导入 Spring 项目所依赖的 jar 包。

右击 Spring_ AOP_demo2 项目,从弹出的快捷菜单中选择 Configure Facets→Install Spring Facet。

(3) 建立 com.ascent.aop 和 com.ascent.aopimpl 目录。

(4) 在 src 目录下编写配置文件 applicationContext.xml,内容如下。

```
<?xml version="1.0" encoding="UTF-8"? >
<beans
    xmlns="http://www.springframework.org/schema/beans"
    xmlns:xsi="http://www.w3.org/2001/XMLSchema-instance"
    xmlns:p="http://www.springframework.org/schema/p"
    xsi:schemaLocation="http://www.springframework.org/schema/beans
    http://www.springframework.org/schema/beans/spring-beans-3.0.xsd
    http://www.springframework.org/schema/aop
    http://www.springframework.org/schema/aop/spring-aop-3.0.xsd"
    xmlns:aop="http://www.springframework.org/schema/aop"
    >
    <!--配置需要被 Spring 管理的 Bean-->
```

```
    <!--老师 -->
    <bean id="TeacherImpl" class="com.ascent.aopimpl.TeacherImpl"/>

    <!--学生 -->
    <bean id="Student" class="com.ascent.aopimpl.Student"></bean>

    <!--为接口类设置切点 -->
    <aop:config proxy-target-class="true">
<aop:aspect ref="Student">
    <!--之前 -->
        <aop:before pointcut="execution(*
                com.ascent.aopimpl.TeacherImpl.teach(..))" method="sit"/>
            <!--之前 -->
            <aop:before pointcut="execution(*
              com.ascent.aopimpl.TeacherImpl.teach(..))" method="greet"/>

            <!--之后 -->
            <aop:after-returning pointcut="execution(*
                com.ascent.aopimpl.TeacherImpl.teach(..))" method="ask"/>

            <!--之后 -->
            <aop:after-returning pointcut="execution(*
             com.ascent.aopimpl.TeacherImpl.teach(..))" method="dismiss"/>
        </aop:aspect>
    </aop:config>

</beans>
```

(5) 在 com.ascent.aop 目录下编写 Teacher.java,在 com.ascent.aopimpl 目录下编写 TeacherImpl.java、Student.java 和 Test.java,代码如下。

```
//Teacher.java
package com.ascent.aop;

public interface Teacher {
    public void teach();
}

//TeacherImpl.java
package com.ascent.aopimpl;

import com.ascent.aop.Teacher;

public class TeacherImpl implements Teacher{

    @Override
```

```
        public void teach() {
                System.out.println("教师开始教课");
        }
    }

    //Student.java
    package com.ascent.aopimpl;

    public class Student {

        public Student() {

        }
        public void sit()
        {
            System.out.println("学生来到教室");
        }
        public void greet()
        {
            System.out.println("向老师问好");
        }
        public void ask()
        {
            System.out.println("上课提问");
        }
        public void dismiss()
        {
            System.out.println("下课");
        }
    }

    //Test.java
    package com.ascent.aopimpl;

    import org.springframework.context.ApplicationContext;
    import org.springframework.context.support.ClassPathXmlApplicationContext;

    public class Test {

        public static void main(String[] args)
        {
            ApplicationContext apc=new ClassPathXmlApplicationContext
                                                ("applicationContext.xml");
            TeacherImpl teacher=(TeacherImpl) apc.getBean("TeacherImpl");
            teacher.teach();
```

```
    }
}
```

(6) 右击 Test 类,从弹出的快捷菜单中选择 Run As→Java Application。

得到如下结果:
学生来到教室
向老师问好
教师开始教课
上课提问
下课

4.6 事务处理

Spring 框架引人注目的重要因素之一是它全面的事务支持。Spring 框架提供了一致的事务管理抽象,这带来了以下好处:

- 为复杂的事务 API 提供了一致的编程模型,如 JTA、JDBC、MyBatis、JPA 和 JDO。
- 支持声明式事务管理。
- 提供比大多数复杂的事务 API(诸如 JTA)更简单的、更易于使用的编程式事务管理 API。
- 非常好地整合 Spring 的各种数据访问抽象。

4.6.1 声明式事务处理

首先介绍声明式事务处理(Declarative Transactions)。声明式事务处理是由 Spring AOP 实现的。大多数 Spring 用户选择声明式事务管理。这是最少影响应用代码的选择,因而这和非侵入性的轻量级容器的观念一致。如果你的应用中存在大量事务操作,那么声明式事务管理通常是首选方案。它将事务管理与业务逻辑分离,而且在 Spring 中配置也不难。

从考虑 EJB CMT 和 Spring 声明式事务管理的相似以及不同之处出发是很有益的。它们的基本方法相似:都可以指定事务管理到单独的方法;如果需要,可以在事务上下文调用 setRollbackOnly()方法。

不同之处在于:

- 不像 EJB CMT 绑定在 JTA 上,Spring 声明式事务管理可以在任何环境下使用。只更改配置文件,就可以和 JDBC、JDO、MyBatis 或其他的事务机制一起工作。
- Spring 的声明式事务管理可以应用到任何类(以及那个类的实例)上,不仅是像 EJB 那样的特殊类。
- Spring 提供了声明式的回滚规则:EJB 没有对应的特性,我们将在下面讨论。回滚可以声明式地控制,不仅仅是编程式地。
- Spring 允许通过 AOP 定制事务行为。例如,如果需要,可以在事务回滚中插入定制的行为。也可以增加任意的通知,就像事务通知一样。使用 EJB CMT,除了使用 setRollbackOnly(),没有办法能够影响容器的事务管理。

Spring 不提供高端应用服务器提供的跨越远程调用的事务上下文传播。如果需要这些特性,推荐使用 EJB。然而,不要轻易使用这些特性。通常我们并不希望事务跨越远程调用。

回滚规则的概念比较重要:它使我们能够指定什么样的异常(和 throwable)将导致自动回滚。在配置文件中声明式地指定,无须在 Java 代码中。同时,仍旧可以通过调用 TransactionStatus 的 setRollbackOnly()方法编程式地回滚当前事务。通常,定义一条规则,声明 MyApplicationException 必须总是导致事务回滚。这种方式带来了显著的好处,它使你的业务对象不必依赖于事务设施。典型的例子是不必在代码中导入 Spring API、事务等。

对 EJB 来说,默认的行为是 EJB 容器在遇到系统异常(通常指运行时异常)时自动回滚当前事务。EJB CMT 遇到应用异常(例如,除 java.rmi.RemoteException 外别的 checked exception)时并不会自动回滚。默认式 Spring 处理声明式事务管理的规则遵守 EJB 习惯(只在遇到 unchecked exceptions 时自动回滚),但通常定制这条规则会更有用。

Spring 的事务管理是通过 AOP 代理实现的。其中的事务通知由元数据(目前基于 XML 或注解)驱动。代理对象与事务元数据结合产生了一个 AOP 代理,它使用一个 PlatformTransactionManager 实现配合 TransactionInterceptor,在方法调用前后实施事务。从概念上来说,在事务代理上调用方法的工作过程看起来如图 4-1 所示。

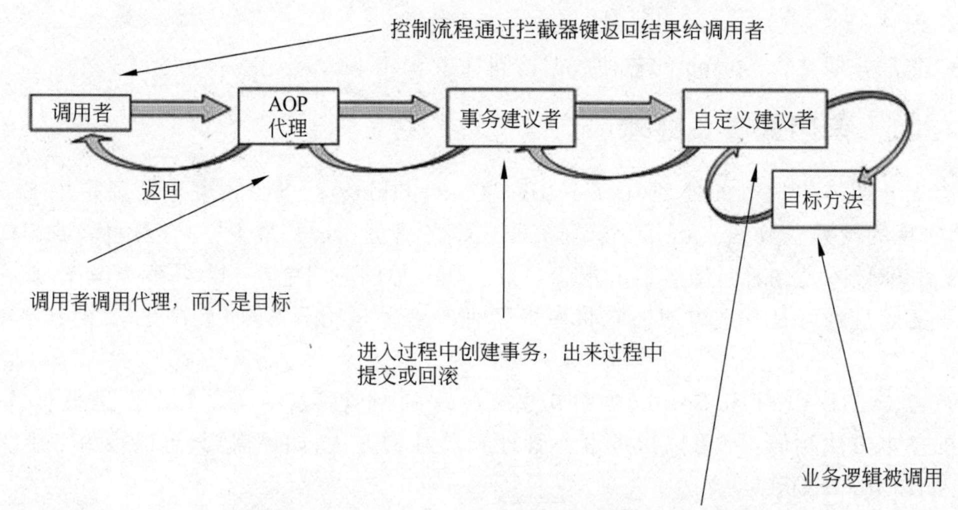

图 4-1 Spring 的事务代理模型

类似于 EJB 的容器管理事务(Container Managed Transaction),可以在配置文件中声明对事务的支持,可以精确到单个方法的级别,这通常通过 TransactionProxyFactoryBean 设置 Spring 事务代理。我们需要一个目标对象包装在事务代理中。这个目标对象一般是一个普通 Java 对象的 bean。当定义 TransactionProxyFactoryBean 时,必须提供一个相关的 PlatformTransactionManager 的引用和事务属性。事务属性含有上面描述的事务定义。

例如,可以使用以下配置:

```
<bean id="orderService" class="org.springframework.transaction.
```

```
    interceptor.TransactionProxyFactoryBean">

  <property name="transactionManager">
    <ref local="myTransactionManager"/>
  </property>

  <property name="target"><ref local="orderTarget"/></property>

  <property name="transactionAttributes">
    <props>
  <prop key="find * ">
   PROPAGATION_REQUIRED,readOnly,-OrderException
    </prop>
    <prop key="save * ">
   PROPAGATION_REQUIRED,-OrderMinimumAmountException
    </prop>
    <prop key="update * ">
   PROPAGATION_REQUIRED,-OrderException
    </prop>
  </props>
  </property>
</bean>
```

通过以上配置声明,Spring会自动帮助我们处理事务。也就是说,对于orderTarget类中的所有以find、save和update开头的方法,自动增加事务管理服务。

这里的 transaction attributes 属性定义在 org. springframework. transaction. interceptor.NameMatchTransactionAttributeSource中的属性格式设置。这个包括通配符的方法名称映射是很直观的。注意,save * 的映射的值包括回滚规则。添加的-OrderMinimumAmountException 指定如果方法抛出 OrderMinimumAmountException或它的子类,事务将会自动回滚。可以用逗号分隔定义多个回滚规则。一前缀强制回滚,+前缀指定提交(这允许即使抛出unchecked异常时,也可以提交事务。当然,你自己要明白自己在做什么)。

TransactionProxyFactoryBean允许通过preInterceptors和postInterceptors属性设置“前”或“后”通知提供额外的拦截行为。可以设置任意数量的“前”和“后”通知,它们的类型可以是Advisor(可以包含一个切入点)、MethodInterceptor或被当前Spring配置支持的通知类型(例如 ThrowAdvice、AfterReturningtAdvice 或 BeforeAdvice,这些都是默认支持的)。这些通知必须支持实例共享模式。如果需要高级AOP特性使用事务,最好使用通用的org.springframework.aop.framework.ProxyFactoryBean,而不是TransactionProxyFactoryBean实用代理创建者。

也可以设置自动代理:配置AOP框架,不需要单独的代理定义类就可以生成类的代理。

Spring 2.0及以后的版本中声明式事务的配置与之前的版本有相当大的不同,主要差异在于不再需要配置TransactionProxyFactoryBean。当然,Spring 2.0之前的旧版本风格的配置仍然是有效的。

4.6.2 编程式事务处理

当只有很少的事务操作时,编程式事务处理通常比较合适。例如,如果有一个 Web 应用,其中只有特定的更新操作有事务要求,你可能不愿使用 Spring 或其他技术设置事务代理。

Spring 提供了以下两种方式的编程式事务处理:

- 使用 TransactionTemplate。
- 直接使用一个 PlatformTransactionManager 实现。

推荐采用第一种方法(即使用 TransactionTemplate)。

1. 使用 TransactionTemplate

TransactionTemplate 采用与 Spring 中别的模板同样的方法,如 JdbcTemplate 和 MyBatisTemplate。它使用回调机制,将应用代码从样板式的资源获取和释放代码中解放出来,不再有大量的 try/catch/finally/try/catch 代码块。同样,和别的模板类一样,TransactionTemplate 类的实例是线程安全的。

必须在事务上下文中执行的应用代码看起来像如下的代码:(注意,使用 TransactionCallback 可以有返回值)

```
Object result=tt.execute(new TransactionCallback() {
    public Object doInTransaction(TransactionStatus status) {
        updateOperation1();
        return resultOfUpdateOperation2();
    }
});
```

如果不需要返回值,更方便的方式是创建一个 TransactionCallbackWithoutResult 的匿名类,代码如下:

```
tt.execute(new TransactionCallbackWithoutResult() {
    protected void doInTransactionWithoutResult(TransactionStatus status) {
        updateOperation1();
        updateOperation2();
    }
});
```

回调方法内的代码可以通过调用 TransactionStatus 对象的 setRollbackOnly()方法回滚事务。

想使用 TransactionTemplate 的应用类,必须能访问一个 PlatformTransactionManager(典型情况下通过依赖注入提供)。这样的类很容易做单元测试,只需要引入一个 PlatformTransactionManager 的伪类或桩类。这里没有 JNDI 查找,没有静态轨迹,它是一个如此简单的接口。像往常一样,使用 Spring 给你的单元测试带来极大的简化。

2. 使用 PlatformTransactionManager

也可以直接使用 org.springframework.transaction.PlatformTransactionManager 的实现管理事务。只需通过 Bean 引用简单地传入一个 PlatformTransactionManager 实现,然

后使用 TransactionDefinition 和 TransactionStatus 对象，就可以启动一个事务，提交或回滚。

```
DefaultTransactionDefinition def=new DefaultTransactionDefinition();
def.setPropagationBehavior(TransactionDefinition.PROPAGATION_REQUIRED);

TransactionStatus status=txManager.getTransaction(def);
try {
    //execute your business logic here
}
catch (MyException ex) {
    txManager.rollback(status);
    throw ex;
}
txManager.commit(status);
```

4.7 项目案例

4.7.1 学习目标

本章案例介绍了 Spring AOP 概念，包括 Spring 的切入点、通知类型和 Advisor，以及如何使用声明式事务处理。

4.7.2 案例描述

本章案例仍为一般用户浏览新闻信息的头版头条新闻和综合新闻数据。在第 3 章中，NewsDAO 使用了 DefaultTransactionDefinition、DataSourceTransactionManager、TransactionStatus 3 个对象，在 NewsDAO 的 findHeaderNews 和 findIndexNews 中分别用来处理数据库查询事务，这种方法可以用于任何数据库 SQL 操作，包括增、删、改，如增加新闻、编辑新闻、添加权限等。本章案例将使用 Spring 提供的事务处理标签和切点表达式完成 AOP 配置，简化类似 NewsDAO 的数据库 SQL 开发。

4.7.3 案例要点

第 3 章的注册功能使用 Spring IoC 管理 IAuthorizationService、NewsDAO 以及 JdbcTemplate 模拟类的依赖关系，将 JdbcTemplate 注入给 NewsDAO，再将 NewsDAO 注入给 IAuthorizationService 接口实现类 IAuthorizationServiceImpl。本案例在此基础上给 NewsDAO 使用 AOP 事务处理功能，主要使用 Spring 的声明式事务处理完成事务操作。

4.7.4 案例实施

1. 事务代理需要代理对象具有接口，为 NewsDAO 增加 INewsDAO 接口

```
INewsDAO.java
package com.ascent.jdbc.dao;
```

```
import java.util.List;
import com.ascent.jdbc.rowmapper.NewsRowMapper;
import com.ascent.po.News;
public interface INewsDAO{
    /*
     * @paramsql 执行查询头版头条新闻的 SQL 语句
     * @paramrowMapper 告知 JdbcTemplate 用什么 RowMapper 封装查询结果
     * @return News 返回 News 新闻对象
     */
    News findHeaderNews(String sql, NewsRowMapperrowMapper);

    /**
     *
     * @paramsql 执行查询综合新闻的 SQL 语句
     * @return 返回所有符合查询条件的 News 新闻集合 List 对象
     */
    List<News> findIndexNews(String sql);
}
```

2. 去掉 NewsDAO 中的事务处理代码

```
NewsDAO.java
package com.ascent.jdbc.dao;

import java.util.List;

import org.springframework.jdbc.core.support.JdbcDaoSupport;

import com.ascent.jdbc.rowmapper.NewsRowMapper;
import com.ascent.po.News;
/**
 * 该类使用 Spring 提供的 JdbcTempate 完成 SQL 操作
 * @author gary
 *
 */
public class NewsDAO extends JdbcDaoSupport implements INewsDAO{

    /*
     * @paramsql 执行查询头版头条新闻的 SQL 语句
     * @paramrowMapper 告知 JdbcTemplate 用什么 RowMapper 封装查询结果
     * @return News 返回 News 新闻对象
     */

    @Override
    public News findHeaderNews(String sql, NewsRowMapperrowMapper){
```

```
    return this.getJdbcTemplate().queryForObject(sql, rowMapper);

    }

    @Override
    public List<News>findIndexNews(String sql) {

        return this.getJdbcTemplate().query(sql,new NewsRowMapper());

    }

}
```

3. 修改applicationContext.xml,追加以下内容

```
<!--配置事务处理器 -->
    <bean id="txManager" class="org.springframework.jdbc.datasource.
        DataSourceTransactionManager">
        <property name="dataSource" ref="dataSource"/>
    </bean>
<!--事务Advice配置-->
    <tx:advice id="txAdvice" transaction-manager="txManager">
        <!--事务内容属性定义什么样的方法使用什么策略 -->
        <tx:attributes>
            <!--所有用find开头的方法都是只读的,并且事务隔离特性使用避免脏读策略-->
                <tx:method name="find*" read-only="true" isolation="READ_
                COMMITTED"/>
    <!--所有save开头的方法都是具有事务的,事务隔离特性使用避免脏读策略 -->
                    <!--rollback-for是指该方法抛出SQLException异常时回滚 -->
            <tx:method name="save*" isolation="READ_COMMITTED"
                                    rollback-for="java.sql.SQLException" />
        <!--其他方法使用默认的事务配置-->
        <tx:method name="*"/>
    </tx:attributes>
</tx:advice>

<!--使得上面的事务配置对NewsDAO接口的所有操作有效 -->
<aop:config>
    <!--定义切面为com.ascent.dao.INewsDAO,接口下的*表示所有方法-->
    <aop:pointcut id="newsDaoPointcut"
    expression="execution(* com.ascent.jdbc.dao.INewsDAO.*(..))"/>
    <!--当上述切面方法被调用时,使用上述事务处理建议advice -->
    <aop:advisor advice-ref="txAdvice" pointcut-ref="newsDaoPointcut"/>
</aop:config>
```

4. 运行 IndexTestCase.java,结果如图 4-2 所示

```
Problems  Tasks  Console  Servers  Workspace Migration
<terminated> IndexTestCase [Java Application] D:\Tools\myeclipse2017ci10\binary\com.sun.java.jdk8.win32.x86_64_1.8.0.v112\bin\javaw.exe
log4j:WARN No appenders could be found for logger (org.springframework.core.env.StandardEnvironment).
log4j:WARN Please initialize the log4j system properly.
log4j:WARN See http://logging.apache.org/log4j/1.2/faq.html#noconfig for more info.
Wed Apr 08 14:07:44 CST 2020 WARN: Establishing SSL connection without server's identity verification is not recom
Wed Apr 08 14:07:44 CST 2020 WARN: Establishing SSL connection without server's identity verification is not recom
[News [id=21, title=JavaScript, author=测试1, deptid=0, content=Javascript是一种由Netscape的LiveScript发展而来的脚本
    JavaScript 的正式名称是 "ECMAScript"。这个标准由 ECMA 组织发展和维护。ECMA-262 是正式的 JavaScript 标准。这个标准基
    在 1998 年，该标准成为了国际 ISO 标准 (ISO/IEC 16262)。这个标准仍然处于发展之中。, type=2, checkopinion=, checkopir
[News [id=22, title=Ajax, author=测试1, deptid=0, content=Ajax（Asynchronous JavaScript + XML）的定义
    基于web标准（standards-based presentation）XHTML+CSS的表示；
    使用 DOM（Document Object Model）进行动态显示及交互；
    使用 XML 和 XSLT 进行数据交换及相关操作；
    使用 XMLHttpRequest 进行异步数据查询、检索；
    使用 JavaScript 将所有的东西绑定在一起。英文参见Ajax的提出者Jesse James Garrett的原文,原文题目(Ajax: A New Approac
    类似于DHTML或LAMP，AJAX不是指一种单一的技术，而是有机地利用了一系列相关的技术。事实上，一些基于AJAX的“派生/合成”式（
    AJAX的应用使用支持以上技术的web浏览器作为运行平台。这些浏览器目前包括：Mozilla、Firefox、Internet Explorer、Opera、
    2、Java Server Faces (JSF)
    JSF全称Java Server Faces ，是一种用于构建Java Web 应用程序的标准框架(是Java Community Process 规定的JSR-127标准)。
```

图 4-2　运行结果

运行成功表示事务配置完成。

4.7.5　特别提示

该案例为模拟案例,上述配置针对的是 INewsDAO 接口下的所有方法,也可以将 INewsDAO 替换为 *,表示某个包下所有接口的方法。本例依然模拟前端,都是采用 Java 测试类实现,Spring 的 AOP 代理采用面向接口产生代理,所以 DAO 类要有其对应的接口,如 INewsDAO、IUsrDAO、IDepartmentDAO 等。

4.7.6　拓展与提高

模拟该功能案例,开发并配置查询更多头版头条新闻及综合新闻功能的 DAO 接口的事务处理。

习题

1. Spring AOP 的基本概念有哪些?
2. Spring 的通知类型有哪些?
3. 如何使用 ProxyFactoryBean 创建 AOP 代理?
4. 如何使用 Spring AOP 代理类,而不是代理接口?
5. 如何使用 Spring 的声明式事务处理?

第5章 Spring MVC基础

学习目的与学习要求

学习目的：了解Spring MVC框架，掌握Spring MVC框架的工作流程及组件，熟悉Spring MVC原理，包括核心控制器DispatcherServlet、Controller控制器、ModelAndView和视图解析ViewResolver等。

学习要求：熟练搭建和开发基于Spring MVC框架的应用系统，顺利完成本章案例的开发，并能够开发项目的其他功能。

本章主要内容

本章主要内容包括MVC模式的概述、MVC与Spring MVC的映射、Spring MVC框架的工作流程和组件、Spring MVC原理(包括核心控制器DispatcherServlet、Controller控制器、ModelAndView和视图解析ViewResolver等)以及Spring MVC开发步骤。

Spring MVC框架是MVC设计模式的一个具体实现，所以先介绍MVC模式。

5.1 MVC模式概述

计算机软件工程领域常常提到设计模式(Design Pattern)。那么，什么是模式(Pattern)呢？一般来说，模式是指一种从一个一再出现的问题背景中抽象出的解决问题的固定方案，而这个问题背景不应该是绝对的，或者说是不固定的。很多时候看来不相关的问题，会有相同的问题背景，从而需要应用相同的模式解决。

模式的概念开始时出现在城市建筑领域中。后来，这个概念逐渐被计算机科学所采纳，并在一本广为接受的经典书籍的推动下流行起来。这本书就是 *Design Patterns*: *Elements of Reusable Object-Oriented Software*(设计模式:可复用面向对象软件元素),由 4 位软件大师合写而成(很多时候直接用 GoF 指这 4 位作者,GoF 的意思是 Gangs of Four,即四人帮)。

设计模式指的是在软件的建模和设计过程中运用到的模式。设计模式中的很多种方法其实很早就出现了,并且应用得也比较多。但是,直到 GoF 的书出来之前,并没有一种统一的认识。或者说,那时对模式并没有形成一个概念,这些方法还仅处在经验阶段,并没有被系统地整理,形成一种理论。

每个设计模式都系统地命名、解释和评价了面向对象系统中一个重要的和重复出现的设计。这样,只要清楚这些设计模式,就可以完全或者说很大程度上吸收那些蕴含在模式中的宝贵经验,对面向对象的系统能够有更完善的了解。更重要的是,这些模式都可以直接用来指导面向对象系统中至关重要的对象建模问题。如果有相同的问题背景,直接套用这些模式就可以了,这可以省去很多工作量。

MVC 模式是一种很常见的设计模式。所谓的 MVC 模式,即模型-视图-控制器(Model-View-Controller)模式。MVC 架构图如图 5-1 所示。

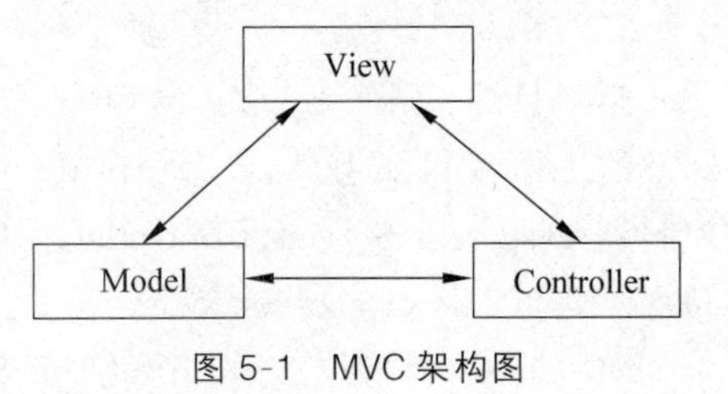

图 5-1　MVC 架构图

1. Model 端

在 MVC 中,模型是执行某些任务的代码,而这部分代码并没有任何逻辑决定用户端的表示方法。Model 只有纯粹的功能性接口,也就是一系列的公共方法,通过这些公共方法,可以取得模型端的所有功能。

2. View 端

在 MVC 模式里,一个 Model 可以有几个 View 端,而实际上多个 View 端是使用 MVC 的原始动机。使用 MVC 模式可以允许多于一个的 View 端存在,并可以在需要的时候动态注册需要的 View。

3. Controller 端

MVC 模式的视图端是与 MVC 的控制器结合使用的。当用户端与相应的视图发生交互时,用户可以通过视窗更新模型的状态,而这种更新是通过控制器端进行的。控制器端通过调用模型端的方法更改其状态值。与此同时,控制器端会通知所有注册了的视图刷新用户界面。

那么,使用 MVC 模式有哪些优点呢?MVC 通过以下 3 种方式消除与用户接口和面向对象的设计有关的绝大部分困难:

(1) 控制器通过一个状态机跟踪和处理面向操作的用户事件。这允许控制器在必要时创建和破坏来自模型的对象,并且将面向操作的拓扑结构与面向对象的设计隔离开。这个隔离有助于防止面向对象的设计走向歧途。

(2) MVC 将用户接口与面向对象的模型分开。这允许同样的模型不用修改,就可使用许多不同的界面显示方式。除此之外,如果模型更新由控制器完成,那么界面就可以跨应用再使用。

(3) MVC 允许应用的用户接口进行大的变化而不影响模型。每个用户接口的变化将只需要对控制器进行修改,但是控制器包含很少的实际行为,它是很容易修改的。

面向对象的设计人员在将一个可视化接口添加到一个面向对象的设计中时必须非常小心,因为可视化接口的面向操作的拓扑结构可以大大增加设计的复杂性。

MVC 设计允许一个开发者将一个好的面向对象的设计与用户接口隔离开,允许在同样的模型中容易地使用多个接口,并且允许在实现阶段对接口做大的修改,而不需要对相应的模型进行修改。

5.2 Spring MVC 概述

Spring MVC 属于 Spring Framework 的后续产品,已经融合在 Spring Web Flow 里。Spring MVC 分离了控制器、模型对象、分派器以及处理程序对象的角色,这让它们更容易进行定制。

Spring 的模型-视图-控制器(MVC)框架核心是围绕一个 DispatcherServlet 设计的,这个 Servlet 会将来自客户端的 HTTP 请求分发给各个用户自定义开发的处理器,同时支持可配置的处理器映射关系、页面视图的渲染、消息本地化和国际化,以及时区设置与页面风格主题渲染等。

处理器被称为 Controller,是在开发应用中注解了@Controller 和@RequestMapping 的 Java 类和方法。由于 Spring MVC 作为 Spring 的 Web MVC 开发组件内置在 Spring 中,所以 Spring 为 Spring MVC 的处理器方法提供了极其多样灵活的配置。Spring 3.0 以后提供了@Controller 注解机制、@PathVariable 注解,以及一些其他的特性,使得开发更加灵活、简洁,同时也可以使用 Spring MVC 进行 RESTful Web 站点和应用的开发。

在 Spring Web MVC 中,可以使用任何 Java 对象作为执行请求处理的命令对象或表单数据处理的返回对象,而无须实现与框架相关的接口或基类。Spring 的数据绑定非常灵活:例如,它会把数据类型不匹配当成可由应用自行处理的运行时验证错误,而非系统错误。之前你可能会为了避免非法的类型转换,在表单对象中使用字符串存储数据,但无类型的字符串无法描述业务数据的真正含义,并且还需要把它们转换成对应的业务对象类型。有了 Spring 的验证机制,再也不必这么做了,直接将业务对象绑定到表单对象上通常是更好的选择。

Spring 的视图解析设计得也异常灵活。控制器一般负责准备一个 Map 模型、填充数据、返回一个合适的视图名等,同时它也可以直接将数据写到响应流中。视图名的解析高度灵活,支持多种配置,包括通过文件扩展名、Accept 内容头、Bean、配置文件等的配置,甚至还可以自己实现一个视图解析器 ViewResolver。模型(Model)其实是一个 Map 类型的接口,彻底地把数据从视图技术中抽象分离了出来。可以与基于模板的渲染技术直接整合,如 JSP、Velocity 和 Freemarker 等,也可以直接生成 XML、JSON、Atom 以及其他多种类型的内容。Map 模型会简单地被转换成合适的格式,如 JSP 的请求属性(attribute)、一个 Velocity 模板的模型等。

5.3 MVC 组件和流程

Spring MVC 体系结构实现了 Model-View-Controller 设计模式的概念，它将这些概念映射到 Web 应用程序的组件和概念中。

Spring MVC 对于 URL 请求的处理框架如图 5-2 所示。

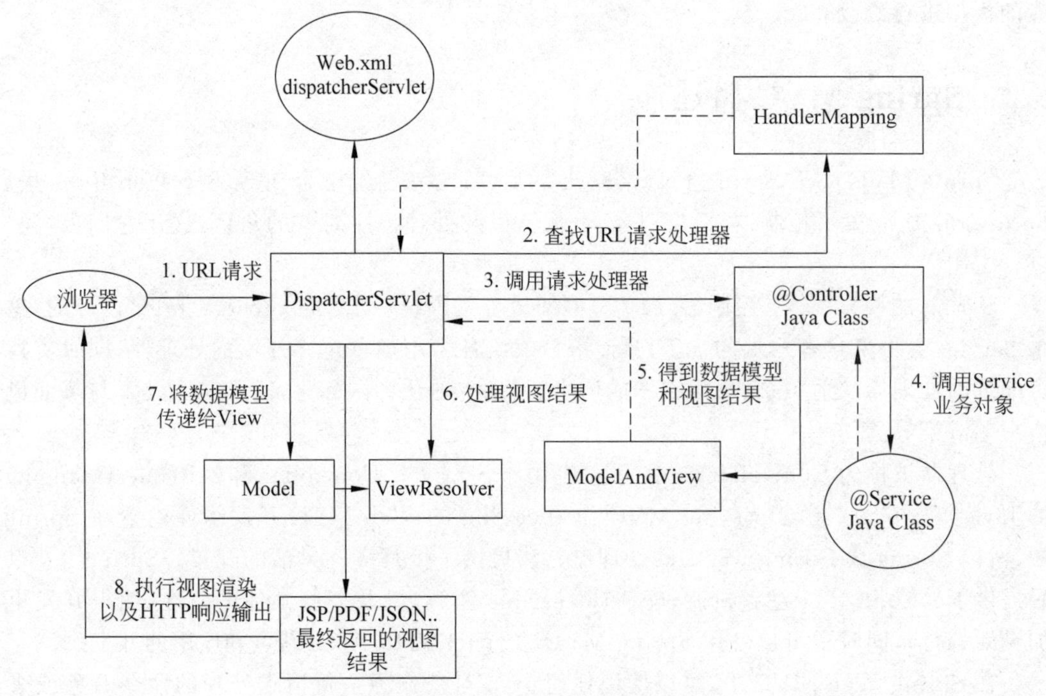

图 5-2 Spring MVC 的体系概图

上述的架构图中包括了 Spring MVC 中的核心功能组件的定义。

(1) DispatcherServlet：Spring MVC 提供的前端控制器，所有的请求都经过它识别，以及负责统一请求分发和响应处理。如 login.action 这样的 URL 请求信息，首先需要 DispatcherServlet 识别请求，然后根据配置信息在 HandlerMapping 中查找处理该 URL 的具体的 Java 类和相应的处理方法。

(2) HandlerMapping：保存了 URL 与 Controller Java 类和方法的映射关系，DispatcherServlet 需要借助该对象定位具体的 Java 对象和方法，以便处理 URL 请求。

(3) Controller：被标记和实现为 Controller 的类通过内部定义的 Java 方法，为并发发出同样 URL 请求的用户处理每个请求，因此实现 Controller 接口时，须保证线程安全并且可重用。Controller 类将调用对应 URL 的 Java 方法处理用户请求，并由开发人员决定是否使用 ModelAndView。一旦 Controller 中的 Java 方法处理完用户请求，则由开发人员决定使用怎样的方式返回 ModelAndView 对象给 DispatcherServlet 前端控制器。

(4) ModelAndView：包含了模型(Model)和视图(View)。模型是处理之后需要展示或者传递到 View 中的数据，如用户信息、登录消息等。View 负责展示动态数据的各种页面技术，如 JSP、PDF、XML、JSON 格式(JavaScript 中的数据对象)等。

(5) ViewResolver：Spring 提供的视图解析器(ViewResolver)在 Web 应用中查找 View(数据展示视图)对象,从而将相应模型中的数据与视图进行渲染(解析处理),最后将渲染后的结果返回给客户端。

从框架设计整体来说,DispatcherServlet 是整个 Web 应用的控制器,需要在 web.xml 中配置；Controller 是针对具体某个 URL 的单个 Http 请求处理过程中的控制器,是需要开发与配置的 Java 类;而 ModelAndView 是包含了处理 Http 请求过程中的需要返回客户端的数据模型(Model)和在客户端需要展现数据方式的视图(View)两部分内容。

这些对象在 Spring MVC 框架中相互配合,用于处理来自客户端的请求,其主要工作流程如下：

① 客户端请求提交到 DispatcherServlet。

② 由 DispatcherServlet 控制器查询一个或多个 HandlerMapping,找到处理请求的 Controller。

③ DispatcherServlet 将请求提交到 Controller。

④ Controller 调用业务逻辑处理后,返回 ModelAndView。

⑤ DispatcherServlet 查询一个或多个 ViewResoler 视图解析器,找到 ModelAndView 指定的视图。

⑥ 视图负责将结果显示到客户端。

先看一个用 MyEclipse 开发的 Spring MVC Web 应用 helloworld 实例,接下来将详细讲解相关理论内容。

(1) 新建一个 Web 项目,项目名称为 Spring MVC_HelloWorld_demo。

选择 File→New→Web Project 命令,如图 5-3 所示。

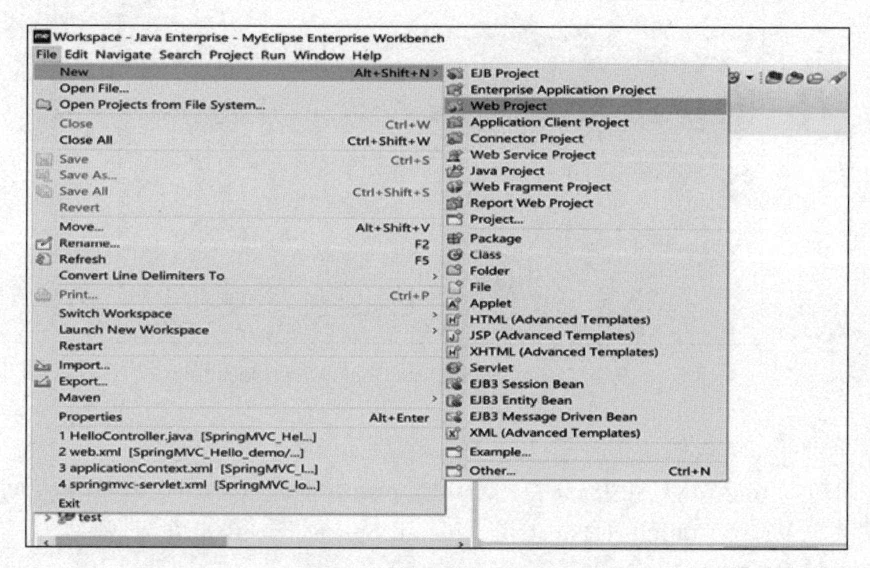

图 5-3 新建 Web Project

在 Project name 中命名为 Spring MVC_HelloWorld_demo,如图 5-4 所示。

单击 Next 按钮,保持默认配置,之后再单击 Next 按钮,勾选 Generate web.xml deployment descriptor,如图 5-5 所示。

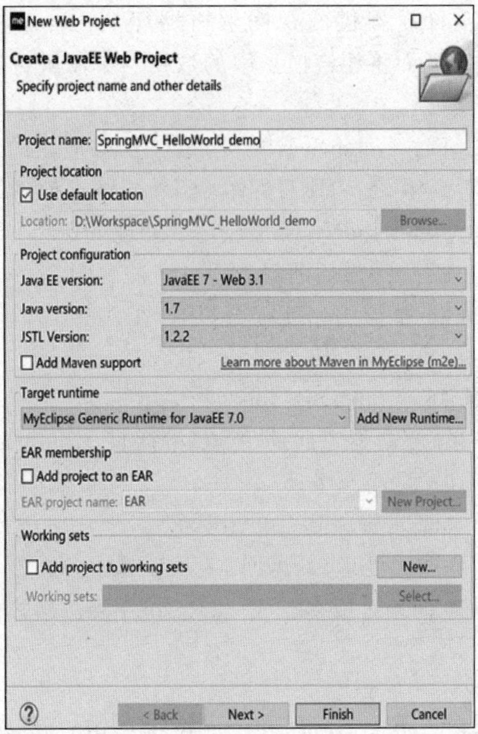

图 5-4　命名 Web Project

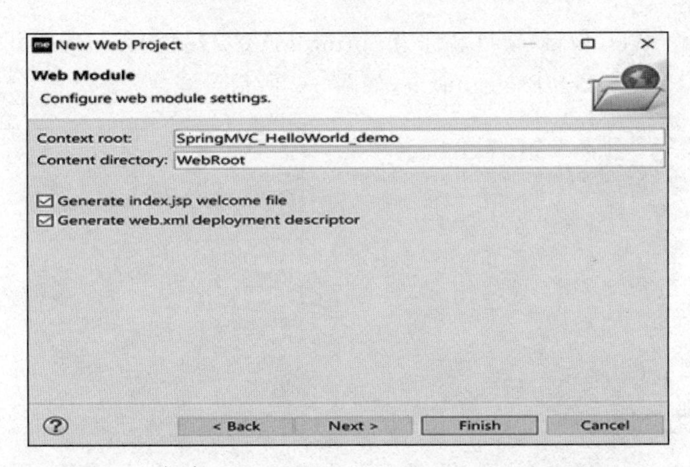

图 5-5　勾选 Generate web.xml deployment descriptor

最后单击 Finish 按钮。

(2) 添加 Spring MVC 相关 jar 包。右击 Spring MVC_HelloWorld_demo 项目，从弹出的快捷菜单中选择 Configure Facets...→Install Spring Facet，如图 5-6 所示。

之后出现图 5-7。

保持默认设置，单击 Next 按钮，出现图 5-8。

保持默认设置，单击 Next 按钮，出现图 5-9。

不需修改(这里需要 Spring 4.1.0 Libraries 中的 Core、Facets 和 Spring Web jar 包)，单击 Finish 按钮。

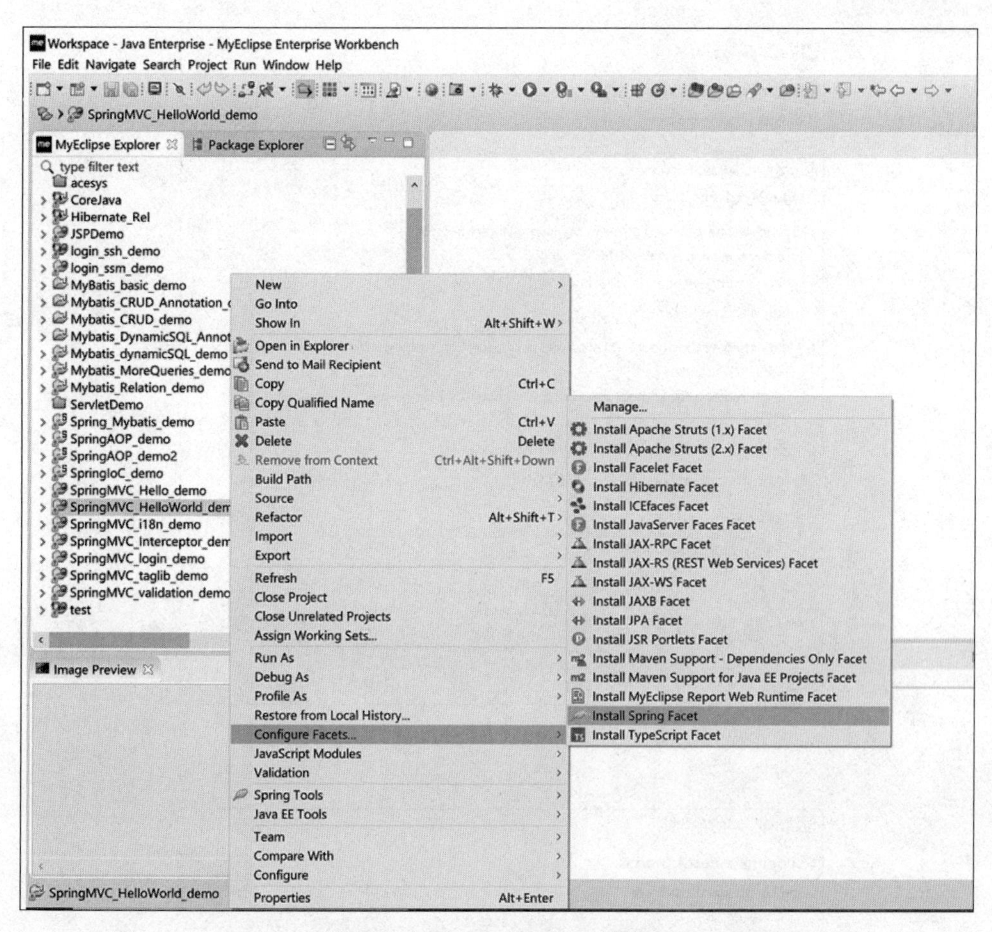

图 5-6 Install Spring Facet 界面 1

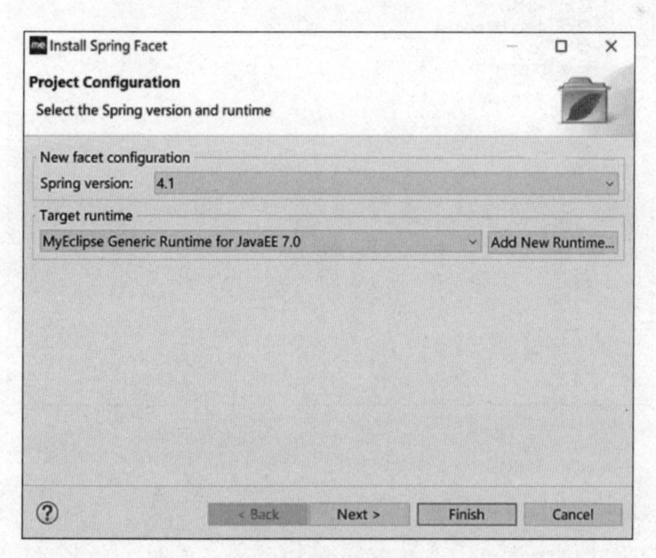

图 5-7 Install Spring Facet 界面 2

Install Spring Facet

Add Spring Capabilities

Configure Spring project

Configure Spring libraries

Type: MyEclipse Library

Libraries will be provided through MyEclipse Library System

Spring Framework version: 4.1.0

Spring Security version: <none>

Spring Webflow version: <none>

☑ Add Spring-Web support (TLDs copied to project, ContextLoaderListener added to web.xml)

☑ Enable AOP Builder

i Larger projects that do not use Spring AOP may see a performance gain by leaving this disabled.

☑ Specify new or existing Spring bean configuration file?

Bean configuration type: ◉ New ○ Existing

Folder: src Browse

File: applicationContext.xml

< Back Next > Finish Cancel

图 5-8 Install Spring Facet 界面 3

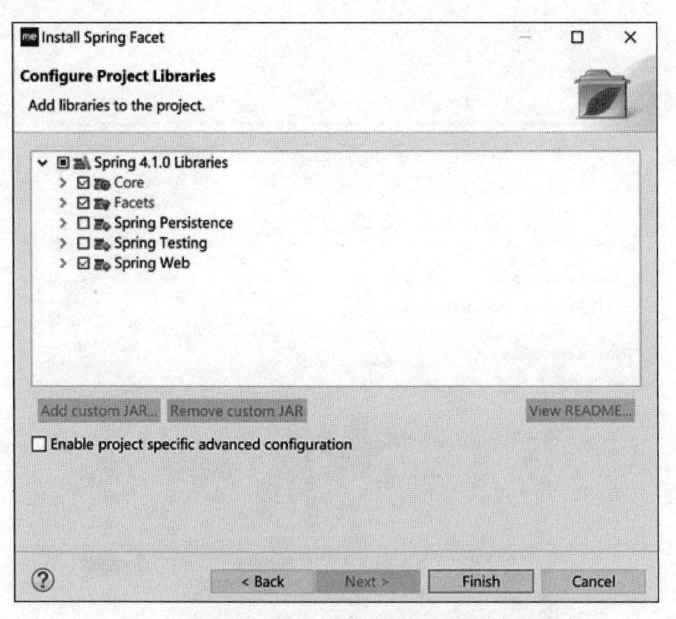

图 5-9 Install Spring Facet 界面 4

(3) 建立项目目录。

右击 src,从弹出的快捷菜单中选择 New→Package,如图 5-10 所示。

命名包名为 com.ascent,如图 5-11 所示,之后单击 Finish 按钮。

我们将在这个包里存放 Java 代码。

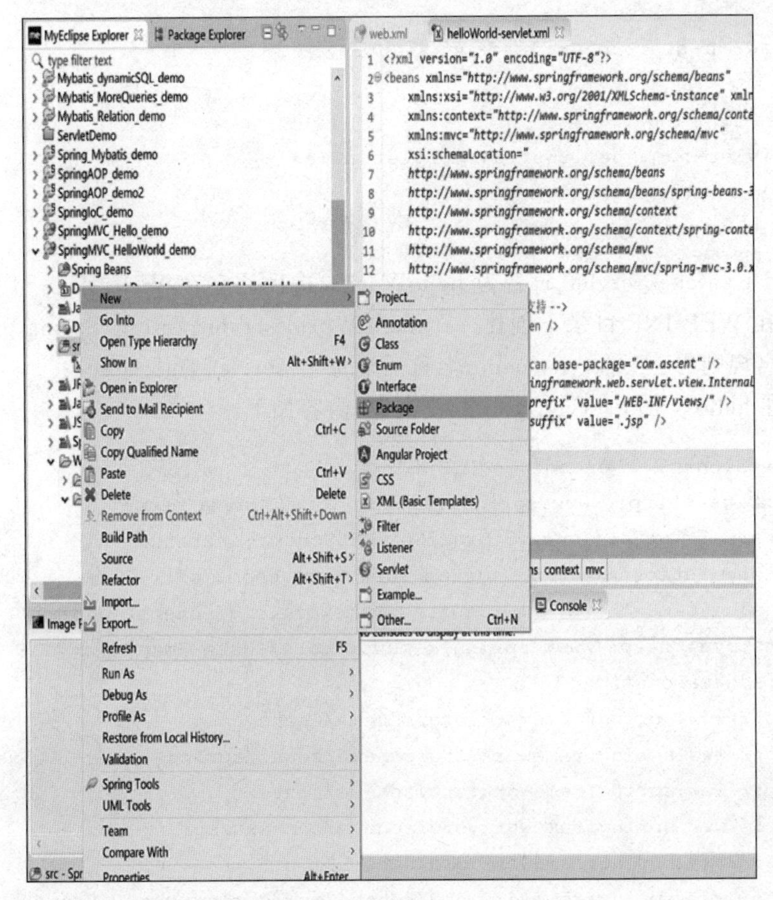

图 5-10 新建 Package

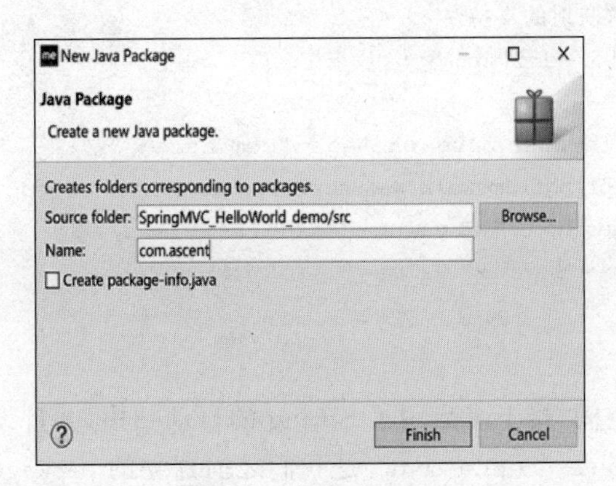

图 5-11 命名 Package

（4）修改 web.xml，添入以下内容。

```
<servlet>
    <servlet-name>helloWorld</servlet-name>
    <servlet-class>
        org.springframework.web.servlet.DispatcherServlet
```

```
        </servlet-class>
        <load-on-startup>1</load-on-startup>
    </servlet>
    <servlet-mapping>
        <servlet-name>helloWorld</servlet-name>
        <url-pattern>/</url-pattern>
    </servlet-mapping>
```

这里将 DispatcherServlet 命名为 helloWorld，并且让它在 Web 项目一启动就加载。接下来需要在 WEB-INF 目录下创建一个 helloWorld-servlet.xml 的 Spring 配置文件，此文件名的命名规则为：在 servlet-name 名称后面加上-servlet.xml。

(5) 添加 helloWorld-servlet.xml 配置文件，内容如下。

```
<?xml version="1.0" encoding="UTF-8"? >
<beans xmlns="http://www.springframework.org/schema/beans"
    xmlns:xsi="http://www.w3.org/2001/XMLSchema-instance"
    xmlns:p="http://www.springframework.org/schema/p"
    xmlns:context="http://www.springframework.org/schema/context"
    xmlns:mvc="http://www.springframework.org/schema/mvc"
    xsi:schemaLocation="
    http://www.springframework.org/schema/beans
    http://www.springframework.org/schema/beans/spring-beans-3.0.xsd
    http://www.springframework.org/schema/context
    http://www.springframework.org/schema/context/spring-context-3.0.xsd
    http://www.springframework.org/schema/mvc
    http://www.springframework.org/schema/mvc/spring-mvc-3.0.xsd">

    <!--默认的注解映射的支持 -->
    <mvc:annotation-driven />
<!--启用自动扫描  -->
<context:component-scan base-package="com.ascent" />
<bean class="org.springframework.web.servlet.view.InternalResourceViewResolver">
    <property name="prefix" value="/WEB-INF/views/" />
    <property name="suffix" value=".jsp" />
</bean>
</beans>
```

添加了 mvc 名称空间，接下来启用了 Spring 的自动扫描，并且设置了默认的注解映射支持。其中 base-package="com.ascent" 是我们之前建立的 package 名。

配置文件中的 Bean 的类型是 Spring MVC 中最常用的一种视图解析器。prefix 属性是指视图前缀，suffix 是视图后缀，这里配置的是.jsp，我们在控制器的方法 sayHello()中返回的是 hello，结合这里的配置，对应的完整的视图是/WEB-INF/views/hello.jsp。

(6) 编写 HelloWorldController 文件。

右击 com.ascent 包，从弹出的快捷菜单中选择 New→Class，如图 5-12 所示。

在 Name 处填入 HelloWorldController，单击 Finish 按钮，如图 5-13 所示。

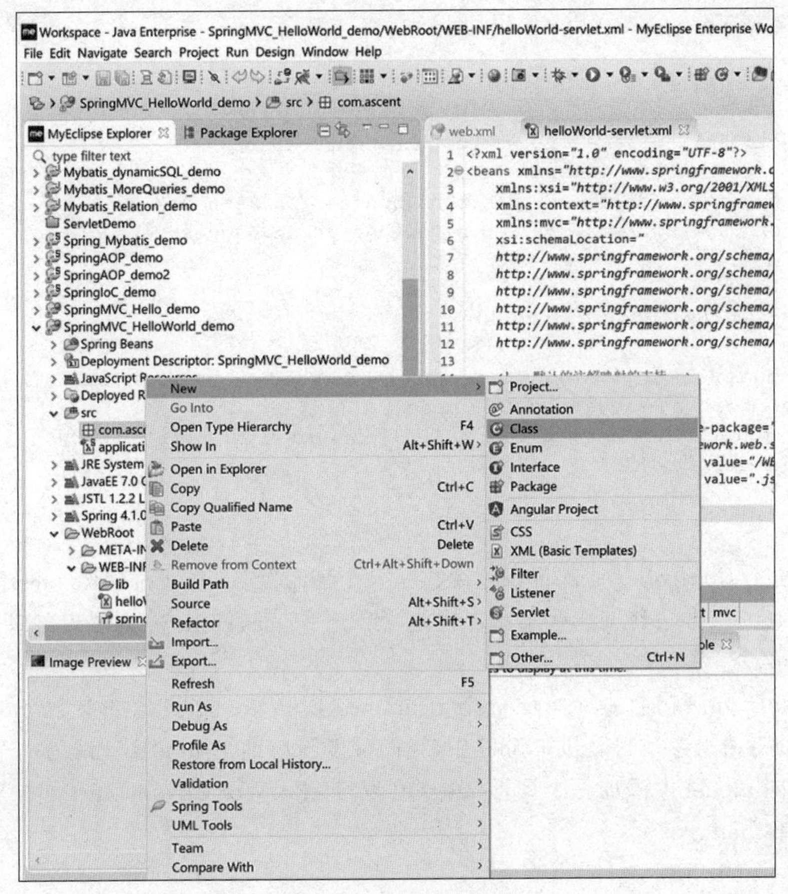

图 5-12 新建 Class

New Java Class
Java Class
Create a new Java class.
Source folder: SpringMVC_HelloWorld_demo/src Browse...
Package: com.ascent Browse...
Enclosing type: Browse...
Name: HelloWorldController
Modifiers: public package private protected
abstract final static
Superclass: java.lang.Object Browse...
Interfaces: Add... Remove
Which method stubs would you like to create?
public static void main(String[] args)
Constructors from superclass
Inherited abstract methods
Do you want to add comments? (Configure templates and default value here)
Generate comments
Finish Cancel

图 5-13 命名 Class

编写 HelloWorldController 代码如下：

```
package com.ascent;

import org.springframework.stereotype.Controller;
import org.springframework.ui.Model;
import org.springframework.web.bind.annotation.RequestMapping;
import org.springframework.web.bind.annotation.RequestMethod;

@Controller
public class HelloWorldController {

    @RequestMapping(value="/helloWorld",method=RequestMethod.GET)
    public String sayHello(Model model) {
        model.addAttribute("msg", "Hello World from Spring MVC!");
        return "hello";
    }
}
```

首先通过@Controller 注解标识这个类是一个控制器，接着通过@RequestMapping 注解指定 sayHello()方法处理哪些请求。在这个例子中，sayHello()方法仅处理 GET 类型的/hello 请求。

sayHello()方法接收一个 org.springframework.ui.Model 类型的参数 model，Spring MVC 会自动将请求参数封装进 model 中，我们可以简单地把 model 理解为一个 Map。在方法中首先向 model 中添加一个属性 msg，值为"Hello World from Spring MVC!"，然后返回视图名称 hello。

(7) 编写 hello JSP 文件。

根据 helloWorld-servlet.xml，对应的完整的视图是/WEB-INF/views/hello.jsp。

右击 WEB-INF，从弹出的快捷菜单中选择 New→Folder，如图 5-14 所示。

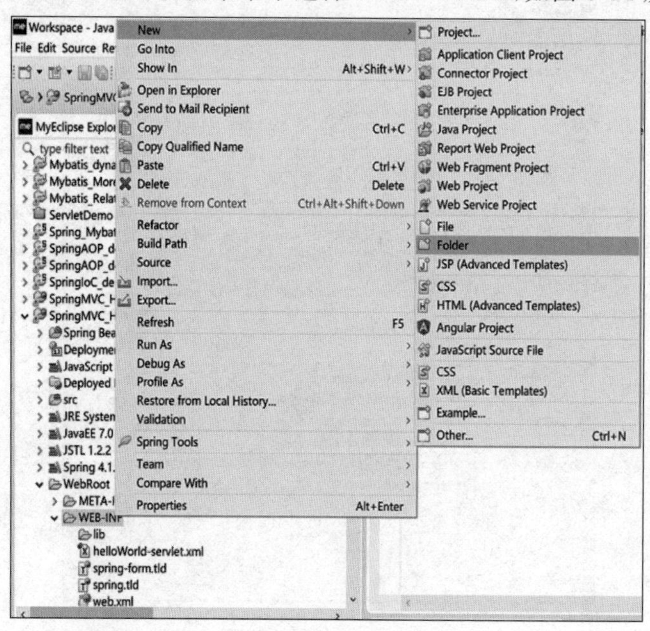

图 5-14 新建 Folder

命名 Folder name 为 views，如图 5-15 所示。

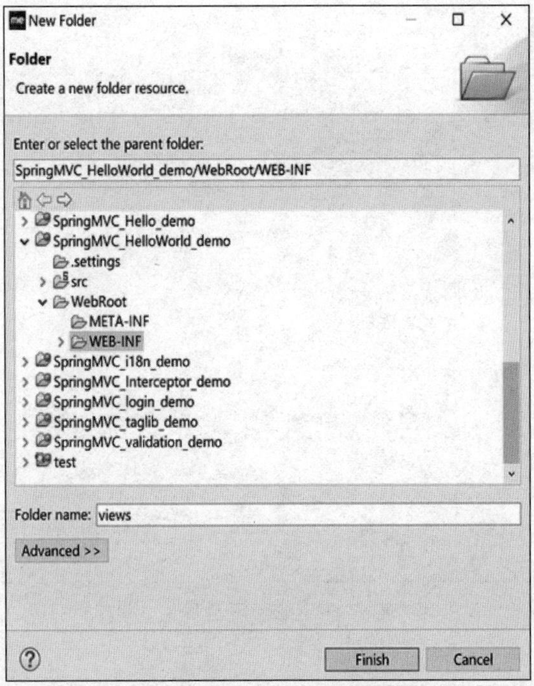

图 5-15　命名 Folder

单击 Finish 按钮，之后在 WEB-INF views 文件夹下建立 hello.jsp，内容如下：

```
<%@ page language="java" contentType="text/html; charset=UTF-8"
    pageEncoding="UTF-8"%>
<!DOCTYPE html PUBLIC "-//W3C//DTD HTML 4.01 Transitional//EN" "http://www.w3.
org/TR/html4/loose.dtd">
<html>
<head>
<meta http-equiv="Content-Type" content="text/html; charset=UTF-8">
<title>hello.jsp</title>
</head>
<body>
    ${msg}
</body>
</html>
```

(8) 部署和运行项目。

选中 SpringMVC_HelloWorld_demo 项目，单击红色处的部署图标，部署在 MyEclipse 自带的 Tomcat v8.5 服务器上，如图 5-16 所示。

单击 Next 按钮，保持默认设置，最后单击 Finish 按钮，出现图 5-17。

单击 OK 按钮，之后在弹出的对话框中单击红色处的启动 Tomcat 图标，如图 5-18 所示。

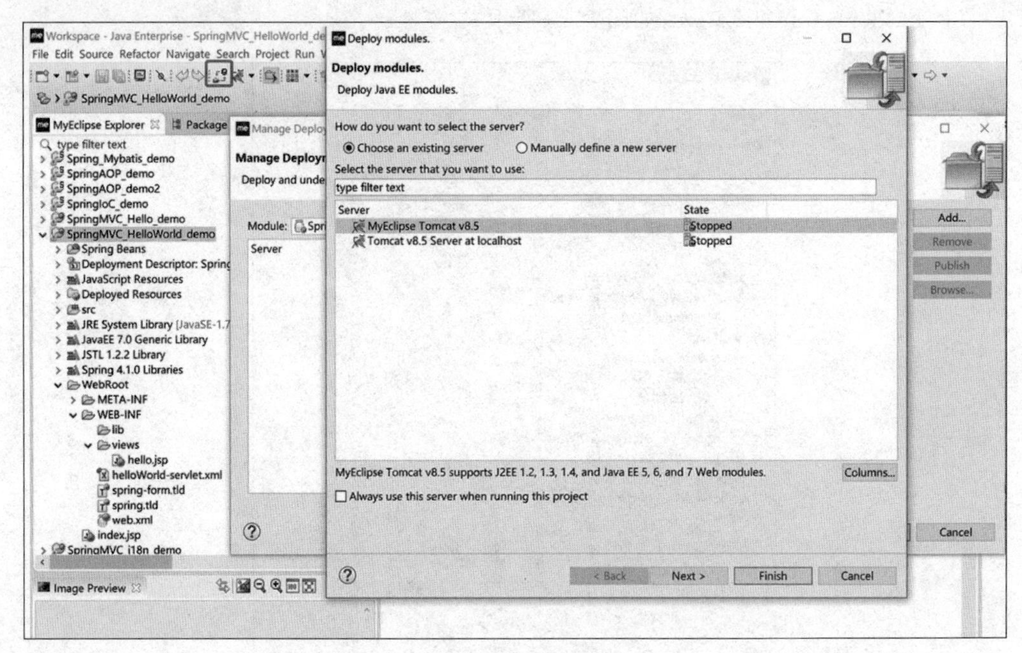

图 5-16　部署项目界面 1

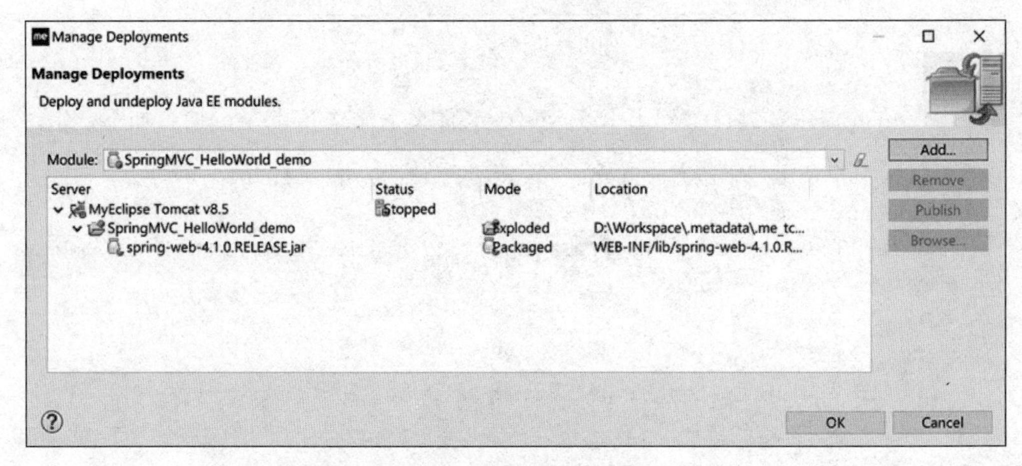

图 5-17　部署项目界面 2

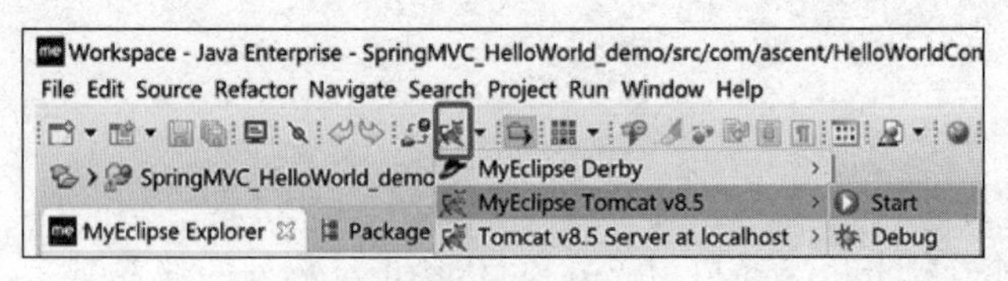

图 5-18　启动项目

Tomcat 启动成功后，在浏览器中输入 http://localhost:8080/Spring MVC_HelloWorld_demo/helloWorld，浏览器显示 Hello World from Spring MVC!，应用开发完成。

接下来具体讲解相关知识点。

5.4 Spring MVC 原理

5.4.1 核心控制器 DispatcherServlet

Spring MVC 框架,与其他很多 Web 应用的 MVC 框架一样是基于请求驱动的,所有设计都围绕一个中央 Servlet 展开。它负责把所有请求分发到控制器,同时提供其他 Web 应用开发所需要的功能。不过,Spring 的中央处理器 DispatcherServlet 能做的事情比这更多。它与 Spring IoC 容器做到了无缝集成,这意味着,Spring 提供的任何特性在 Spring MVC 中你都可以使用。熟悉设计模式的朋友会发现,DispatcherServlet 应用的其实就是一个"前端控制器"的设计模式(其他很多优秀的 Web 框架也都使用了这个设计模式)。

DispatcherServlet 是整个 Spring MVC 的核心。它负责接收 HTTP 请求,组织协调 Spring MVC 的各个组成部分。其主要工作有以下 3 项。

- 截获符合特定格式的 URL 请求。
- 初始化 DispatcherServlet 上下文对应的 WebApplicationContext,并将其与业务层、持久化层的 WebApplicationContext 建立关联。
- 初始化 Spring MVC 的各个组成组件,并装配到 DispatcherServlet 中。

1. URL Mapping

DispatcherServlet 其实就是一个 Servlet(它继承自 HttpServlet 基类),同样也需要在 Web 应用的 web.xml 配置文件下声明。通常需要在 web.xml 文件中把希望 DispatcherServlet 处理的请求映射到对应的 URL 上。下面的代码展示了对 DispatcherServlet 和路径映射的声明:

```
<web-app>
    <servlet>
        <servlet-name>example</servlet-name>
        <servlet-class>org.springframework.web.servlet.DispatcherServlet
            </servlet-class>
        <load-on-startup>1</load-on-startup>
    </servlet>
    <servlet-mapping>
        <servlet-name>example</servlet-name>
        <url-pattern>*.action</url-pattern>
    </servlet-mapping>
</web-app>
```

在上面的例子中,所有路径以.action 结尾的 URL 请求都会被名字为 example 的 DispatcherServlet 处理。在 Servlet 3.0 以上的环境下,还可以用编程的方式配置 Servlet 容器。下面是一个基于代码配置的例子,它与上面定义的 web.xml 配置文件等效。

```
public class MyWebApplicationInitializer implements WebApplicationInitializer {
@Override
  public void onStartup(ServletContext container) {
```

```
        ServletRegistration.Dynamic registration=container.addServlet("dispatcher",
    new DispatcherServlet());
        registration.setLoadOnStartup(1);
        registration.addMapping("*.action");
      }
    }
```

这两种方式二选一，可以使用 web.xml 声明配置，也可以使用 Java 代码替代 web.xml。

2. WebApplicationContext

WebApplicationInitializer 是 Spring MVC 提供的一个接口，它会查找所有基于代码的配置，并应用它们初始化 Servlet 3 版本以上的 Web 容器。它有一个抽象的实现 AbstractDispatcherServletInitializer，用以简化 DispatcherServlet 的注册工作：只指定其 servlet 映射(mapping)即可。上面只是配置 Spring Web MVC 的第一步，接下来需要配置其他一些 Bean(除 DispatcherServlet 外的其他 Bean)，它们也会被 Spring Web MVC 框架使用到。

Spring 中的 ApplicationContext 实例是可以有范围(scope)的。在 Spring MVC 中，每个 DispatcherServlet 都持有一个自己的上下文对象 WebApplicationContext，它又继承了根对象 Root WebApplicationContext 中已经定义的所有 Bean。这些继承的 Bean 可以在具体的 Servlet 实例中被重载，在每个 Servlet 实例中也可以定义其 scope 下的新 bean，如图 5-19 所示。

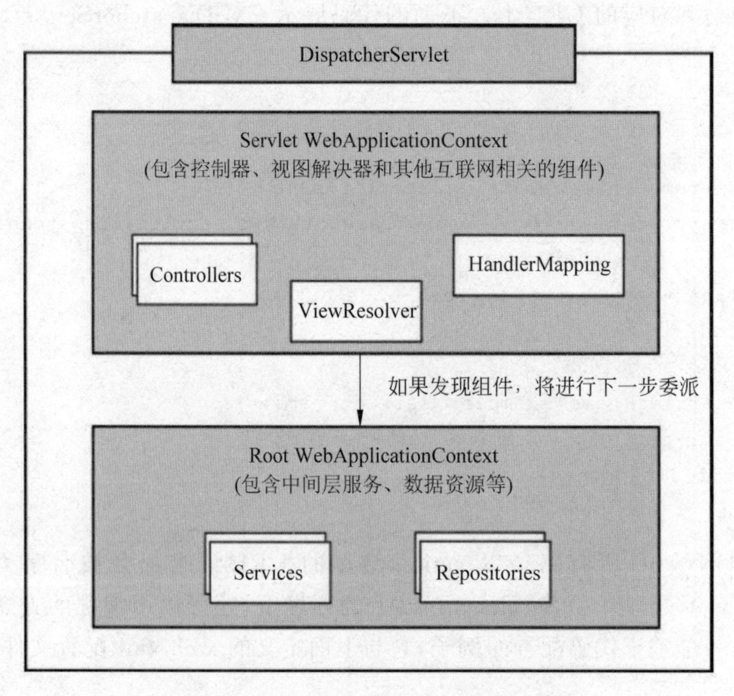

图 5-19 WebApplicationContext 结构

在 DispatcherServlet 的初始化过程中,Spring MVC 会在 Web 应用的 WEB-INF 目录下查找一个名为[servlet-name]-servlet.xml 的配置文件,并创建其中定义的 Bean。如果在全局上下文中存在相同名字的 Bean,则它们将被新定义的同名 Bean 覆盖。也可以在 WEB-INF 下的 classes 子目录查找用户自定义的配置文件。

看下面这个 DispatcherServlet 的 Servlet 配置(定义于 web.xml 文件中):

```
<web-app>
    <servlet>
        <servlet-name>helloWorld</servlet-name>
        <servlet-class>org.springframework.web.servlet.DispatcherServlet
            </servlet-class>
        <load-on-startup>1</load-on-startup>
    </servlet>
    <servlet-mapping>
        <servlet-name>helloWorld</servlet-name>
        <url-pattern>/</url-pattern>
    </servlet-mapping>
</web-app>
```

有了以上 Servlet 配置文件,还需要在应用中的 /WEB-INF/ 路径下创建一个 helloWorld -servlet.xml 文件,在该文件中定义所有 Spring MVC 相关的组件(如 Bean 等)。可以通过 servlet 初始化参数为这个配置文件指定其他的路径。

当应用中只需要一个 DispatcherServlet 时,只配置一个根 context 对象也是可行的,如图 5-20 所示。

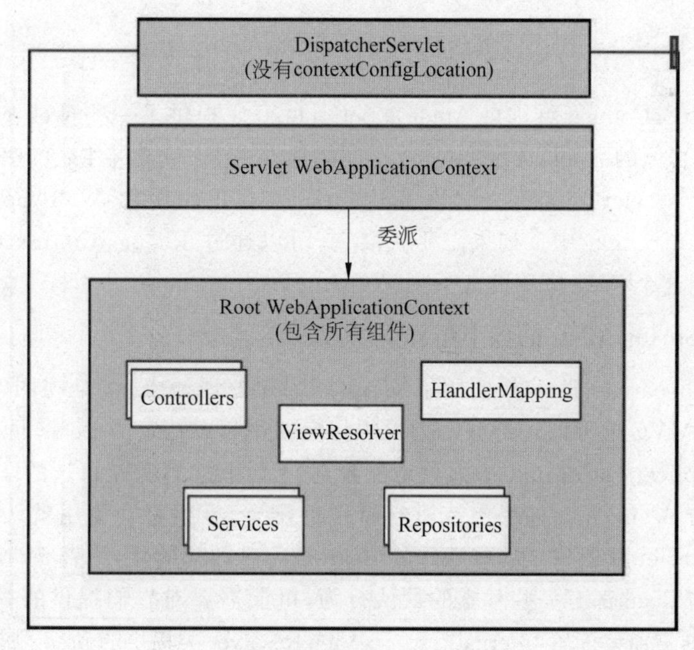

图 5-20 唯一的根 context 对象

要配置一个唯一的根 context 对象，可以通过在 servlet 初始化参数中配置一个空的 contextConfigLocation 做到，如下所示。

```
<web-app>
    <context-param>
        <param-name>contextConfigLocation</param-name>
        <param-value>/WEB-INF/root-context.xml</param-value>
    </context-param>
    <servlet>
        <servlet-name>dispatcher</servlet-name>
        <servlet-class>org.springframework.web.servlet.DispatcherServlet
            </servlet-class>
        <init-param>
            <param-name>contextConfigLocation</param-name>
            <param-value></param-value>
        </init-param>
        <load-on-startup>1</load-on-startup>
    </servlet>
    <servlet-mapping>
        <servlet-name>dispatcher</servlet-name>
        <url-pattern>/*</url-pattern>
    </servlet-mapping>
    <listener>
        <listener-class>org.springframework.web.context.ContextLoaderListener
            </listener-class>
    </listener>
</web-app>
```

WebApplicationContext 继承自 ApplicationContext，它提供了一些 Web 应用经常需要用到的特性。它与普通的 ApplicationContext 不同的地方在于：它支持主题的解析，并且知道它关联到的是哪个 servlet（它持有一个该 ServletContext 的引用）。WebApplicationContext 被绑定在 ServletContext 中。如果需要获取它，可以通过 RequestContextUtils 工具类中的静态方法拿到这个 Web 应用的上下文 WebApplicationContext。

3. 初始化 Spring MVC 的各个组件

Spring 的 DispatcherServlet 使用了特殊的 Bean 处理请求、渲染视图等，这些特定的 Bean 是 Spring MVC 框架的一部分。如果想指定使用哪个特定的 Bean，可以在 Web 应用上下文 WebApplicationContext 中简单地配置它们。当然，这只是可选的，Spring MVC 维护了一个默认的 Bean 列表，如果没有进行特别的配置，框架将会使用默认的 Bean。看一下 DispatcherServlet 都依赖于哪些特殊的 Bean 进行它的初始化，见表 5-1。

这些特殊的 Bean 都有一些基本的默认行为，可能需要对它们提供的一些默认配置进行定制。例如，通常需要配置 InternalResourceViewResolver 类提供的 prefix 属性，使其指向视图文件所在的目录。这里需要说明的是，一旦在 Web 应用上下文 WebApplicationContext 中配置了某个特殊 Bean 后（如 InternalResourceViewResolver），实际上也重写了该 Bean

的默认实现。例如，如果配置了 InternalResourceViewResolver，那么框架就不会再使用 bean ViewResolver 的默认实现了。

表 5-1 DispatcherServlet 依赖的 Bean

Bean 类型	作 用
HandlerMapping	处理器映射。它会根据某些规则将进入容器的请求映射到具体的处理器以及一系列前处理器和后处理器(即处理器拦截器)上。具体的规则视 HandlerMapping 类的实现不同而有所不同。其最常用的一个实现支持在控制器上添加注解，配置请求路径。当然，也存在其他的实现
HandlerAdapter	处理器适配器。拿到请求对应的处理器后，适配器将负责调用该处理器，这使得 DispatcherServlet 无须关心具体的调用细节。例如，要调用的是一个基于注解配置的控制器，那么调用前还需要从许多注解中解析出一些相应的信息。因此，HandlerAdapter 的主要任务就是对 DispatcherServlet 屏蔽这些具体的细节
HandlerExceptionResolver	处理器异常解析器。它负责将捕获的异常映射到不同的视图上。此外，它还支持更复杂的异常处理代码
ViewResolver	视图解析器。它负责将一个代表逻辑视图名的字符串(String)映射到实际的视图类型 View 上
LocaleResolver & LocaleContextResolver	地区解析器和地区上下文解析器。它们负责解析客户端所在的地区信息甚至时区信息，为国际化的视图定制提供了支持
ThemeResolver	主题解析器。它负责解析 Web 应用中可用的主题，如提供一些个性化定制的布局等
MultipartResolver	解析 multi-part 的传输请求，如支持通过 HTML 表单进行的文件上传等
FlashMapManager	FlashMap 管理器。它能够存储并取回两次请求之间的 FlashMap 对象。后者可用于在请求之间传递数据，通常在请求重定向的情境下使用

4. DispatcherServlet 的处理流程

配置好 DispatcherServlet 后，开始有请求会经过这个 DispatcherServlet。此时，DispatcherServlet 会依照以下次序对请求进行处理。

(1) 首先，搜索应用的上下文对象 WebApplicationContext 并把它作为一个属性(attribute)绑定到该请求上，以便控制器和其他组件能够使用它。属性的键名默认为 DispatcherServlet.WEB_APPLICATION_CONTEXT_ATTRIBUTE。

(2) 将地区(locale)解析器绑定到请求上，以便其他组件在处理请求(如渲染视图、准备数据等)时可以获取区域相关的信息。如果应用不需要解析区域相关的信息，忽略它即可。

(3) 将主题(theme)解析器绑定到请求上，以便其他组件(如视图等)能够了解要渲染哪个主题文件。同样，如果不需要使用主题相关的特性，忽略它即可。

(4) 如果配置了 multipart 文件处理器，那么框架将查找该文件是不是 multipart(分为多个部分连续上传)的。若是，则将该请求包装成一个 MultipartHttpServletRequest 对象，以便处理链中的其他组件对它做进一步的处理。

(5) 为该请求查找一个合适的处理器。如果可以找到对应的处理器,则与该处理器关联的整条执行链(如前处理器、后处理器、控制器等)都会被执行,以完成相应模型的准备或视图的渲染。

(6) 如果处理器返回的是一个模型(model),那么框架将渲染相应的视图。若没有返回任何模型(可能是因为前后的处理器出于某些原因拦截了请求等,如安全问题),则框架不会渲染任何视图,此时认为对请求的处理可能已经由处理链完成了。

(7) 如果在处理请求的过程中抛出了异常,那么上下文 WebApplicationContext 对象中定义的异常处理器将会负责捕获这些异常。通过配置自己的异常处理器,可以定制自己处理异常的方式。

Spring 的 DispatcherServlet 也允许处理器返回一个 Servlet API 规范中定义的最后修改时间戳(last-modification-date)的值。决定请求最后修改时间的方式很直接:DispatcherServlet 会先查找合适的处理器映射找到请求对应的处理器,然后检测它是否实现了 LastModified 接口。如果是,则调用接口的 long getLastModified(request) 方法,并将该返回值返回给客户端。可以定制 DispatcherServlet 的配置,具体的做法是在 web.xml 文件中 Servlet 的声明元素上添加一些 Servlet 的初始化参数(通过 init-param 元素)。该元素可选的参数列表见表 5-2。

表 5-2　Servlet 的初始化参数

可选参数	说　明
contextClass	任意实现了 WebApplicationContext 接口的类。这个类会初始化该 servlet 需要用到的上下文对象。默认情况下,框架会使用一个 XmlWebApplicationContext 对象
contextConfigLocation	一个指定了上下文配置文件路径的字符串,该值会传给 contextClass 指定的上下文实例对象。该字符串内可以包含多个字符串,字符串之间以逗号分隔,以此支持进行多个上下文的配置。在多个上下文中重复定义的 Bean,以最后加载的 Bean 定义为准
namespace	WebApplicationContext 的命名空间,默认是[servlet-name]-servlet

5.4.2　Controller 控制器

控制器作为应用程序逻辑的处理入口,它会负责调用已经实现的一些服务。通常,一个控制器会接收并解析用户的请求,然后把它转换成一个模型交给视图,由视图渲染出页面最终呈现给用户。Spring 对控制器的定义非常宽松,这意味着在实现控制器时非常灵活。

Spring 2.5 以后引入了基于注解的编程模型,可以在控制器实现上添加 @RequestMapping、@RequestParam、@ModelAttribute 等注解。注解特性既支持基于 Servlet 的 MVC,也支持基于 Portlet 的 MVC。通过此种方式实现的控制器既无须继承某个特定的基类,也无须实现某些特定的接口。而且,它通常也不会直接依赖于 Servlet 或 Portlet 的 API 进行编程,而你仍然可以很容易地获取 Servlet 或 Portlet 相关的变量、特性和设施等。

例如：

```
@Controller
public class HelloWorldController {

    @RequestMapping(value="/helloWorld",method=RequestMethod.GET)
    public String sayHello(Model model) {
        model.addAttribute("msg", "Hello World from Spring MVC!");
        return "hello";
    }
}
```

可以看到，@Controller 注解和 @RequestMapping 注解支持多样的方法名和方法签名。在上面这个例子中，方法接受一个 Model 类型的参数并返回一个字符串 String 类型的视图名。但事实上，方法支持的参数和返回值有非常多的选择。@Controller 和 @RequestMapping 及其他的一些注解，共同构成了 Spring MVC 框架的基本实现。

1. 使用@Controller 注解定义一个控制器

@Controller 注解表明了一个类是作为控制器的角色而存在的。Spring 不要求继承任何控制器基类，也不要求实现 Servlet 的那套 API。当然，如果需要，也可以使用任何与 Servlet 相关的特性和设施。

@Controller 注解可以认为是被标注类的原型(stereotype)，表明了这个类所承担的角色。DispatcherServlet 会扫描所有注解了 @Controller 的类，检测其中通过 @RequestMapping 注解配置的方法。

当然，也可以不使用 @Controller 注解而显式地定义被注解的 bean，这点通过标准的 Spring bean 的定义方式，在 dispather 的上下文属性下配置即可做到。但是，@Controller 原型是可以被框架自动检测的，Spring 支持 classpath 下组件类的自动检测，以及对已定义 bean 的自动注册。通常需要在配置中加入组件扫描的配置代码开启框架对注解控制器的自动检测，例如下面的 XML 代码。

```
<?xml version="1.0" encoding="UTF-8"?>
<beans xmlns="http://www.springframework.org/schema/beans"
  xmlns:xsi="http://www.w3.org/2001/XMLSchema-instance"
  xmlns:p="http://www.springframework.org/schema/p"
  xmlns:context="http://www.springframework.org/schema/context"
  xsi:schemaLocation="
    http://www.springframework.org/schema/beans
    http://www.springframework.org/schema/beans/spring-beans.xsd
    http://www.springframework.org/schema/context
    http://www.springframework.org/schema/context/spring-context.xsd">
  <context:component-scan base-package="com.ascent"/>
  <!-- ... -->
</beans>
```

2. 使用@RequestMapping 注解映射请求路径

1）@RequestMapping 注解概述

可以使用 @RequestMapping 注解将请求 URL（如/helloWorld 等）映射到整个类上或某个特定的处理器方法上。一般来说，类级别的注解负责将一个特定（或符合某种模式）的请求路径映射到一个控制器上，同时通过方法级别的注解细化映射，即根据特定的 HTTP 请求方法（如 GET()、POST()方法等）、HTTP 请求中是否携带特定参数等条件，将请求映射到匹配的方法上。类级别的 @RequestMapping 注解并不是必需的，若不配置，则所有的路径都是绝对路径，而非相对路径。

例如：

```
@RequestMapping(value="/helloWorld",method=RequestMethod.GET)
public String sayHello(Model model) {
    model.addAttribute("msg", "Hello World from Spring MVC!");
    return "hello";
}
```

这里的@RequestMapping 注解将 URL 为/helloWorld 和 HTTP 方法为 GET 的请求映射到了 sayHello()方法上。

下面这段代码示例来自 Petcare，它展示了在 Spring MVC 中如何在控制器上使用 @RequestMapping 注解。

```
@Controller
@RequestMapping("/appointments")
public class AppointmentsController {
  private final AppointmentBookappointmentBook;
  @Autowired
  public AppointmentsController(AppointmentBookappointmentBook) {
    this.appointmentBook=appointmentBook;
  }
  @RequestMapping(method=RequestMethod.GET)
  public Map<String, Appointment>get() {
    return appointmentBook.getAppointmentsForToday();
  }
  @RequestMapping(path="/{day}", method=RequestMethod.GET)
  public Map<String, Appointment>getForDay(@PathVariable @DateTimeFormat
    (iso=ISO.DATE) Date day, Model model) {
    return appointmentBook.getAppointmentsForDay(day);
  }
  @RequestMapping(path="/new", method=RequestMethod.GET)
  public AppointmentFormgetNewForm() {
    return new AppointmentForm();
  }
  @RequestMapping(method=RequestMethod.POST)
  public String add(@Valid AppointmentForm appointment, BindingResult result) {
```

```
        if (result.hasErrors()) {
            return "appointments/new";
        }
        appointmentBook.addAppointment(appointment);
        return "redirect:/appointments";
    }
}
```

在上面的示例中,许多地方都使用到@RequestMapping 注解。第一次使用是作用于类级别的,它指示所有以/appointments 开头的路径都会被映射到控制器下。get()方法上的@RequestMapping 注解对请求路径进行了进一步细化:它仅接受 GET()方法的请求。这样,一个请求路径为/appointments、HTTP()方法为 GET 的请求,将会最终进入这个方法被处理。add()方法也做了类似的细化,getNewForm()方法则同时注解了能够接受的请求的 HTTP()方法和路径。这种情况下,一个路径为 appointments/new、HTTP()方法为 GET 的请求将会被这个方法处理。

getForDay()方法则展示了使用@RequestMapping 注解的另一个技巧:URI 模板。

注意:String add(@Valid AppointmentForm appointment,BindingResult result)中的@Valid 和 BindingResult 是和 Spring MVC 数据验证相关的,第 6 章将会讲解。

2) Spring MVC 3.1 中新增了支持@RequestMapping 的一些类

Spring 3.1 中新增了一组类,用以增强@RequestMapping,分别是 RequestMappingHandlerMapping 和 RequestMappingHandlerAdapter。在 MVC 命名空间和 MVC Java 编程配置方式下,这组类及其新特性默认是开启的。但如果使用其他配置方式,则该特性必须手动配置才能使用。新类相比之前的类有一些重要变化。

在 Spring 3.1 之前,框架会在两个不同的阶段分别检查类级别和方法级别的请求映射——DefaultAnnotationHanlderMapping 首先会在类级别上选中一个控制器,然后再通过 AnnotationMethodHandlerAdapter 定位到具体要调用的方法。

现在有了在 Spring 3.1 之后引入的这组新类,RequestMappingHandlerMapping 成为这两个决策实际发生的唯一的地方。可以把控制器中的一系列处理方法当成一系列独立的服务节点,从类级别和方法级别的 @RequestMapping 注解中获取足够匹配请求路径的映射信息。这种新的处理方式带来了新的可能性。HandlerInterceptor 或 HandlerExceptionResolver 现在可以确定获得的这个处理器肯定是 HandlerMethod 类型,因此它能够精确地了解这个方法的所有信息,包括它的参数、应用于其上的注解等。这样,对于一个 URL 的处理流程,再也不需要分隔到不同的控制器里执行了。

同时,也有其他一些变化,如以下事情就没法做了。

(1) 首先通过 SimpleUrlHandlerMapping 或 BeanNameUrlHandlerMapping 获得负责处理请求的控制器,然后通过 @RequestMapping 注解配置的信息定位到具体的处理方法。

(2) 依靠方法名称作为选择处理方法的标准。例如,两个注解了 @RequestMapping 的方法除了方法名称外都拥有完全相同的 URL 映射和 HTTP 请求方法。在新版本下,@RequestMapping注解的方法必须具有唯一的请求映射。

(3) 定义一个默认方法(即没有声明路径映射),在请求路径无法被映射到控制器下更精确的方法上时,为该请求提供默认处理。在新版本中,如果无法为一个请求找到合适的处理方法,那么一个 404 错误将被抛出。

如果使用原来的类,以上功能还是可以做到的。但是,如果要利用 Spring MVC 3.1 版本带来的方便特性,就需要使用新的类。

3) URI 模板

URI 模板可以为快速访问 @RequestMapping 中指定的 URL 的一个特定的部分提供很大的便利。URI 模板是一个类似于 URI 的字符串,只不过其中包含了一个或多个变量名。当使用实际的值填充这些变量名的时候,模板就转化成一个 URI。URI 模板的 RFC 提议中定义了一个 URI 是如何进行参数化的。例如,这个 URI 模板“http://www.example.com/users/{userId}”就包含了一个变量名 userId。将值 fred 赋给这个变量名后,它就变成一个 URI,即 http://www.example.com/users/fred。

在 Spring MVC 中,可以在方法参数上使用 @PathVariable 注解,将其与 URI 模板中的参数绑定起来:

```
@RequestMapping(path="/owners/{ownerId}", method=RequestMethod.GET)
public String findOwner(@PathVariable String ownerId, Model model) {
  Owner owner=ownerService.findOwner(ownerId);
  model.addAttribute("owner", owner);
  return "displayOwner";
}
```

URI 模板"/owners/{ownerId}"指定了一个变量,名为 ownerId。当控制器处理这个请求的时候,ownerId 的值就会被 URI 模板中对应部分的值所填充。例如,如果请求的 URI 是/owners/fred,此时变量 ownerId 的值就是 fred。为了处理@PathVariables 注解,Spring MVC 必须通过变量名找到 URI 模板中对应的变量。可以在注解中直接声明:

```
@RequestMapping(path="/owners/{ownerId}}", method=RequestMethod.GET)
public String findOwner (@ PathVariable ( " ownerId") String theOwner, Model
model) {
  //具体的方法代码…
}
```

或者,如果 URI 模板中的变量名与方法的参数名相同,就不必再指定一次。只要在编译的时候留下 debug 信息,Spring MVC 就可以自动匹配 URL 模板中与方法参数名相同的变量名。

```
@RequestMapping(path="/owners/{ownerId}", method=RequestMethod.GET)
public String findOwner(@PathVariable String ownerId, Model model) {
    //具体的方法代码…
}
```

一个方法可以拥有任意数量的 @PathVariable 注解:

```
@RequestMapping(path="/owners/{ownerId}/pets/{petId}", method=RequestMethod.
GET)
public String findPet(@PathVariable String ownerId, @PathVariable String petId,
```

```
Modelmodel) {
    Owner owner=ownerService.findOwner(ownerId);
    Pet pet=owner.getPet(petId);
    model.addAttribute("pet", pet);
    return "displayPet";
}
```

当@PathVariable 注解被应用于 Map<String,String>类型的参数上时，框架会使用所有的 URI 模板变量填充这个 Map。URI 模板可以从类级别和方法级别的@RequestMapping 注解中获取数据。因此，像这样的 findPet()方法可以被类似于/owners/42/pets/21 这样的 URL 获得并调用到。

```
@Controller
@RequestMapping("/owners/{ownerId}")
public class RelativePathUriTemplateController {
  @RequestMapping("/pets/{petId}")
  public void findPet(_@PathVariable_ String ownerId, _@PathVariable_ String
petId,Model model) {
      //这里忽略了方法实现体
  }
}
```

@PathVariable 可以应用于所有简单类型的参数上，如 int、long、Date 等类型。Spring 会自动把参数转换成合适的类型，如果转换失败，就抛出一个 TypeMismatchException。如果需要处理其他数据类型的转换，也可以注册自己的类。

4）带正则表达式的 URI 模板

有时可能需要更准确地描述一个 URI 模板的变量，比如说 URL："/springweb/spring-web-3.0.5.jar。怎么把它分解成几个有意义的部分呢？@RequestMapping 注解支持在 URI 模板变量中使用正则表达式。语法是{varName:regex}，其中第一部分定义了变量名，第二部分就是所要应用的正则表达式。例如下面的代码实例：

```
@RequestMapping("/spring-web/{symbolicName:[a-z-]+}-{version:\\d\\.\\d\\.\\
d}{extension:\\.[a-z]+}")
  public void handle(@PathVariable String version, @PathVariable String
extension) {
    //代码部分省略...
  }
}
```

5）路径模式

@RequestMapping 注解还支持 Ant 风格的路径模式(如/myPath/*.do 等)。不仅如此，还可以把 URI 模板变量和 Ant 风格的 glob 组合起来使用(如/owners/*/pets/{petId}这样的用法等)。

6）路径样式的匹配

当一个 URL 同时匹配多个模板(pattern)时，将需要一个算法决定其中最匹配的一个。URI 模板变量的数目和通配符数量的总和最少的那个路径模板更加准确。举一个例子,/hotels/{hotel}/* 这个路径拥有一个 URI 变量和一个通配符,而/hotels/{hotel}/** 这个路径则拥有一个 URI 变量和两个通配符,因此,前者是更准确的路径模板。如果两个模板的 URI 模板数量和通配符数量总和一致,则路径更长的那个模板更准确。举一个例子,/foo/bar* 被认为比/foo/* 更准确,因为前者的路径更长。如果两个模板的数量和长度均一致,则那个具有更少通配符的模板更加准确。例如,/hotels/{hotel}比/hotels/* 更精确。除此之外,还有一些其他的规则:默认的通配模式/**比其他所有的模式都更“不准确”。例如,/api/{a}/{b}/{c}就比默认的通配模式/**要准确,前缀通配(如/public/**)被认为比其他任何不包括双通配符的模式更不准确。例如,/public/path3/{a}/{b}/{c}就比/public/**更准确。更多的细节,请参考这两个类:AntPatternComparator 和 AntPathMatcher。值得一提的是,PathMatcher 类是可以配置的。

7)带占位符的路径模式

@RequestMapping 注解支持在路径中使用占位符,以取得一些本地配置、系统配置、环境变量等。这个特性有时很有用,比如说控制器的映射路径需要通过配置定制的场景。如果想了解更多关于占位符的信息,可以参考 PropertyPlaceholderConfigure 这个类的文档。

8)后缀模式匹配

Spring MVC 默认采用".*"的后缀模式匹配进行路径匹配,因此,一个映射到/person 路径的控制器也会隐式地被映射到/person.*。这使得通过 URL 请求同一资源文件的不同格式变得更简单(如/person.pdf,/person.xml)。

可以关闭默认的后缀模式匹配,或者显式地将路径后缀限定到一些特定格式上。我们推荐这样做,它可以减少映射请求时带来的一些二义性,例如,请求以下路径/person/{id}时,路径中的点号后面带的可能不是描述内容格式,如/person/joe@email.com vs /person/joe@email.com.json。而且后缀模式通配以及内容协商有时可能会被黑客用来进行攻击,因此,对后缀通配进行有意义的限定是有好处的。

9)后缀模式匹配与 RFD

RFD(Reflected File Download)攻击最先于 2014 年在 Trustwave 的一篇论文中被提出。它与 XSS 攻击有些相似,因为这种攻击方式也依赖于某些特征,即需要你的输入(如查询参数、URI 变量等)等也在输出(response)中以某种形式出现。不同的是,RFD 攻击并不通过在 HTML 中写入 JavaScript 代码进行,而是依赖于浏览器跳转到下载页面,并把特定格式(如.bat,.cmd 等)的 response 当成可执行脚本,双击它就会执行。Spring MVC 的@ResponseBody 和 ResponseEntity 方法是有风险的,因为它们会根据客户的请求包括 URL 的路径后缀,渲染不同的内容类型。因此,禁用后缀模式匹配或者禁用仅为内容协商开启的路径文件后缀名携带,都是防范 RFD 攻击的有效方式。

若要开启对 RFD 更高级的保护模式,可以在 Spring MVC 渲染开始请求正文之前,在请求头中增加一行配置 Content-Disposition:inline;filename=f.txt,指定固定的下载文件的文件名。这仅在 URL 路径中包含一个文件符合以下特征的拓展名时适用:该扩展名既不在信任列表(白名单)中,也没有显式地注册于内容协商时使用,并且这种做法还可以有一些副作用,例如,当 URL 是通过浏览器手动输入的时候,很多常用的路径文件后缀默认是被信任的。另外，REST 的 API 一般不应该直接用作 URL。不过，可以自己定制 HttpMessageConverter 的实现，然后显式地注册用于内容协商的文件类型，这种情形下 Content-Disposition 头将不会被加入到请求头中。

10）矩阵变量

原来的 URI 规范 RFC 3986 中允许在路径段落中携带键值对，但规范没有明确给这样的键值对定义术语。有人叫“URI 路径参数”，也有人叫“矩阵 URI”。后者是 Tim Berners-Lee 首先在其博客中提到的术语，使用得要频繁一些，知名度也更高一些。而在 Spring MVC 中，我们称这样的键值对为矩阵变量。矩阵变量可以在任何路径段落中出现，每对矩阵变量之间使用一个分号“;”隔开。例如这样的 URI："/cars;color=red;year=2012"。多个值可以用逗号隔开，如"color=red,green,blue"，或者重复变量名多，如"color=red;color=green;color=blue"。如果一个 URL 需要包含矩阵变量，那么在请求路径的映射配置上就需要使用 URI 模板体现这一点，这样才能确保请求可以被正确地映射，而不管矩阵变量在 URI 中是否出现、出现的次序怎样等。

下面是一个例子，展示了如何从矩阵变量中获取到变量“q”的值。

```
//GET /pets/42;q=11;r=22
@RequestMapping(path="/pets/{petId}", method=RequestMethod.GET)
public void findPet(@PathVariable String petId, @MatrixVariable int q) {
  //petId==42
  //q==11
}
```

由于任意路径段落中都可以含有矩阵变量，在某些场景下，需要用更精确的信息指定一个矩阵变量的位置：

```
//GET /owners/42;q=11/pets/21;q=22
@RequestMapping(path="/owners/{ownerId}/pets/{petId}", method=
RequestMethod.GET)
public void findPet(
  @MatrixVariable(name="q", pathVar="ownerId") int q1,
  @MatrixVariable(name="q", pathVar="petId") int q2) {
  //q1==11
  //q2==22
}
```

也可以声明一个矩阵变量不是必须出现的，并给它赋一个默认值：

```
//GET /pets/42
@RequestMapping(path="/pets/{petId}", method=RequestMethod.GET)
public void findPet(@MatrixVariable(required=false, defaultValue="1") int q) {
    //q==1
}
```

可以通过一个 Map 存储所有的矩阵变量：

```
//GET /owners/42;q=11;r=12/pets/21;q=22;s=23
```

```
@RequestMapping(path="/owners/{ownerId}/pets/{petId}", method=
RequestMethod.GET)
public void findPet(
  @MatrixVariable Map<String, String>matrixVars,
  @MatrixVariable(pathVar="petId") Map<String, String>petMatrixVars){
  //matrixVars:["q" :[11,22], "r" : 12, "s" : 23]
  //petMatrixVars:["q" : 11, "s" : 23]
}
```

如果允许使用矩阵变量,必须把 RequestMappingHandlerMapping 类的 removeSemicolonContent 属性设置为 false。该值默认为 true。

MVC 的 Java 编程配置和命名空间配置都提供了启用矩阵变量的方式。如果是使用 Java 编程的方式,请使用 RequestMappingHandlerMapping 进行定制。

使用 MVC 的命名空间配置时,可以把<mvc:annotation-driven>元素下的 enablematrix-variables 属性设置为 true。该值默认情况下是配置为 false 的。

```
<?xml version="1.0" encoding="UTF-8"?>
<beans xmlns="http://www.springframework.org/schema/beans"
  xmlns:mvc="http://www.springframework.org/schema/mvc"
  xmlns:xsi="http://www.w3.org/2001/XMLSchema-instance"
  xsi:schemaLocation="
    http://www.springframework.org/schema/beans
    http://www.springframework.org/schema/beans/spring-beans.xsd
    http://www.springframework.org/schema/mvc
    http://www.springframework.org/schema/mvc/spring-mvc.xsd">
  <mvc:annotation-driven enable-matrix-variables="true"/>
</beans>
```

11)可消费的媒体类型

可以指定一组可消费的媒体类型,以缩小映射的范围。这样,只有请求头中 Content-Type 的值与指定可消费的媒体类型的值相同时,请求才会被匹配。例如下面这个例子:

```
@Controller
@RequestMapping(path="/pets", method=RequestMethod.POST, consumes="application/
json")
public void addPet(@RequestBody Pet pet, Model model) {
    //方法实现省略
}
```

指定可消费媒体类型的表达式中还可以使用否定,例如,可以使用! text/plain 匹配所有请求头 Content-Type 中不含 text/plain 的请求。同时,在 MediaType 类中还定义了一些常量,如 APPLICATION_JSON_VALUE、APPLICATION_JSON_UTF8_VALUE 等,更多的时候推荐使用它们。

consumes 属性提供的是方法级的类型支持。与其他属性不同,当在类型级使用时,方法级的消费类型将覆盖类型级的配置,而非继承关系。

12）可生产的媒体类型

可以指定一组可生产的媒体类型，以缩小映射的范围。这样，只有请求头中 Accept 的值与指定可生产的媒体类型中有相同值的时候，请求才会被匹配。而且，使用 produces 条件可以确保用于生成响应（response）的内容与指定的可生产的媒体类型是相同的。举一个例子：

```
@Controller
@RequestMapping(path="/pets/{petId}", method=RequestMethod.GET, produces=
MediaType.APPLICATION_JSON_UTF8_VALUE)
@ResponseBody
public Pet getPet(@PathVariable String petId, Model model) {
    //方法实现省略
}
```

需要注意的是，通过 condition 条件指定的媒体类型也可以指定字符集。例如，在上面的代码中，我们还是重写 MappingJackson2HttpMessageConverter 类中默认配置的媒体类型，同时还指定了使用 UTF-8 的字符集。与 consumes 条件类似，可生产的媒体类型表达式也可以使用否定。例如，可以使用！text/plain 匹配所有请求头 Accept 中不含 text/plain 的请求。同时，在 MediaType 类中还定义了一些常量，如 APPLICATION_JSON_VALUE、APPLICATION_JSON_UTF8_VALUE 等，更多的时候推荐使用它们。

produces 属性提供的是方法级的类型支持。与其他属性不同，当在类型级使用时，方法级的消费类型将覆盖类型级的配置，而非继承关系。

13）请求参数与请求头的值

可以筛选请求参数的条件，以缩小请求匹配的范围，如"myParam" "!myParam"及"myParam=myValue"等。前两个条件用于筛选存在/不存在某些请求参数的请求，第三个条件用于筛选具有特定参数值的请求。下面的例子展示了如何使用请求参数值的筛选条件：

```
@Controller
@RequestMapping("/owners/{ownerId}")
public class RelativePathUriTemplateController {
  @RequestMapping(path="/pets/{petId}", method=RequestMethod.GET,
    params="myParam=myValue")
  public void findPet(@PathVariable String ownerId, @PathVariable String petId,
Model model) {
    //实际实现省略
  }
}
```

同样，可以用相同的条件筛选请求头的出现与否，或者筛选出一个具有特定值的请求头：

```
@Controller
@RequestMapping("/owners/{ownerId}")
public class RelativePathUriTemplateController {
```

```
    @RequestMapping(path="/pets", method=RequestMethod.GET, headers=
      "myHeader=myValue")
    public void findPet(@PathVariable String ownerId, @PathVariable String petId,
Model model) {
        //方法体实现省略
    }
}
```

尽管可以使用媒体类型的通配符(如"content-type=text/ * ")匹配请求头 Content-Type 和 Accept 的值,但更推荐独立使用 consumes 和 produces 条件筛选各自的请求,因为它们就是专门为区分这两种不同的场景而生的。

3. 定义@RequestMapping 注解的处理方法

使用 @RequestMapping 注解的处理方法可以拥有非常灵活的方法签名,它支持的方法参数及返回值类型将在接下来的内容中讲述。大多数参数都可以任意次序出现,除 BindingResult 参数外。

1) 支持的方法参数类型

下面列出所有支持的方法参数类型:

- 请求或响应对象(Servlet API)。可以是任何具体的请求或响应类型的对象,如 ServletRequest 或 HttpServletRequest 对象等。
- HttpSession 类型的会话对象(Servlet API)。使用该类型的参数要求存在一个 session,因此这样的参数永不为 null。存取 session 可能不是线程安全的,特别是在一个 Servlet 的运行环境中。如果应用可能有多个请求同时并发存取一个 session 场景,请考虑将 RequestMappingHandlerAdapter 类中的"synchronizeOnSession"标志设置为"true"。
- org. springframework. web. context. request. WebReques 或 org. springframework. web.context.request.NativeWebRequest。允许存取一般的请求参数和请求/会话范围的属性(attribute),同时无须绑定使用 Servlet/Portlet 的 API。
- 当前请求的地区信息 java.util.Locale,由已配置的最相关的地区解析器解析得到。在 MVC 的环境下,就是应用中配置的 LocaleResolver 或 LocaleContextResolver。
- 与当前请求绑定的时区信息 java.util.TimeZone(java 6 以上的版本)/java.time.ZoneId(java 8),由 LocaleContextResolver 解析得到。
- 用于存取请求正文的 java.io.InputStream 或 java.io.Reader。该对象与通过 Servlet API 拿到的输入流/Reader 是一样的。
- 用于生成响应正文的 java.io.OutputStream 或 java.io.Writer。该对象与通过 ServletAPI 拿到的输出流/Writer 是一样的。
- org.springframework.http.HttpMethod。可以拿到 HTTP 请求方法。
- 包装了当前被认证用户信息的 java.security.Principal。
- 带@PathVariable 注解的方法参数,其存放了 URI 模板变量中的值。
- 带@MatrixVariable 注解的方法参数,其存放了 URI 路径段中的键值对。
- 带@RequestParam 注解的方法参数,其存放了 Servlet 请求中指定的参数。参数的值会被转换成方法参数所声明的类型。

- 带@RequestHeader 注解的方法参数，其存放了 Servlet 请求中指定的 HTTP 请求头的值。
- 带@RequestBody 注解的参数，提供了对 HTTP 请求体的存取。参数的值通过 HttpMessageConverter 转换成方法参数所声明的类型。
- 带@RequestPart 注解的参数，提供了对一个 multipart/form-data 请求块(request part)内容的存取。
- HttpEntity<?>类型的参数，其提供了对 HTTP 请求头和请求内容的存取。请求流是通过 HttpMessageConverter 转换成 entity 对象的。
- java.util.Map/org.springframework.io.Model/org.springframework.ui.ModelMap 类型的参数，用以增强默认暴露给视图层的模型(model)的功能。
- org.springframework.web.servlet.mvc.support.RedirectAttributes 类型的参数，用以指定重定向下使用到的属性集以及添加的 flash 属性(暂存在服务端的属性，它们会在下次重定向请求的范围中有效)。
- 命令或表单对象，用于将请求参数直接绑定到 bean 字段(可能通过 setter()方法)。可以通过@InitBinder 注解和/或 HanderAdapter 的配置定制这个过程的类型转换。具体请参考 RequestMappingHandlerAdapter 类 webBindingInitializer 属性的文档。这样的命令对象，以及其上的验证结果，默认会被添加到模型 model 中，键名默认是该命令对象类的类名——例如，some.package.OrderAddress 类型的命令对象就使用属性名 orderAddress 类获取。ModelAttribute 注解可以应用在方法参数上，用以指定该模型所用的属性名。
- org.springframework.validation.Errors/org.springframework.validation.BindingResult 验证结果对象，用于存储前面的命令或表单对象的验证结果(紧接其前的第一个方法参数)。
- org.springframework.web.bind.support.SessionStatus 对象，用以标记当前的表单处理已结束。这将触发一些清理操作：@SessionAttributes 在类级别注解的属性将被移除。
- org.springframework.web.util.UriComponentsBuilder 构造器对象，用于构造当前请求 URL 相关的信息，如主机名、端口号、资源类型(scheme)、上下文路径、servlet 映射中的相对部分(literal part)等。

在参数列表中，Errors 或 BindingResult 参数必须紧跟在其绑定的验证对象后面。这是因为，在参数列表中允许有多于一个的模型对象，Spring 会为它们创建不同的 BindingResult 实例。因此，下面这样的代码是不能工作的：

```
BindingResult 与@ModelAttribute 错误的参数次序
@RequestMapping(method=RequestMethod.POST)
public String processSubmit (@ ModelAttribute ( "pet") Pet pet, Model model,
BindingResultresult) { ... }
```

上例中，因为在模型对象 Pet 和验证结果对象 BindingResult 中间还插了一个 Model 参数，这是不行的。要达到预期的效果，必须调整参数的次序如下：

```
@RequestMapping(method=RequestMethod.POST)
public String processSubmit (@ ModelAttribute ( "pet") Pet pet, BindingResult
result, Model model) { ... }
```

对于一些带有 required 属性的注解（如@RequestParam、@RequestHeader 等），JDK1.8 的 java.util.Optional 可以作为被它们注解的方法参数。在这种情况下，使用 java.util.Optional 与 required=false 的作用是相同的。

2）支持的方法返回类型

以下是 handler()方法允许的所有返回类型。

- ModelAndView 对象，其中 model 隐含填充了命令对象，以及注解了@ModelAttribute 字段的存取器被调用所返回的值。
- Model 对象，其中视图名称默认由 RequestToViewNameTranslator 决定，model 隐含填充了命令对象以及注解了@ModelAttribute 字段的存取器被调用所返回的值。
- Map 对象，用于暴露 model，其中视图名称默认由 RequestToViewNameTranslator 决定，model 隐含填充了命令对象以及注解了@ModelAttribute 字段的存取器被调用所返回的值。
- View 对象。其中 model 隐含填充了命令对象，以及注解了@ModelAttribute 字段的存取器被调用所返回的值。handler 方法也可以增加一个 Model 类型的方法参数增强 model。
- String 对象，其值会被解析成一个逻辑视图名。其中，model 默认取填充了命令对象以及注解了@ModelAttribute 字段的存取器被调用所返回的值。handler()方法也可以增加一个 Model 类型的方法参数增强 model。
- void。如果处理器方法中已经对 response 响应数据进行了处理（如在方法参数中定义一个 ServletResponse 或 HttpServletResponse 类型的参数，并直接向其响应体中写东西），那么方法可以返回 void。handler()方法也可以增加一个 Model 类型的方法参数增强 model。
- 如果处理器方法注解了 ResponseBody，那么返回类型将被写到 HTTP 的响应体中，而返回值会被 HttpMessageConverters 转换成该方法声明的参数类型。
- HttpEntity＜?＞或 ResponseEntity＜?＞对象，用于提供对 Servlet HTTP 响应头和响应内容的存取。对象体会被 HttpMessageConverters 转换成响应流。
- HttpHeaders 对象，返回一个不含响应体的 response。
- Callable＜?＞对象，当应用希望异步地返回方法值时使用，这个过程由 Spring MVC 自身的线程管理。
- DeferredResult＜?＞对象，当应用希望方法的返回值交由线程自身决定时使用。
- ListenableFuture＜?＞对象，当应用希望方法的返回值交由线程自身决定时使用。
- ResponseBodyEmitter 对象，可用它异步地向响应体中同时写多个对象，也可作为 ResponseEntity 中的主体。
- SseEmitter 对象，可用它异步地向响应体中写服务器端事件。
- StreamingResponseBody 对象，可用它异步地向响应对象的输出流中写内容。
- 其他任何返回类型，都会被处理成 model 的一个属性并返回给视图，该属性的名称为方法级的@ModelAttribute 所注解的字段名(或者以返回类型的类名作为默认的属性名)。model 隐含填充了命令对象以及注解了@ModelAttribute 字段的存取器被调用所返回的值。

3) 使用@RequestParam 将请求参数绑定至方法参数

可以使用 @RequestParam 注解将请求参数绑定到控制器的方法参数上。

下面这段代码展示了它的用法:

```
@Controller
@RequestMapping("/pets")
@SessionAttributes("pet")
public class EditPetForm{
  //...
  @RequestMapping(method=RequestMapping.GET)
  public String setupForm(@RequestParam("petId") int petId, ModelMap model) {
    Pet pet=this.clinic.loadPet(petId);
    model.addAttribute("pet", pet);
    return "petForm";
  }
  //...
}
```

若参数使用了该注解,则该参数默认是必须提供的,但也可以把该参数标注为非必需的:只将@RequestParam 注解的 required 属性设置为 false 即可(例如,@RequestParam(path="id",required=false))。

若所注解的方法参数类型不是 String,则类型转换会自动发生。若@RequestParam 注解的参数类型是 Map＜String,String＞或者 MultiValueMap＜String,String＞,则该 Map 中会自动填充所有的请求参数。

4) 使用@RequestBody 注解映射请求体

方法参数中的@RequestBody 注解暗示了方法参数被绑定了 HTTP 请求体的值。举一个例子:

```
@RequestMapping(path="/something", method=RequestMethod.PUT)
public void handle(@RequestBody String body, Writer writer)
  throws IOException{
    writer.write(body);
}
```

请求体到方法参数的转换是由 HttpMessageConverter 完成的。HttpMessageConverter 负责将 HTTP 请求信息转换成对象,以及将对象转换成一个 HTTP 响应体。对于@RequestBody 注解,RequestMappingHandlerAdapter 提供了以下 4 种默认的 HttpMessageConverter 支持。

- ByteArrayHttpMessageConverter 用以转换字节数组。
- StringHttpMessageConverter 用以转换字符串。
- FormHttpMessageConverter 用以将表格数据转换成 MultiValueMap＜String,

String>或从 MultiValueMap<String,String>中转换出表格数据。

- SourceHttpMessageConverter 用于 javax.xml.transform.Source 类的互相转换。

另外,如果使用的是 MVC 命名空间或 Java 编程的配置方式,会有更多默认注册的消息转换器。若倾向阅读和编写 XML 文件,就需要配置一个 MarshallingHttpMessageConverter 并为其提供 org.springframework.oxm 包下的一个 Marshaller 和 Unmarshaller 实现。下面的示例就展示了如何直接在配置文件中配置它。

```
<bean class="org.springframework.web.servlet.mvc.method.annotation.
RequestMappingHandlerAdapter">
  <property name="messageConverters">
    <util:list id="beanList">
      <ref bean="stringHttpMessageConverter"/>
      <ref bean="marshallingHttpMessageConverter"/>
    </util:list>
  </property
</bean>
<bean id="stringHttpMessageConverter"
  class="org.springframework.http.converter.StringHttpMessageConverter"/>
  <bean id="marshallingHttpMessageConverter"
    class="org.springframework.http.converter.xml.MarshallingHttpMessageConverter">
  <property name="marshaller" ref="castorMarshaller"/>
  <property name="unmarshaller" ref="castorMarshaller"/>
</bean>
<bean id="castorMarshaller" class="org.springframework.oxm.castor.
                                          CastorMarshaller"/>
```

注解了@RequestBody 的方法参数还可以被@Valid 注解,这样框架会使用已配置的 Validator 实例对该参数进行验证。若应用是使用 MVC 命令空间或 MVC Java 编程的方式配置的,框架会假设在 classpath 路径下存在一个符合 JSR-303 规范的验证器,并自动将其作为默认配置。

与@ModelAttribute 注解的参数一样,Errors 也可以被传入为方法参数,用于检查错误。如果没有声明这样一个参数,那么程序会抛出一个 MethodArgumentNotValidException 异常。该异常默认由 DefaultHandlerExceptionResolver 处理,处理程序会返回一个 400 错误给客户端。

5)使用@ResponseBody 注解映射响应体

@ResponseBody 注解与@RequestBody 注解类似。@ResponseBody 注解可应用于方法上,标识该方法的返回值被直接写回到 HTTP 响应体中(而不会被放置到 Model 中或被解释为一个视图名)。举一个例子:

```
@RequestMapping(path="/something", method=RequestMethod.PUT)
@ResponseBody
public String helloWorld() {
  return "Hello World"
}
```

执行上面代码的结果是文本 Hello World 将被写入 HTTP 的响应流中。与@RequestBody注解类似，Spring 使用了一个 HttpMessageConverter 将返回对象转换到响应体中。

6）使用@RestController 注解创建 REST 控制器

让控制器实现一个 REST API 是非常常见的，这种场景下控制器只提供 JSON、XML 或其他自定义的媒体类型内容即可。不需要在每个@RequestMapping 方法上都增加一个@ResponseBody 注解，更简明的做法是，给控制器加上一个@RestController 的注解。@RestController是一个原生内置的注解，它结合了@ResponseBody 与@Controller 注解的功能。与普通的@Controller 一样，@RestController 也可以与@ControllerAdvice bean 配合使用。

7）使用 HTTP 实体 HttpEntity

HttpEntity 与@RequestBody 和@ResponseBody 相似，除了能获得请求体和响应体中的内容之外，HttpEntity（以及专门负责处理响应的 ResponseEntity 子类）还可以存取请求头和响应头，像下面这样：

```
@RequestMapping("/something")
public ResponseEntity<String> handle(HttpEntity<byte[]> requestEntity) throws
UnsupportedEncodingException {
  String requestHeader=requestEntity.getHeaders().getFirst("MyRequestHeader");
  byte[] requestBody=requestEntity.getBody();
  //do something with request header and body
  HttpHeadersresponseHeaders=new HttpHeaders();
  responseHeaders.set("MyResponseHeader", "MyValue");
  return new ResponseEntity<String>("Hello World", responseHeaders,
HttpStatus.CREATED);
}
```

上面这段示例代码先是获取了 MyRequestHeader 请求头的值，然后读取请求体的主体内容。读完以后向响应头中添加了一个自己的响应头 MyResponseHeader，然后向响应流中写了字符串 Hello World，最后把响应状态码设置为 201（创建成功）。

与@RequestBody 与@ResponseBody 注解一样，Spring 使用了 HttpMessageConverter 对请求流和响应流进行转换。

8）对方法使用@ModelAttribute 注解

@ModelAttribute 注解可被应用在方法或方法参数上。这里先介绍其被注解于方法上时的用法，接下来介绍其被用于注解方法参数的用法。

注解在方法上的@ModelAttribute 说明了方法的作用是添加一个或多个属性到 model 上。这样的方法能接受与@RequestMapping 注解相同的参数类型，只不过不能直接映射到具体的请求上。在同一个控制器中，注解了@ModelAttribute 的方法实际上会在@RequestMapping 方法之前被调用。以下是几个例子：

```
//Add one attribute
//The return value of the method is added to the model under the name "account"
//You can customize the name via @ModelAttribute("myAccount")
```

```
@ModelAttribute
public Account addAccount(@RequestParam String number) {
  return accountManager.findAccount(number);
}
//Add multiple attributes
@ModelAttribute
public void populateModel(@RequestParam String number, Model model) {
  model.addAttribute(accountManager.findAccount(number));
  //add more ...
}
```

@ModelAttribute 方法通常用来填充一些公共需要的属性或数据，例如一个下拉列表预设的几种状态，或者宠物的几种类型，或者取得一个 HTML 表单渲染所需要的命令对象，如 Account 等。

注意：@ModelAttribute 方法的两种风格。在第一种写法中，方法通过返回值的方式默认将添加一个属性；在第二种写法中，方法接收一个 Model 对象，然后可以向其中添加任意数量的属性。可以根据需要在两种风格中选择一种。

一个控制器可以拥有数量不限的@ModelAttribute 方法。同一个控制器内的所有这些方法，都会在@RequestMapping 方法之前被调用。@ModelAttribute 方法也可以定义在@ControllerAdvice 注解的类中，并且这些@ModelAttribute 可以同时对许多控制器生效。属性名没有被显式指定的时候又如何呢？在这种情况下，框架将根据属性的类型给予一个默认名称。举一个例子，若方法返回一个 Account 类型的对象，则默认的属性名为“account”。可以通过设置@ModelAttribute 注解的值改变默认值。当向 Model 中直接添加属性时，请使用合适的重载方法 addAttribute(..)，即带或不带属性名的方法。@ModelAttribute注解也可以用在@RequestMapping 方法上。这种情况下，@RequestMapping 方法的返回值将会被解释为 model 的一个属性，而非一个视图名。此时视图名将以视图命名约定的方式决定，与返回值为 void 的方法所采用的处理方法类似，请求与视图名的对应。

9）在方法参数上使用@ModelAttribute 注解

@ModelAttribute 注解既可以被用在方法上，也可以被用在方法参数上。注解在方法参数上的@ModelAttribute 说明了该方法参数的值将从 model 中取得。如果从 model 中找不到，那么该参数会先被实例化，然后被添加到 model 中。在 model 中存在以后，请求中所有名称匹配的参数都会填充到该参数中。这在 Spring MVC 中被称为数据绑定，一个非常有用的特性，节约了每次都需要手动从表格数据中转换这些字段数据的时间。

```
@RequestMapping(path="/owners/{ownerId}/pets/{petId}/edit", method=
RequestMethod.POST)
public String processSubmit(@ModelAttribute Pet pet) { }
```

以上面的代码为例，这个 Pet 类型的实例可能来自哪里呢？有几种可能：

它可能因为@ SessionAttributes 注解的使用已经存在于 model 中，用@ SessionAttributes 注解，它可能因为在同一个控制器中使用了@ModelAttribute 方法已经存在于 model 中，它可能是从 URI 模板变量和类型转换中取得的，它可能是调用了自身的默认构造器被实例化出来的。

@ModelAttribute 方法常用于从数据库中取一个属性值，该值可能通过 @SessionAttributes 注解在请求中间传递。在一些情况下，使用 URI 模板变量和类型转换的方式取得一个属性是较方便的方式。这里有一个例子：

```
@RequestMapping(path="/accounts/{account}", method=RequestMethod.PUT)
public String save(@ModelAttribute("account") Account account) {
}
```

上面这个例子中，model 属性的名称("account")与 URI 模板变量的名称相匹配。如果配置了一个可以将 String 类型的账户值转换成 Account 类型实例的转换器 Converter <String，Account>，那么上面这段代码就可以工作得很好，而不需要再额外写一个 @ModelAttribute 方法。

下一步是数据的绑定。WebDataBinder 类能将请求参数——包括字符串的查询参数和表单字段等——通过名称匹配到 model 的属性上。成功匹配的字段在需要的时候会进行一次类型转换(从 String 类型到目标字段的类型)，然后被填充到 model 对应的属性中。进行数据绑定后，可能会出现一些错误，如没有提供必需的字段、类型转换过程的错误等。若想检查这些错误，可以在注解了@ModelAttribute 的参数后紧跟着声明一个 BindingResult 参数：

```
@RequestMapping(path="/owners/{ownerId}/pets/{petId}/edit", method=
RequestMethod.POST)
public String processSubmit (@ ModelAttribute ( "pet") Pet pet, BindingResult
result) {
  if(result.hasErrors()) {
    return "petForm";
  }
  //...
}
```

拿到 BindingResult 参数后，可以检查是否有错误。有时可以通过 Spring 的<errors>表单标签在同一个表单上显示错误信息。BindingResult 用于记录数据绑定过程的错误，因此，除了数据绑定外，还可以把该对象传给自己定制的验证器调用验证。这使得数据绑定过程和验证过程出现的错误可以被收集到一处，然后一并返回给用户。

```
@RequestMapping(path="/owners/{ownerId}/pets/{petId}/edit", method=
RequestMethod.POST)
public String processSubmit (@ ModelAttribute ( "pet") Pet pet, BindingResult
result) {
  new PetValidator().validate(pet, result);
  if(result.hasErrors()) {
    return "petForm";
  }
  //...
}
```

或者,可以添加一个 JSR-303 规范的@Valid 注解,这样验证器就会自动被调用。

```
@RequestMapping(path="/owners/{ownerId}/pets/{petId}/edit", method=
RequestMethod.POST)
public String processSubmit(@Valid @ModelAttribute("pet") Pet pet,
BindingResult result) {
  if(result.hasErrors()) {
    return "petForm";
  }
  //...
}
```

10) 在请求之间使用@SessionAttributes 注解,使用 HTTP 会话保存模型数据

类型级别的@SessionAttributes 注解声明了某个特定处理器所使用的会话属性。通常它会列出该类型希望存储到 session 或 converstaion 中的 model 属性名或 model 的类型名,一般用于在请求之间保存一些表单数据的 Bean。

以下的代码段演示了该注解的用法,它指定了模型属性的名称。

```
@Controller
@RequestMapping("/editPet.do")
@SessionAttributes("pet")
public class EditPetForm{
    //...
}
```

11) 使用"application/x-www-form-urlencoded"数据

对于不是使用的浏览器的客户端,推荐使用这个注解处理请求。但是,当请求是一个 HTTP PUT 方法的请求时,有一个事情需要注意。浏览器可以通过 HTTP 的 GET()方法或 POST()方法提交表单数据,非浏览器的客户端还可以通过 HTTP 的 PUT()方法提交表单。这样设计是一个挑战,因为在 Servlet 规范中明确规定,ServletRequest.getParameter*()系列的方法只能支持通过 HTTP POST()方法的方式提交表单,而不支持 HTTP PUT 的方式。

为了支持 HTTP 的 PUT 类型和 PATCH 类型的请求,Spring 的 spring-web 模块提供了一个过滤器 HttpPutFormContentFilter。可以在 web.xml 文件中配置它:

```
<filter>
  <filter-name>httpPutFormFilter</filter-name>
  <filter-class>org.springframework.web.filter.HttpPutFormContentFilter
    </filterclass>
</filter>
<filter-mapping>
  <filter-name>httpPutFormFilter</filter-name>
  <servlet-name>dispatcherServlet</servlet-name>
</filter-mapping>
<servlet>
```

```
    <servlet-name>dispatcherServlet</servlet-name>
    <servlet-class>org.springframework.web.servlet.DispatcherServlet
      </servlet-class>
</servlet>
```

上面的过滤器将会拦截内容类型（content type）为 application/x-www-form-urlencoded、HTTP 方法为 PUT 或 PATCH 类型的请求，然后从请求体中读取表单数据，把它们包装在 ServletRequest 中。这是为了使表单数据能够通过 ServletRequest.getParameter * ()系列的方法获取。

因为 HttpPutFormContentFilter 会消费请求体的内容，因此，它不应该用于处理那些依赖于其他 application/x-www-form-urlencoded 转换器的 PUT 和 PATCH 请求，这包括了@ RequestBodyMultiValueMap＜String，String＞和 HttpEntity＜MultiValueMap＜String，String＞＞。

12）使用@CookieValue 注解映射 cookie 值

@CookieValue 注解能将一个方法参数与一个 HTTP cookie 的值进行绑定。看一个这样的场景：以下的这个 cookie 存储在一个 HTTP 请求中：

```
JSESSIONID=415A4AC178C59DACE0B2C9CA727CDD84
```

下面的代码演示了拿到 JSESSIONID 这个 cookie 值的方法：

```
@RequestMapping("/displayHeaderInfo.do")
public void displayHeaderInfo(@CookieValue("JSESSIONID") String cookie) {
//...
}
```

若注解的目标方法参数不是 String 类型，则类型转换会自动进行。这个注解可以注解到处理器方法上，在 Servlet 环境和 Portlet 环境都能使用。

13）使用@RequestHeader 注解映射请求头属性

@RequestHeader 注解能将一个方法参数与一个请求头属性进行绑定。以下是一个请求头的例子：

```
Host localhost:8080
Accept text/html,application/xhtml+xml,application/xml;q=0.9
Accept-Language fr,en-gb;q=0.7,en;q=0.3
Accept-Encoding gzip,deflate
Accept-Charset ISO-8859-1,utf-8;q=0.7,*;q=0.7
Keep-Alive 300
```

以下的代码片段展示了如何取得 Accept-Encoding 请求头和 Keep-Alive 请求头的值：

```
@RequestMapping("/displayHeaderInfo.do")
public void displayHeaderInfo (@ RequestHeader ( " Accept - Encoding ") String
encoding,
  @RequestHeader("Keep-Alive") long keepAlive){
  //...
}
```

若注解的目标方法参数不是 String 类型，则类型转换会自动进行。如果@RequestHeader 注解应用在 Map＜String，String＞、MultiValueMap＜String，String＞或 HttpHeaders 类型的参数上，那么所有的请求头属性值都会被填充到 map 中。Spring 内置支持将一个逗号分隔的字符串（或其他类型转换系统所能识别的类型）转换成一个 String 类型的列表/集合。举一个例子，一个注解了@RequestHeader（"Accept"）的方法参数可以是一个 String 类型，也可以是 String[]或 List＜String＞类型。这个注解可以注解到处理器方法上，在 Servlet 环境和 Portlet 环境都能使用。

14）方法参数与类型转换

从请求参数、路径变量、请求头属性或者 cookie 中抽取出的 String 类型的值，可能需要被转换成其所绑定的目标方法参数或字段的类型（例如，通过@ModelAttribute 将请求参数绑定到方法参数上）。如果目标类型不是 String，Spring 会自动进行类型转换。所有的简单类型（如 int、long、Date）都有内置的支持。如果想进一步定制这个转换过程，可以通过 WebDataBinder，或者为 Formatters 配置一个 FormattingConversionService 做到。

15）定制 WebDataBinder 的初始化

如果想通过 Spring 的 WebDataBinder 在属性编辑器中做请求参数的绑定，可以使用在控制器内使用@InitBinder 注解的方法、在注解了@ControllerAdvice 的类中使用@InitBinder 注解的方法，或者提供一个定制的 WebBindingInitializer。

（1）数据绑定的定制：使用@InitBinder。

使用@InitBinder 注解控制器的方法，可以直接在控制器类中定制应用的数据绑定。@InitBinder 用来标记一些方法，这些方法会初始化一个 WebDataBinder，并用以为处理器方法填充命令对象和表单对象的参数。除了命令/表单对象以及相应的验证结果对象，这样的"绑定器初始化"方法能够接收@RequestMapping 所支持的所有参数类型。"绑定器初始化"方法不能有返回值，因此，一般将它们声明为 void 返回类型。特别地，当 WebDataBinder 与 WebRequest 或 java.util.Locale 一起作为方法参数时，可以在代码中注册上下文相关的编辑器。下面的代码示例演示了如何使用@InitBinder 配置一个 CustomerDateEditor，后者会对所有 java.util.Date 类型的表单字段进行操作：

```
@Controller
public class MyFormController{
  @InitBinder
  public void initBinder(WebDataBinder binder) {
    SimpleDateFormatdateFormat=new SimpleDateFormat("yyyy-MM-dd");
    dateFormat.setLenient(false);
    binder.registerCustomEditor(Date.class, new CustomDateEditor(dateFormat,
false));
  }
  //...
}
```

或者，可以使用 Spring 4.2 提供的 addCustomFormatter 指定 Formatter 的实现，而非通过 PropertyEditor 实例。这在拥有一个需要 Formatter 的 setup()方法，并且该方法位于一个共享的 FormattingConversionService 中时非常有用。这样，对于控制器级别的绑定规则的定制,代码更容易被复用。

```
@Controller
public class MyFormController{
  @InitBinder
  public void initBinder(WebDataBinder binder) {
    binder.addCustomFormatter(new DateFormatter("yyyy-MM-dd"));
}
  //...
}
```

(2) 配置定制的 WebBindingInitializer。

为了 externalize 数据绑定的初始化过程,可以为 WebBindingInitializer 接口提供一个自己的实现,在其中可以为 AnnotationMethodHandlerAdapter 提供一个默认的配置 Bean,以此重写默认的配置。

以下的代码来自 PetClinic 的应用,它展示了为 WebBindingInitializer 接口提供一个自定义实现:org.springframework.samples.petclinic.web.ClinicBindingInitializer 完整的配置过程。后者中配置了 PetClinic 应用中许多控制器需要的属性编辑器 PropertyEditors。

```
<bean class="org.springframework.web.servlet.mvc.method.annotation.
    RequestMappingHandlerAdapter">
  <property name="cacheSeconds" value="0"/>
  <property name="webBindingInitializer">
    <bean class="org.springframework.samples.petclinic.web.
      ClinicBindingInitializer"/>
  </property>
</bean>
```

@InitBinder 方法也可以定义在@ControllerAdvice 注解的类上,这样,配置可以为许多控制器所共享。这提供了除使用 WebBindingInitializer 外的另外一种方法。

16) 使用@ControllerAdvice 辅助控制器

@ControllerAdvice 是一个组件注解,它使得其实现类能够被 classpath 扫描自动发现。若应用是通过 MVC 命令空间或 MVCJava 编程方式配置的,那么该特性默认是自动开启的。注解@ControllerAdvice 的类可以拥有@ExceptionHandler、@InitBinder 或@ModelAttribute注解的方法,并且这些方法会被应用至控制器类层次的所有@RequestMapping 方法上。

也可以通过@ControllerAdvice 的属性指定其只对一个子集的控制器生效:

```
@ControllerAdvice(annotations=RestController.class)
public class AnnotationAdvice {}
@ControllerAdvice("org.example.controllers")
public class BasePackageAdvice {}
@ControllerAdvice(assignableTypes={ControllerInterface.class, AbstractController.
class})
public class AssignableTypesAdvice {}
```

4. 异步请求处理

1）异步请求处理概述

Spring MVC 3.2 开始引入了基于 Servlet 3 的异步请求处理。相比以前，控制器方法已经不一定需要返回一个值，而是可以返回一个 java.util.concurrent.Callable 的对象，并通过 Spring MVC 所管理的线程产生返回值。与此同时，Servlet 容器的主线程则可以退出并释放其资源了，同时也允许容器处理其他的请求。通过一个 TaskExecutor，Spring MVC 可以在另外的线程中调用 Callable。当 Callable 返回时，请求再携带 Callable 返回的值，再次被分配到 Servlet 容器中恢复处理流程。以下代码给出一个这样的控制器方法作为例子：

```
@RequestMapping(method=RequestMethod.POST)
public Callable<String>processUpload(final MultipartFile file) {
  return new Callable<String>() {
    public String call() throws Exception {
    //...
    return "someView";
    }
  };
}
```

另一个选择是让控制器方法返回一个 DeferredResult 的实例。这种场景下，返回值可以由任何一个线程产生，也包括那些不是由 Spring MVC 管理的线程。举一个例子，返回值可能是为了响应某些外部事件所产生的，如一条 JMS 的消息、一个计划任务等。以下代码给出一个这样的控制器作为例子：

```
@RequestMapping("/quotes")
@ResponseBody
public DeferredResult<String>quotes() {
  DeferredResult<String>deferredResult=new DeferredResult<String>();
  //Save the deferredResult somewhere..
  return deferredResult;
}
//In some other thread...
deferredResult.setResult(data);
```

如果对 Servlet 3.0 的异步请求处理特性没有了解，理解这个特性可能会有一点困难。以下给出这个机制运作背后的一些原理：

- 一个 servlet 请求 ServletRequest 可以通过调用 request.startAsync()方法而进入异步模式。这样做的主要结果是该 servlet 以及所有的过滤器都可以结束，但其响应(response)会留待异步处理结束后再返回异步请求的处理。
- 调用 request.startAsync()方法会返回一个 AsyncContext 对象，可用它对异步处理进行进一步的控制和操作。例如，说它也提供了一个与转发(forward)很相似的 dispatch()方法，只不过它允许应用恢复 Servlet 容器的请求处理进程。

- ServletRequest 提供了获取当前 DispatherType 的方式，后者可以用来区别当前处理的是原始请求、异步分发请求、转向，或是其他类型的请求分发类型。

有了上面的知识，下面可以看一下 Callable 的异步请求被处理时所发生的事件：

- 控制器先返回一个 Callable 对象。
- Spring MVC 开始进行异步处理，并把该 Callable 对象提交给另一个独立线程的执行器 TaskExecutor 处理。
- DispatcherServlet 和所有过滤器都退出 Servlet 容器线程，但此时方法的响应对象仍未返回。
- Callable 对象最终产生一个返回结果，此时 Spring MVC 会重新把请求分派回 Servlet 容器，恢复处理。
- DispatcherServlet 再次被调用，恢复对 Callable 异步处理所返回结果的处理。

对 DeferredResult 异步请求的处理顺序也非常类似，区别仅在于应用可以通过任何线程来计算返回一个结果。

- 控制器先返回一个 DeferredResult 对象，并把它存取在内存(队列或列表等)中，以便存取。
- Spring MVC 开始进行异步处理。
- DispatcherServlet 和所有过滤器都退出 Servlet 容器线程，但此时方法的响应对象仍未返回。
- 由处理该请求的线程对 DeferredResult 进行设值，然后 Spring MVC 会重新把请求分派回 Servlet 容器，恢复处理。
- DispatcherServlet 再次被调用，恢复对该异步返回结果的处理。

2) 异步请求的异常处理

若控制器返回的 Callable 在执行过程中抛出了异常，又会发生什么事情？简单来说，这与一般的控制器方法抛出异常一样。它会被正常的异常处理流程捕获处理。更具体地说，当 Callable 抛出异常时，Spring MVC 会把一个 Exception 对象分派给 Servlet 容器进行处理，而不是正常返回方法的返回值，然后容器恢复对此异步请求异常的处理。若方法返回的是一个 DeferredResult 对象，可以选择调用 Exception 实例的 setResult()方法或 setErrorResult()方法。

3) 拦截异步请求

处理器拦截器 HandlerInterceptor 可以实现 AsyncHandlerInterceptor 接口拦截异步请求，因为在异步请求开始时，被调用的回调方法是该接口的 afterConcurrentHandlingStarted()方法，而非一般的 postHandle()和 afterCompletion()方法。如果需要与异步请求处理的生命流程有更深入的集成，如需要处理 timeout 的事件等，则 HandlerInterceptor 需要注册一个 CallableProcessingInterceptor 或 DeferredResultProcessingInterceptor 拦截器。DeferredResult 类还提供了 onTimeout(Runnable)和 onCompletion(Runnable)等方法，Callable 需要请求过期(timeout)和完成后的拦截时，可以把它包装在一个 WebAsyncTask 实例中，后者提供了相关的支持。

4) HTTP 内容流

控制器可以使用 DeferredResult 或 Callable 对象异步地计算其返回值，这可以用于实现一些有用的技术,如 long polling 技术,让服务器可以尽可能快地向客户端推送事件。

如果想在一个 HTTP 响应中同时推送多个事件,怎么办?这样的技术已经存在,与“Long Polling”相关,叫“HTTP Streaming”。Spring MVC 支持这项技术,可以通过让方法返回一个 ResponseBodyEmitter 类型对象实现,该对象可用于发送多个对象。通常使用的@ResponseBody 只能返回一个对象,它是通过 HttpMessageConverter 写到响应体中的。

下面是一个实现该技术的例子:

```
@RequestMapping("/events")
public ResponseBodyEmitter handle() {
  ResponseBodyEmitter emitter=new ResponseBodyEmitter();
  //Save the emitter somewhere..
  return emitter;
}
//In some other thread
emitter.send("Hello once");
//and again later on
emitter.send("Hello again");
//and done at some point
emitter.complete();
```

ResponseBodyEmitter 也可以被放到 ResponseEntity 体里使用,这可以对响应状态和响应头做一些定制。

5) 使用“服务器端事件推送”的 HTTP 内容流

SseEmitter 是 ResponseBodyEmitter 的一个子类,提供了对服务器端事件推送的技术的支持。服务器端事件推送其实只是一种 HTTP Streaming 的类似实现,只不过服务器端所推送的事件遵循了 W3C Server-Sent Events 规范中定义的事件格式。“服务器端事件推送”技术正如其名,是用于由服务器端向客户端进行的事件推送。这在 Spring MVC 中很容易做到,只需要方法返回一个 SseEmitter 类型的对象。需要注意的是,Internet Explorer 并不支持这项服务器端事件推送的技术。另外,对于更大型的 Web 应用及更精致的消息传输场景(如在线游戏、在线协作、金融应用等)来说,使用 Spring 的 WebSocket(包含 SockJS 风格的实时 WebSocket)更成熟一些,因为它支持的浏览器范围非常广,并且,对于一个以消息为中心的架构,它为服务器端-客户端间的事件发布-订阅模型的交互提供了更高层级的消息模式(messaging patterns)的支持。

6) 直接写回输出流的 HTTP 内容流

ResponseBodyEmitter 也允许通过 HttpMessageConverter 向响应体中支持写事件对象。这可能是最常见的情形,如写返回的 JSON 数据的时候。但有时跳过消息转换的阶段,直接把数据写回响应的输出流 OutputStream 可能更有效,如文件下载这样的场景。这可以通过返回一个 StreamingResponseBody 类型的对象实现。

以下是一个实现的例子:

```
@RequestMapping("/download")
public StreamingResponseBody handle() {
```

```
    return new StreamingResponseBody() {
      @Override
      public void writeTo(OutputStreamoutputStream) throws IOException{
        //write...
      }
    };
  }
```

ResponseBodyEmitter 也可以被放到 ResponseEntity 体里使用，这可以对响应状态和响应头做一些定制。

7）异步请求处理的相关配置

（1）Servlet 容器配置。

对于使用 web.xml 配置文件的应用，请确保 web.xml 的版本更新到 3.0：

```
<web-app xmlns="http://java.sun.com/xml/ns/javaee"
  xmlns:xsi="http://www.w3.org/2001/XMLSchema-instance
  http://java.sun.com/xml/ns/javaeehttp://java.sun.com/xml/ns/javaee/web-app
_3_0.xsd" version="3.0">
...
</web-app>
```

异步请求必须在 web.xml 中将 DispatcherServlet 下的子元素<async-supported>true</asyncsupported>设置为 true。此外，所有可能参与异步请求处理的过滤器 Filter 都必须配置为支持 ASYNC 类型的请求分派。在 Spring 框架中为过滤器启用支持 ASYNC类型的请求分派应是安全的，因为这些过滤器一般都继承了基类 OncePerRequestFilter，后者在运行时会检查该过滤器是否需要参与到异步分派的请求处理中。下面的例子展示了 web.xml 的配置：

```
<web-app xmlns="http://java.sun.com/xml/ns/javaee"
  xmlns:xsi=http://www.w3.org/2001/XMLSchema-instance
  xsi:schemaLocation="http://java.sun.com/xml/ns/javaee
    http://java.sun.com/xml/ns/javaee/web-app_3_0.xsd" version="3.0">
  <filter>
    <filter-name>Spring OpenEntityManagerInViewFilter</filter-name>
    <filter-class>org.springframework.~ .OpenEntityManagerInViewFilter
      </filterclass>
    <async-supported>true</async-supported>
  </filter>
  <filter-mapping>
    <filter-name>Spring OpenEntityManagerInViewFilter</filter-name>
    <url-pattern>/*</url-pattern>
    <dispatcher>REQUEST</dispatcher>
    <dispatcher>ASYNC</dispatcher>
  </filter-mapping>
</web-app>
```

如果应用使用的是 Servlet 3 规范基于 Java 编程的配置方式，如通过 WebApplicationInitializer，那么也需要设置"asyncSupported"标志和 ASYNC 分派类型的支持，就像在 web.xml 中所配置的一样。可以直接继承 AbstractDispatcherServletInitializer 或 AbstractAnnotationConfigDispatcherServletInitializer 来简化配置，它们都自动设置了这些配置项，并使得注册 Filter 过滤器实例变得非常简单。

(2) Spring MVC 配置。

MVC Java 编程配置和 MVC 命名空间配置方式都提供了配置异步请求处理支持的选择。WebMvcConfigurer 提供了 configureAsyncSupport() 方法，而<mvc: annotation-driven>有一个子元素<async-support>，它们都用以为此提供支持。这些配置允许重写异步请求默认的超时时间，在未显式设置时，它们的值与所依赖的 Servlet 容器是相关的(例如，Tomcat 设置的超时时间是 10s)。也可以配置用于执行控制器返回值 Callable 的执行器 AsyncTaskExecutor。Spring 强烈推荐配置这个选项，因为 Spring MVC 默认使用的是普通的执行器 SimpleAsyncTaskExecutor。MVC Java 编程配置及 MVC 命名空间配置的方式都允许注册自己的 CallableProcessingInterceptor 和 DeferredResultProcessingInterceptor 拦截器实例。若需要为特定的 DeferredResult 重写默认的超时时间，可以选用合适的构造方法实现。类似地，对于 Callable 返回，可以把它包装在一个 WebAsyncTask 对象中，并使用合适的构造方法定义超时时间。WebAsyncTask 类的构造方法同时也能接受一个任务执行器 AsyncTaskExecutor 类型的参数。

5. 处理器映射

在 Spring 的前期版本中，用户需要在 Web 应用的上下文中定义一个或多个 HandlerMapping bean，用以将进入容器的 Web 请求映射到合适的处理器方法上。允许在控制器上添加注解后，通常就不必这么做了，因为 RequestMappingHandlerMapping 类会自动查找所有注解了@RequestMapping 的@Controller 控制器 Bean。同时也应知道，所有继承自 AbstractHandlerMapping 的处理器方法映射 HandlerMapping 类都拥有下列的属性，可以对它们进行定制：

- 一个 interceptors 列表，指示了应用其上的一个拦截器列表。
- defaultHandler，生效的默认处理器，根据 order 属性的值，Spring 会对上下文可用的所有处理器映射进行排序，并应用第一个匹配成功的处理器。
- alwaysUseFullPath(总是使用完整路径)。若设置为 true，Spring 将在当前 Servlet 上下文中使用完整路径查找合适的处理器。若设置为 false(默认就为 false)，则使用当前 Servlet 的 mapping 路径。举一个例子，若一个 Servlet 的 mapping 路径是/testing/*，并且 alwaysUseFullPath 属性被设置为 true，此时用于查找处理器的路径将是/testing/viewPage.html；而若 alwaysUseFullPath 属性的值为 false，则此时查找路径是/viewPage.html。
- urlDecode，默认设置为 true(也是 Spring 2.5 的默认设置)。若需要比较加密过的路径，则把此标志设为 false。需要注意的是，HttpServletRequest 永远以未加密的方式存储 Servlet 路径。此时，该路径将无法匹配到加密过的路径。

下面的代码展示了配置一个拦截器的方法：

```
<beans>
  <bean id="handlerMapping" class="org.springframework.web.servlet.mvc.
    method.annotation.RequestMappingHandlerMapping">
    <property name="interceptors">
      <bean class="example.MyInterceptor"/>
    </property>
  </bean>
<beans>
```

5.4.3 ModelAndView

Spring MVC 使用 ModelAndView 类存储处理后的结果数据，以及显示该数据的视图。从名字上看，ModelAndView 中的 Model 代表模型，View 代表视图，这个名字就很好地解释了该类的作用。业务处理器调用模型层处理完用户请求后，把结果数据存储在该类的 model 属性中，把要返回的视图信息存储在该类的 view 属性中，然后 ModelAndView 返回该 Spring MVC 框架。框架通过调用配置文件中定义的视图解析器，对该对象进行解析，最后把结果数据显示在指定的页面上。

1. ModelAndView()方法

1）添加模型数据

将控制器方法中处理的结果数据传递到结果页面，也就是把在结果页面上需要的数据放到 ModelAndView 对象中，其作用类似于 request 对象的 setAttribute()方法的作用，用来在一个请求过程中传递处理的数据。通过以下方法添加模型数据：

```
public ModelAndView addObject(String attributeName, Object
attributeValue){...}
public ModelAndView addObject(Object attributeValue){...}
public ModelAndView addAllObjects(Map modelMap)
```

在页面中可以获取并展示 ModelAndView 中的数据。

2）返回指定页面

ModelAndView()方法可以指定返回的页面名称：

```
ModelAndView view=new ModelAndView("xxx");
```

也可以通过 setViewName()方法跳转到指定的页面：

```
public void setViewName(String viewName){...}
```

接下来看一个使用 ModelAndView 的实例(Spring MVC_ModelAndView_demo)。具体步骤如下：

(1) 新建一个 Web 项目，项目名称为 Spring MVC_ ModelAndView _demo。

(2) 添加 Spring MVC 相关 jar 包，右击 Spring MVC_ ModelAndView _demo 项目，从弹出的快捷菜单中选择 Configure Facets...→Install Spring Facet。

(3) 建立项目目录。

在 src 下创建包 com.ascent.controller，并存放在 ModelAndViewController 控制类中。

在 src 下创建包 com.ascent.po,并存放在 Student 实体类中。

(4) 修改 web.xml,添入以下内容:

```
<?xml version="1.0"encoding="UTF-8"?>
<web-app xmlns:xsi="http://www.w3.org/2001/XMLSchema-instance"
xmlns="http://xmlns.jcp.org/xml/ns/javaee"
xsi:schemaLocation="http://xmlns.jcp.org/xml/ns/javaee
http://xmlns.jcp.org/xml/ns/javaee/web-app_3_1.xsd"id="WebApp_ID"version=
"3.1">
  <display-name>Spring MVC_login_demo</display-name>

  <listener>
<listener-class>org.springframework.web.context.ContextLoaderListener
</listener-class>
  </listener>
  <context-param>
    <param-name>contextConfigLocation</param-name>
    <param-value>classpath:applicationContext.xml</param-value>
  </context-param>
  <servlet>
    <servlet-name>Spring MVC</servlet-name>
    <servlet-class>
        org.springframework.web.servlet.DispatcherServlet
    </servlet-class>
    <load-on-startup>1</load-on-startup>
  </servlet>
  <servlet-mapping>
    <servlet-name>Spring MVC</servlet-name>
    <url-pattern>/</url-pattern>
  </servlet-mapping>

  <filter>
    <filter-name>CharacterEncodingFilter</filter-name>
<filter-class>org.springframework.web.filter.CharacterEncodingFilter
</filter-class>
    <init-param>
      <param-name>encoding</param-name>
      <param-value>UTF-8</param-value>
    </init-param>
  </filter>
  <filter-mapping>
    <filter-name>CharacterEncodingFilter</filter-name>
    <url-pattern>/*</url-pattern>
  </filter-mapping>
```

</web-app>这里将 DispatcherServlet 命名为 Spring MVC,并且让它在 Web 项目一

启动就加载。接下来需要在 WEB-INF 目录下创建一个 Spring MVC -servlet. xml 的 Spring 配置文件，此文件名的命名规则为：在 servlet-name 名称后面加上-servlet.xml。

（5）添加 Spring MVC-servlet.xml 配置文件，内容如下。

```
<?xml version="1.0"encoding="UTF-8"?>
<beans xmlns="http://www.springframework.org/schema/beans"
    xmlns:xsi="http://www.w3.org/2001/XMLSchema-instance"
    xmlns:p="http://www.springframework.org/schema/p"
    xmlns:context="http://www.springframework.org/schema/context"
    xmlns:mvc="http://www.springframework.org/schema/mvc"
    xsi:schemaLocation="
    http://www.springframework.org/schema/beans
    http://www.springframework.org/schema/beans/spring-beans-3.0.xsd
    http://www.springframework.org/schema/context
    http://www.springframework.org/schema/context/spring-context-3.0.xsd
    http://www.springframework.org/schema/mvc
    http://www.springframework.org/schema/mvc/spring-mvc-3.0.xsd">

<!--默认的注解映射的支持 -->
<mvc:annotation-driven/>
<!--启用自动扫描  -->
<context:component-scan base-package="com.ascent"/>
<bean class="org.springframework.web.servlet.view.InternalResourceViewResolver">
    <property name="prefix"value="/WEB-INF/views/"/>
    <property name="suffix"value=".jsp"/>
</bean>
</beans>
```

（6）在 com.ascent po 包中编写 Student 类。

```
package com.ascent.po;

public class Student {
    private int id;
    private String name;            //姓名
    private int age;                //年龄
    private String description;     //描述

    public int getId() {
      return id;
    }
    public void setId(int id) {
      this.id=id;
    }
    public String getName() {
      return name;
```

```
    }
    public void setName(String name) {
      this.name=name;
    }
    public int getAge() {
      return age;
    }
    public void setAge(int age) {
      this.age=age;
    }
    public String getDescription() {
      return description;
    }
    public void setDescription(String description) {
      this.description=description;
    }
}
```

(7) 在 com.ascent controller 包中编写 ModelAndViewController 类。

```
package com.ascent.controller;

import org.springframework.stereotype.Controller;
import org.springframework.web.bind.annotation.RequestMapping;
import org.springframework.web.servlet.ModelAndView;

import com.ascent.po.Student;

@Controller
@RequestMapping(value="handler")
public class ModelAndViewController{
@RequestMapping(value="testModelAndView")
    public ModelAndView testModelAndView() {
        ModelAndView mv=new ModelAndView("success");
        //设置返回页面 views/success.jsp

        Student student=new Student() ;
        student.setId(1);
        student.setName("Lixin");
        student.setAge(25);

        mv.addObject("student", student);
        //相当于 request.setAttribute("student", student);
        return mv;       //返回数据和页面
    }
}
```

(8) 编写 JSP 文件。

① 在 WebRoot 下编写 index.jsp。

```
<%@ page language="java" contentType="text/html; charset=UTF-8"
    pageEncoding="UTF-8"%>

<!DOCTYPE html PUBLIC "-//W3C//DTD HTML 4.01 Transitional//EN"
"http://www.w3.org/TR/html4/loose.dtd">
<html>
    <head>
    <meta http-equiv="Content-Type" content="text/html; charset=UTF-8">
    <title>Spring MVC_ModelAndView_demo Test</title>
    </head>
    <body>
        <br/>
            <a href="handler/testModelAndView">testModelAndView</a>
        <br/>
    </body>
</html>
```

② 根据 Spring MVC-servlet.xml,在/WEB-INF/views/下编写 success.jsp。

```
<%@ page language="java" contentType="text/html; charset=UTF-8"
    pageEncoding="UTF-8"%>

<!DOCTYPE html PUBLIC "-//W3C//DTD HTML 4.01 Transitional//EN"
"http://www.w3.org/TR/html4/loose.dtd">
<html>
    <head>
    <meta http-equiv="Content-Type" content="text/html; charset=UTF-8">
    <title>Spring MVC_ModelAndView_demo Test </title>
    </head>
    <body>
            ${ requestScope. student. id } - ${ requestScope. student. name } -
${requestScope.student.age }<br/>
    </body>
</html>
```

(9) 部署和运行项目。

选中 Spring MVC_login_demo 项目,部署在 Tomcat 8 服务器上,之后启动 Tomcat。

在浏览器中输入 http://localhost:8080/Spring MVC_ModelAndView_demo 进入 index.jsp 页面,如图 5-21 所示。

单击 testModelAndView 链接,进入成功页面,如图 5-22 所示。

2. ModelAndView 和 Model 的区别

这两者之间有很大的区别,具体表现在 Model 只用来传输数据,并不会进行业务的寻

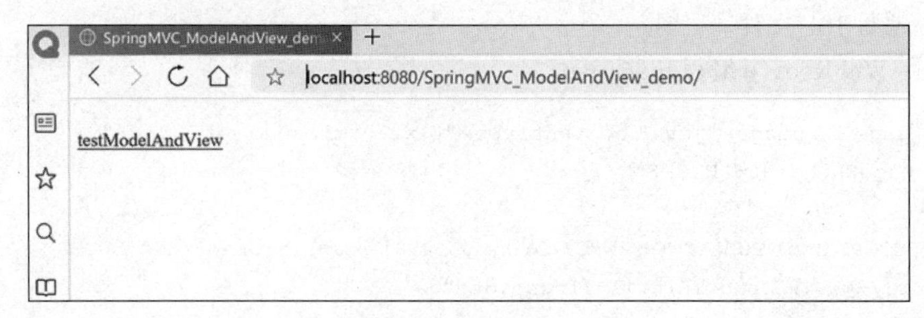

图 5-21　index.jsp 页面

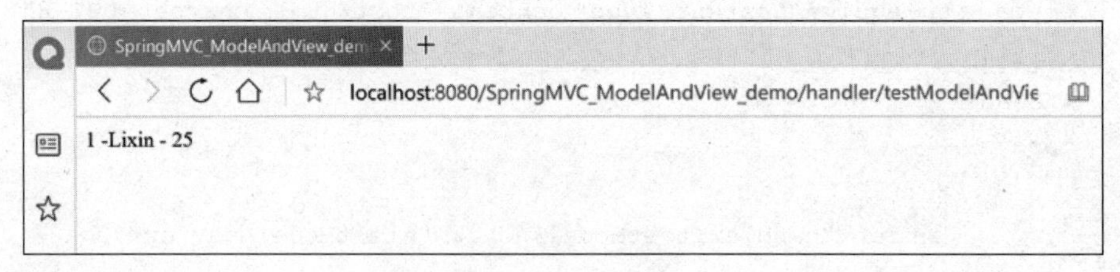

图 5-22　成功页面

址。ModelAndView 却是可以进行业务寻址的，就是设置对应的要请求的视图文件，这里的视图文件指的是类似 jsp 的文件。

其次，两者还有一个最大的区别，就是 Model 是每一次请求都可以自动创建，但是 ModelAndView 是需要我们创建的。

1）Model

Model 是一个接口，其实现类为 ExtendedModelMap，继承了 ModelMap 类。

```
public classExtendedModelMap extends ModelMap implements Model
```

一般来说，可以用 Model 接收各种类型的数据，如果使用 Model 接收一组数据，那么这个时候的 Model 实际上是 ModelMap。

2）ModelMap

ModelMap 对象主要用于传递控制方法处理数据到结果页面，也就是说，把结果页面上需要的数据放到 ModelMap 对象中即可，它的作用类似于 request 对象的 setAttribute()方法的作用：用来在一个请求过程中传递处理的数据。

ModelMap 或者 Model 通过 addAttribute()方法向页面传递参数。

其中 addAttribute()方法的参数有多种方式，常用的有：

```
publicModelMap addAttribute(String attributeName, Object attribute Value){...}
publicModelMap addAttribute(Object attributeValue){...}
publicModelMap addAllAttributes(Collection<?>attribute Values) {...}
publicModelMap addAllAttributes(Map<String, ?>attributes){...}
```

然后在 jsp 页面上可以获取并展示 modelmap 中的数据。modelmap 本身不能设置页面跳转的 url 地址别名或者物理跳转地址。可以通过控制器方法的字符串返回值设置跳转 url 地址别名或者物理跳转地址。

5.4.4 视图解析

所有 Web 应用的 MVC 框架都提供了视图相关的支持。Spring 提供了一些视图解析器，它们让你能够在浏览器中渲染模型，并支持自由选用适合的视图技术，而不必与框架绑定到一起。Spring 原生支持 JSP 视图技术、Velocity 模板技术和 XSLT 视图等。

有两个接口在 Spring 处理视图相关事宜时至关重要，分别是视图解析器接口 ViewResolver 和视图接口 View。视图解析器接口 ViewResolver 负责处理视图名与实际视图之间的映射关系。视图接口 View 负责准备请求，并将请求的渲染交给某种具体的视图技术实现。

1. 处理器映射

Spring MVC 中所有控制器的处理器方法都必须返回一个逻辑视图的名字，无论是显式返回(如返回一个 String、View 或者 ModelAndView)，还是隐式返回(如基于约定的返回)。Spring 中的视图由一个视图名标识，并由视图解析器渲染。Spring 有非常多内置的视图解析器。表 5-3 列出了主要的视图解析器。

表 5-3 主要的视图解析器

视图解析器	描　述
AbstractCachingViewResolver	一个抽象的视图解析器类，提供了缓存视图的功能。通常，视图在能够被使用之前需要经过准备。继承这个基类的视图解析器可以获得缓存视图的能力
XmlViewResolver	视图解析器接口 ViewResolver 的一个实现，该类接受一个 XML 格式的配置文件。该 XML 文件必须与 Spring XML 的 bean 工厂有相同的 DTD。默认的配置文件名是 /WEB-INF/views.xml
ResourceBundleViewResolver	视图解析器接口 ViewResolver 的一个实现，采用 bundle 根路径指定的 ResourceBundle 中的 bean 定义作为配置。一般 bundle 都定义在 classpath 路径下的一个配置文件中。默认的配置文件名为 views.properties
UrlBasedViewResolver	ViewResolver 接口的一个简单实现。它直接使用 URL 解析到逻辑视图名，除此之外，不需要其他任何显式的映射声明。如果逻辑视图名与真正的视图资源名是直接对应的，那么这种直接解析的方式就很方便，不需要再指定额外的映射
InternalResourceViewResolver	UrlBasedViewResolver 的一个好用的子类。它支持内部资源视图(具体来说，是 Servlet 和 JSP)以及诸如 JstlView 和 TilesView 等类的子类
VelocityViewResolver/FreeMarkerViewResolver	UrlBasedViewResolver 下的实用子类，支持 Velocity 视图 VelocityView(Velocity 模板)和 FreeMarker 视图 FreeMarkerView，以及它们对应的子类
ContentNegotiatingViewResolver	视图解析器接口 ViewResolver 的一个实现，它会根据请求的文件名或请求的 Accept 头解析一个视图

假设这里使用的是 JSP 视图技术,那么可以使用一个基于 URL 的视图解析器 UrlBasedViewResolver。这个视图解析器会将 URL 解析成一个视图名,并将请求转交给请求分发器进行视图渲染。

```
<bean id="viewResolver" class="org.springframework.web.servlet.view.UrlBasedViewResolver">
  <property name="viewClass" value="org.springframework.web.servlet.view.JstlView"/>
  <property name="prefix" value="/WEB-INF/jsp/"/>
  <property name="suffix" value=".jsp"/>
</bean>
```

若返回一个 test 逻辑视图名,那么该视图解析器会将请求转发到 RequestDispatcher,后者会将请求交给 /WEB-INF/jsp/test.jsp 视图渲染。

如果需要在应用中使用多种不同的视图技术,可以使用 ResourceBundleViewResolver:

```
<bean id="viewResolver" class="org.springframework.web.servlet.view.ResourceBundleViewResolver">
  <property name="basename" value="views"/>
  <property name="defaultParentView" value="parentView"/>
</bean>
```

ResourceBundleViewResolver 会检索由 bundle 根路径下配置的 ResourceBundle,对于每个视图而言,其视图类由[viewname].(class)属性的值指定,其视图 url 由[viewname].url 属性的值指定。视图还允许有基视图,即 properties 文件中所有视图都"继承"的一个文件。通过继承技术,可以为众多视图指定一个默认的视图基类。

AbstractCachingViewResolver 的子类能够缓存已经解析过的视图实例。关闭缓存特性也是可以的,只将 cache 属性设置为 false 即可。另外,如果实在需要在运行时刷新某个视图(如修改了 Velocity 模板时),可以使用 removeFromCache(String viewName,Locale loc)方法。

2. 视图链

Spring 支持同时使用多个视图解析器。因此,可以配置一个解析器链并做更多的事。例如,在特定条件下重写一个视图等。可以通过把多个视图解析器设置到应用上下文(applicationContext)中的方式串联它们。如果需要指定它们的次序,设置 order 属性即可。记住,order 属性的值越大,该视图解析器在链中的位置越靠后。

在下面的代码例子中,视图解析器链中包含了两个解析器:一个是 InternalResourceViewResolver,它总是自动被放置在解析器链的最后;另一个是 XmlViewResolver,它用来指定 Excel 视图。InternalResourceViewResolver 不支持 Excel 视图。

```
<bean id="jspViewResolver"
class="org.springframework.web.servlet.view.InternalResourceViewResolver">
  <property name="viewClass" value="org.springframework.web.servlet.view.JstlView"/>
```

```
    <property name="prefix" value="/WEB-INF/jsp/"/>
    <property name="suffix" value=".jsp"/>
  </bean>
  <bean id="excelViewResolver"
  class="org.springframework.web.servlet.view.XmlViewResolver">
    <property name="order" value="1"/>
    <property name="location" value="/WEB-INF/views.xml"/>
  </bean>
  <!--in views.xml -->
  <beans>
    <bean name="report" class="org.springframework.example.ReportExcelView"/>
  </beans>
```

如果一个视图解析器不能返回一个视图，那么 Spring 会继续检查上下文中其他的视图解析器。此时如果存在其他的解析器，Spring 会继续调用它们，直到产生一个视图返回为止。如果最后所有视图解析器都不能返回一个视图，Spring 就抛出一个 ServletException。

视图解析器的接口清楚声明了一个视图解析器是可以返回 null 值的，这表示不能找到任何合适的视图。并非所有的视图解析器都这么做，但是也存在不得不如此的场景，即解析器确实无法检测对应的视图是否存在。例如，InternalResourceViewResolver 在内部使用了 RequestDispatcher，并且进入分派过程是检测一个 JSP 视图是否存在的方法，但这个过程仅可能发生一次。类似地，VelocityViewResolver 和部分其他的视图解析器也存在这样的情况。因此，如果不把 InternalResourceViewResolver 放置在解析器链的最后，将可能导致解析器链无法完全执行，因为 InternalResourceViewResolver 会返回一个视图。

3. 视图重定向

控制器通常会返回一个逻辑视图名，然后视图解析器会把它解析到一个具体的视图技术上去渲染。对于一些可以由 Servlet 或 JSP 引擎处理的视图技术，如 JSP 等，这个解析过程通常是由 InternalResourceViewResolver 和 InternalResourceView 协作完成的，而这通常会调用 Servlet 的 API RequestDispatcher.forward(..)方法或 RequestDispatcher.include(..)方法，并发生一次内部的转发（forward）或引用（include）。而对于其他的视图技术，如 Velocity、XSLT 等，视图本身的内容是直接被写回响应流中的。有时，想在视图渲染之前先把一个 HTTP 重定向请求发送回客户端。例如，当一个控制器成功地接收到 POST 过来的数据，而响应仅是委托另一个控制器处理（例如一次成功的表单提交）时，希望发生一次重定向。在这种场景下，如果只是简单地使用内部转发，那么意味着下一个控制器也能看到这次 POST 请求携带的数据，这可能导致一些潜在的问题，如可能与其他期望的数据混淆等。此外，另一种在渲染视图前对请求进行重定向的需求是，防止用户多次提交表单的数据。此时若使用重定向，则浏览器会先发送第一个 POST 请求；请求被处理后，浏览器会收到一个重定向响应，然后浏览器直接被重定向到一个不同的 URL，最后浏览器会使用重定向响应中携带的 URL 发起一次 GET 请求。因此，从浏览器的角度看，当前所见的页面并不是 POST 请求的结果，而是一次 GET 请求的结果。这就防止了用户因刷新等原因意外多次提交同样的数据。此时刷新会重新 GET 一次结果页，而不是把同样的 POST 数据再发送一遍。

1）重定向视图 RedirectView

强制重定向的一种方法是，在控制器中创建并返回一个 Spring 重定向视图 RedirectView 的实例。它会使得 DispatcherServlet 放弃使用一般的视图解析机制，因为你已经返回一个(重定向)视图给 DispatcherServlet 了，所以它会构造一个视图满足渲染的需求。紧接着 RedirectView 会调用 HttpServletResponse.sendRedirect()方法，发送一个 HTTP 重定向响应给客户端浏览器。如果决定返回 RedirectView，并且这个视图实例是由控制器内部创建出来的，那么更推荐在外部配置重定向 URL 然后注入控制器中，而不是写在控制器里。这样它就可以与视图名一起在配置文件中配置了。

2）向重定向目标传递数据

模型中的所有属性默认都会考虑作为 URI 模板变量被添加到重定向 URL 中。剩下的其他属性，如果是基本类型或者基本类型的集合或数组，那它们将被自动添加到 URL 的查询参数中。如果 model 是专门为该重定向准备的，就把所有基本类型的属性添加到查询参数中可能是期望结果的那个。但是，在包含注解的控制器中，model 可能包含了专门作为渲染用途的属性(如一个下拉列表的字段值等)。为了避免把这样的属性也暴露在 URL 中，@RequestMapping 方法可以声明一个 RedirectAttributes 类型的方法参数，用它指定专门供重定向视图 RedirectView 用的属性。如果重定向成功发生，那么 RedirectAttributes 对象中的内容就会被使用；否则使用模型 model 中的数据。

RequestMappingHandlerAdapter 提供了一个"ignoreDefaultModelOnRedirect"标志。它被用来标记默认 Model 中的属性永远不应该用于控制器方法的重定向中。控制器方法应该声明一个 RedirectAttributes 类的参数。如果不声明，就没有参数被传递到重定向的视图 RedirectView 中。在 MVC 命名空间或 MVC Java 编程配置方式中，为了维持向后的兼容性，这个标志仍被保持为 false。但如果你的应用是一个新的项目，那么推荐把它的值设置成 true。注意，当前请求 URI 中的模板变量会在填充重定向 URL 的时候自动对应用可见，而不需要显式地在 Model 或 RedirectAttributes 中再添加属性。请看下面的例子：

```
@RequestMapping(path="/files/{path}", method=RequestMethod.POST)
public String upload(...) {
  //...
  return "redirect:files/{path}";
}
```

另外一种向重定向目标传递数据的方法是通过闪存属性(Flash Attributes)。与其他重定向属性不同，flash 属性是存储在 HTTP session 中的(因此不会出现在 URL 中)。

3）重定向前缀——redirect:

尽管使用 RedirectView 重定向能工作得很好，但如果控制器自身还是需要创建一个 RedirectView，那无疑控制器还是需要了解重定向的发生。这还是有点不尽完美，不同范畴的耦合还是太强。控制器其实不应该关心响应会如何被渲染。一个特别的视图名前缀能完成这个解耦：redirect:。如果返回的视图名中含有 redirect:前缀，那么 UrlBasedViewResolver(及它的所有子类)就会接收到这个信号，意识到这里需要发生重定向。然后视图名剩下的部分会被解析成重定向 URL。这种方式与通过控制器返回一个重定向视图 RedirectView 达到的效果是一样的，不过，这样一来控制器就可以只专注于处理并返回逻辑视图名了。如果逻辑视图名是这样的形式：redirect:/myapp/some/resource，它们重定向的路径将以 Servlet 上下文作为相对路径进行查找，而逻辑视图名如果是这样的形式：redirect:http://myhost.com/some/arbitrary/path，那么重定向 URL 使用的就是绝对路径。

注意：如果控制器方法注解了 @ResponseStatus，那么注解设置的状态码值就会覆盖 RedirectView 设置的响应状态码值。

4) 重定向前缀——forward:

对于最终会被 UrlBasedViewResolver 或其子类解析的视图名，可以使用一个特殊的前缀：forward:。这会导致一个 InternalResourceView 视图对象的创建(它最终会调用 RequestDispatcher.forward()方法)，后者会认为视图名剩下的部分是一个 URL。因此，这个前缀在使用 InternalResourceViewResolver 和 InternalResourceView 时并没有特别的作用(例如，对于 JSP 来说)。但是当主要使用的是其他的视图技术，而又想强制把一个资源转发给 Servlet/JSP 引擎进行处理时，这个前缀可能就很有用了(或者，也可能同时串联多个视图解析器)。

与 redirect:前缀一样，如果控制器中的视图名使用了 forward:前缀，那么控制器本身并不会发觉任何异常，它关注的仍然只是如何处理响应的问题。

5.5 Spring MVC 开发实例

接下来看一个用户登录功能的实例开发(Spring MVC_login_demo)。具体步骤如下：

(1) 新建一个 Web 项目，项目名称为 Spring MVC_login_demo。

(2) 添加 Spring MVC 相关 jar 包，右击 Spring MVC_ login _demo 项目，从弹出的快捷菜单中选择 Configure Facets...→Install Spring Facet。

(3) 建立项目目录。

在 src 下创建包 com.ascent.controller，并存放在 controller 控制类中。

(4) 修改 web.xml，添入以下内容。

```
<?xml version="1.0" encoding="UTF-8"?>
<web-app xmlns:xsi="http://www.w3.org/2001/XMLSchema-instance"
xmlns="http://xmlns.jcp.org/xml/ns/javaee"
xsi:schemaLocation="http://xmlns.jcp.org/xml/ns/javaee http://xmlns.jcp.org/
xml/ns/javaee/web-app_3_1.xsd" id="WebApp_ID" version="3.1">
  <display-name>Spring MVC_login_demo</display-name>

  <listener>
<listener-class>org.springframework.web.context.ContextLoaderListener
</listener-class>
  </listener>
  <context-param>
    <param-name>contextConfigLocation</param-name>
```

```
    <param-value>classpath:applicationContext.xml</param-value>
  </context-param>
  <servlet>
    <servlet-name>Spring MVC</servlet-name>
    <servlet-class>
        org.springframework.web.servlet.DispatcherServlet
    </servlet-class>
    <load-on-startup>1</load-on-startup>
  </servlet>
  <servlet-mapping>
    <servlet-name>Spring MVC</servlet-name>
    <url-pattern>/</url-pattern>
  </servlet-mapping>

  <filter>
    <filter-name>CharacterEncodingFilter</filter-name>
<filter-class>org.springframework.web.filter.CharacterEncodingFilter
</filter-class>
    <init-param>
      <param-name>encoding</param-name>
      <param-value>UTF-8</param-value>
    </init-param>
  </filter>
  <filter-mapping>
    <filter-name>CharacterEncodingFilter</filter-name>
    <url-pattern>/*</url-pattern>
  </filter-mapping>
```

</web-app>这里将 DispatcherServlet 命名为 Spring MVC,并且让它在 Web 项目一启动就加载。接下来需要在 WEB-INF 目录下创建一个 Spring MVC -servlet.xml 的 Spring 配置文件,此文件名的命名规则为:在 servlet-name 名称后面加上-servlet.xml。

(5) 添加 Spring MVC-servlet.xml 配置文件,内容如下。

```
<?xml version="1.0" encoding="UTF-8"?>
<beans xmlns="http://www.springframework.org/schema/beans"
    xmlns:xsi="http://www.w3.org/2001/XMLSchema-instance"
xmlns:p="http://www.springframework.org/schema/p"
    xmlns:context="http://www.springframework.org/schema/context"
    xmlns:mvc="http://www.springframework.org/schema/mvc"
    xsi:schemaLocation="
    http://www.springframework.org/schema/beans
    http://www.springframework.org/schema/beans/spring-beans-3.0.xsd
    http://www.springframework.org/schema/context
    http://www.springframework.org/schema/context/spring-context-3.0.xsd
```

```
    http://www.springframework.org/schema/mvc
http://www.springframework.org/schema/mvc/spring-mvc-3.0.xsd">

<!--默认的注解映射的支持 -->
<mvc:annotation-driven />
<!--启用自动扫描  -->
<context:component-scan base-package="com.ascent" />
<bean class="org.springframework.web.servlet.view.InternalResourceViewResolver">
    <property name="prefix" value="/WEB-INF/views/" />
    <property name="suffix" value=".jsp" />
</bean>
</beans>
```

(6) 在 com.ascent controller 包中编写 Login 类。

```
package com.ascent.controller;

import org.springframework.stereotype.Controller;
import org.springframework.ui.Model;
import org.springframework.web.bind.annotation.RequestMapping;
import org.springframework.web.bind.annotation.RequestParam;

@Controller
public class Login{

    /**
     * @param username  用户名,对应表单的 username
     * @param password  密码,对应表单的 password
     * @param model     用于存储数据值
     */
    @RequestMapping("/login")     //@RequestMapping 注解可以用指定的 URL 路径访问
                                    本控制层
    public String login(@RequestParam("username") String username,
        @RequestParam("password") String password, Model model) {

        if (username.equals("Lixin") && password.equals("123456")){
            model.addAttribute("username", username);
            return "success";
    } else {
            model.addAttribute("username", username);
            return "fail";
        }
    }

}
```

(7) 编写 JSP 文件。

① 在 WebRoot 下编写 login.jsp。

```
<%@ page language="java" contentType="text/html; charset=UTF-8"
pageEncoding="UTF-8"%>
<!DOCTYPE html PUBLIC "-//W3C//DTD HTML 4.01 Transitional//EN"
"http://www.w3.org/TR/html4/loose.dtd">
<html>
<head>
<meta http-equiv="Content-Type" content="text/html; charset=UTF-8">
<title>登录界面</title>
</head>
<body>

    <form action="login" method="post">
          用户名: <input type="text" name="username"><br />
            密码: <input type="password" name="password"><br />
        <input type="submit" value="登录">
    </form>
</body>
</html>
```

② 根据 Spring MVC-servlet.xml,在/WEB-INF/views/下编写 success.jsp 和 fail.jsp。

```
success.jsp 如下:
<%@ page language="java" contentType="text/html; charset=UTF-8"
    pageEncoding="UTF-8"%>
<!DOCTYPE html PUBLIC "-//W3C//DTD HTML 4.01 Transitional//EN"
"http://www.w3.org/TR/html4/loose.dtd">
<html>
<head>
<meta http-equiv="Content-Type" content="text/html; charset=UTF-8">
<title>登录网站</title>
</head>
<body>
<font color="green">${username }</font>欢迎你!

</body>
</html>

fail.jsp 如下:
<%@ page language="java" contentType="text/html; charset=UTF-8"
    pageEncoding="UTF-8"%>
<!DOCTYPE html PUBLIC "-//W3C//DTD HTML 4.01 Transitional//EN"
"http://www.w3.org/TR/html4/loose.dtd">
<html>
```

```
<head>
<meta http-equiv="Content-Type" content="text/html; charset=UTF-8">
<title>登录网站</title>
</head>
<body>

    <font color="red">对不起</font>,没有${username }这个用户!
    <br />
    <a href="login.jsp">重试一下! </a>

</body>
</html>
```

(8) 部署和运行项目。

选中 Spring MVC_login_demo 项目,将其部署在 Tomcat 8 服务器上,之后启动 Tomcat。

在浏览器中输入 http://localhost:8080/Spring MVC_login_demo/login.jsp 进入登录页面,如图 5-23 所示。

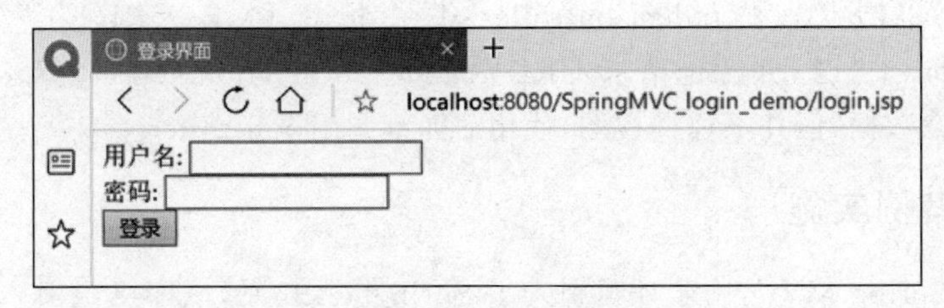

图 5-23 登录页面

在用户名处输入 Lixin,在密码处输入 123456,单击“登录”按钮,进入成功页面,如图 5-24 所示。

图 5-24 成功页面

在用户名或密码处输入其他内容,单击“登录”按钮,进入失败页面,如图 5-25 所示。

图 5-25 失败页面

5.6 项目案例

5.6.1 学习目标

本章学习了 Spring MVC 框架基础，掌握配置 Spring MVC 框架所需的核心知识，以及在 Spring MVC 中如何获取并处理请求，通过控制器产生 ModelAndView 模型给前端视图层，如 JSP。

5.6.2 案例描述

本案例在 eGov 项目首页获取头版头条新闻和综合新闻案例基础上，增加了 Spring MVC 内容，借助 Spring 框架，注意不是结合，是借助 ApplicationContext 中的 IAuthorizationService 对象完成数据获取，并封装到 ModelAndView 中，传递给 Index.jsp。

5.6.3 案例要点

本案例需要配置 Spring MVC，并编写 Controller，通过注解方式实现 Spring MVC 控制器的标注。需要编写 IndexController 类并标记请求 URL 为"/"，编写一个 indexauthAction()方法，并标记请求 URL 为 indexauthAction.action，在该方法中调用查询业务方法获取数据，但并不返回视图，使用 c 标签调用 controller。

5.6.4 案例实施

(1) 配置 Spring MVC，由于项目引入了 Spring 4.1.0 库，该库已经引入了 spring-mvc 核心包，所以不需要额外引入。如需其他扩展功能(如 JSON API)，则另外引入。

在 electrone 项目中新建 config 目录，在该目录下建立 Spring MVC.xml 文件，操作步骤与建立 applicationContext.xml 相同，如图 5-26 所示。

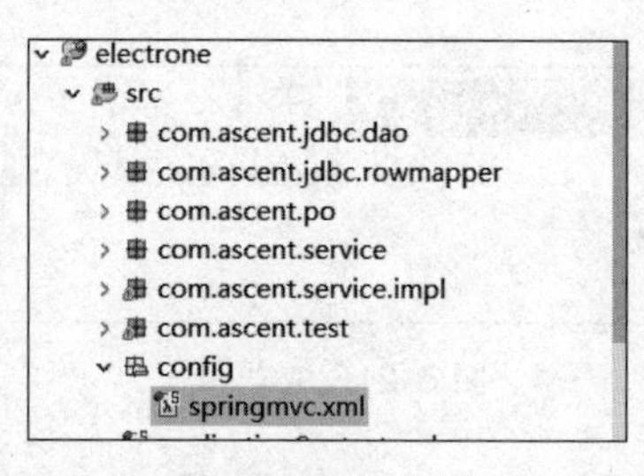

图 5-26 建立 Spring MVC.xml 文件

编写 Spring MVC.xml，忽略 xml 头信息以及 beans 根标签声明，以下为 beans 中的主体内容：

```
<!--spring mvc 框架采用注解注入 Bean,需要配置到包含 IndexController 包的上一层-->
  <context:component-scan
          base-package="com.ascent "/>
  <!--使用注解方式进行控制器装配以及请求 URL 映射 -->
<mvc:annotation-driven />
```

```
<!--配置静态资源映射,包括 js 脚本、css 样式表,以及 image 图片等 -->
<mvc:resources location="/" mapping="/**"></mvc:resources>

<!--配置动态视图解析器 -->
   <bean class="org.springframework.web.servlet.view.
      InternalResourceViewResolver">
      <property name="prefix" value="/"/>
      <!--默认视图起始位置为运行时 Web 项目的根目录 -->
      <property name="suffix" value=".jsp"/>
   </bean>
```

注意：com.ascent.controller 是包名，表示 Spring MVC 自动扫描该包下包括子目录的类，识别注解@Controller、@RequestMapping 等。

(2) 修改 WEB-INF/web.xml。

```
<!--注册编码过滤器 -->
   <filter>
      <filter-name>charset</filter-name>
      <filter-class>org.springframework.web.filter.
         CharacterEncodingFilter</filter-class>
      <init-param>
        <param-name>encoding</param-name>
         <param-value>UTF-8</param-value>
      </init-param>
   </filter>
   <filter-mapping>
      <filter-name>charset</filter-name>
      <url-pattern>/*</url-pattern>
   </filter-mapping>

<!--配置 Spring MVC Servlet  -->
<servlet>
  <servlet-name>Spring MVCServlet</servlet-name>
  <servlet-class>org.springframework.web.servlet.DispatcherServlet
    </servlet-class>
  <init-param>
    <param-name>contextConfigLocation</param-name>
    <param-value>classpath:config/Spring MVC.xml</param-value>
  </init-param>
  <load-on-startup>1</load-on-startup>
</servlet>
  <servlet-mapping>
     <servlet-name>Spring MVCServlet</servlet-name>
     <url-pattern>*.action</url-pattern>
  </servlet-mapping>
```

```
  <welcome-file-list>
  <welcome-file>index.jsp</welcome-file>
</welcome-file-list>
```

注意：servlet-mapping 中 url-pattern 配置的是请求的后缀为.action 时由 Spring MVCServlet 处理。

(3) 编写 IndexController。

```
IndexController.java:
package com.ascent.controller;
import java.util.HashMap;
import java.util.List;

import javax.servlet.http.HttpServletRequest;
import javax.servlet.http.HttpServletResponse;
import javax.servlet.http.HttpSession;

import org.springframework.context.ApplicationContext;
import org.springframework.context.support.ClassPathXmlApplicationContext;
import org.springframework.stereotype.Controller;
import org.springframework.web.bind.annotation.RequestMapping;
import org.springframework.web.servlet.ModelAndView;

import com.ascent.po.News;
import com.ascent.service.IAuthorizationService;

/**
 *所有在 URL/下发出的请求,都由该 index 控制器处理,Class 相当于包含了命名空间的对象,
 可以有多个不同的 URL 映射
 * @author gary
 *
 */
 @Controller
 @RequestMapping(value="/")
 public classIndexController {


   @RequestMapping(value="indexauthAction")
   public void index(HttpServletRequest request,HttpServletResponse response){

       //获取 HttpSession
       HttpSession session=request.getSession(true);
       //获取 applicationContext 对象
       ApplicationContext applicationContext=
```

```
            new ClassPathXmlApplicationContext("applicationContext.xml");
        //获取 IAuthorizationService 实现
        IAuthorizationService authService= (IAuthorizationService)
            applicationContext.getBean("iAuthorizationService");

        //ModelAndView mv=new ModelAndView();

        HashMap<String, List<News>>indexMap=authService.findHeaderNews();

        List<News>headerNews=indexMap.get("headerNews");

        if(headerNews.size()>0){
            News news=(News) headerNews.get(0);
            session.setAttribute("typ1",news);
        } else {
            session.setAttribute("typ1",null);
        }

        List<News>indexNews=indexMap.get("indexNews");
        session.setAttribute("typ2", indexNews);

    }

}
```

注意：该方法使用 jstl:c 标签调用，不需要响应到其他页面，所以该方法为 void，注意 jsp 中对此的调用。

(4) 在 electrone 项目的 WebRoot 下建立 index.jsp 以及其他静态资源包，如图 5-27 所示。

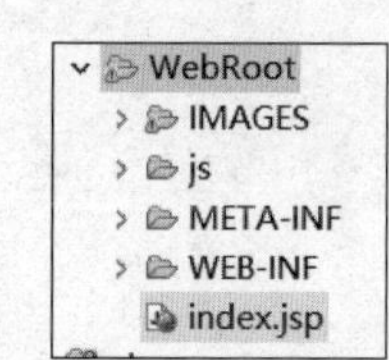

图 5-27 建立 index.jsp 以及其他静态资源包

由于 index.jsp 页面包含登录功能以及其他静态数据，代码较多，所以只选出如下和头版头条以及综合新闻有关的代码：

```
<%@ page language="java" import="java.util.*,com.ascent.po.*"pageEncoding=
"utf8"%>
<%@ taglib uri="http://java.sun.com/jsp/jstl/core" prefix="c"%>
<%
String path=request.getContextPath();
StringbasePath=request.getScheme()+"://"+request.getServerName()+":"+
request.getServerPort()+path+"/";
%>

<!DOCTYPE HTML PUBLIC "-//W3C//DTD HTML 4.01 Transitional//EN">
<html>
```

```
  <head>
    <title>Acsent 电子政务系统</title>
    <meta http-equiv="pragma" content="no-cache">
    <meta http-equiv="cache-control" content="no-cache">
    <meta http-equiv="expires" content="0">
    <meta http-equiv="keywords" content="keyword1,keyword2,keyword3">
    <meta http-equiv="description" content="This is my page">
    <meta http-equiv="Content-Type" content="text/html; charset=gb2312">
    <link rel="stylesheet" href="<%=basePath%>/IMAGES/001.css" type="text/
css">

  </head>
<%--jsp:include 标签调用 Spring MVC 控制器的 indexauthAction URL 所对应的方法--%>
<jsp:include page="${basePath}/indexauthAction.action"></jsp:include>
  <body bgcolor="#FFFFFF" text="#000000"topmargin="0">
                <% //获取头版头条新闻
                    News news=(News)session.getAttribute("typ1");
                    //显示头版头条新闻
                %>
    <table width="555" border="0"cellspacing="0" cellpadding="0">
      <tr>
<td width="333" height="176"valign="top">
        <table width="320" border="0"cellspacing="0" cellpadding="0"
                align="center">
          <tr>
            <td>
    <%String title="";                      //定义新闻标题
        if(news!=null){                     //如果头版头条不为 null
if(news.getTitle().length()>14){            //如果新闻标题长度大于 14 个字符
                title=news.getTitle().substring(0,14)+"……";
                                            //截取新闻标题的前 14 个字符
            }else{ title=news.getTitle(); }
          }else{ title="没有内容";  //如果标题为空,则设置 title 为没有内容。}%>
          <div><span class="p3 p6"><strong><%=title%></strong></span></div>
            </td>
          </tr>
        <tr>
          <td height="50"valign="top">
        <% String str="";                               //保存头版头条新闻内容
        if(news!=null){                                 //如果头条版头新闻不为 null
         if(news.getContent().length()>76){   //如果内容长度大于 76 个字符
          str=news.getContent().substring(0,76)+"……"; //截取前 76 个字符
            }else{ str=news.getContent(); }
          }
```

```
        %>
        <table><tr><td height="40"><div align="left"><br><span class="p6">
          <%=str%>
</span></div></td></tr><tr><td height="10" align="right">
<%
            if(news!=null){
          %>
<a href="viewnewsAction.action?id=<%=news.getId()%>" target="_blank"><font
color="#3333FF">&gt;&gt;详细内容</font></a>
      <%} else {%>  <%}%>
    </td></tr
        </table>
      </td></tr>
  <tr><td height="48" align="center" valign="bottom"><a href="<%=basePath%
  >/listNewsnewsAction.action?typ=1"><font color="#3333FF">&gt;&gt;&gt;更多
  新闻 &lt;&lt;&lt;</font></a></td></tr>
  </table>
  </td></tr></table>

  <div align="center">
 <table width="575" border="0"cellspacing="0" cellpadding="0">
      <tr><td align="center"valign="top">
        <table width="100%" border="0"cellspacing="0" cellpadding="0">
          <tr><td height="24" bgcolor="#99C9FD">
            <table width="382" border="0"cellspacing="0" cellpadding="0">
            <tr><td width="36"> </td>
<td width="346">综合新闻</td>
            </tr></table></td></tr></table>
     <table width="100%" border="0"cellspacing="0" cellpadding="0">
      <tr><td bgcolor="B7B7B7" width="1"><img src="IMAGES/dian_04.gif" width
      ="1" height="1"></td><td bgcolor="#FFFFFF" height="154">
  <table width="100%" border="0"cellspacing="0" cellpadding="0">
  <%--使用 c:forEach 标签获取 session 中的综合新闻集合 List,迭代取出对象数据--%>
      <c:forEach items="${sessionScope.typ2}" var="typ2">
          <tr><td width="35"><div align="center"><img src="IMAGES/dian_03.
              gif" width="3" height="3">
          </div></td>
        <td width="527" height="22">
     <a href ="viewnewsAction.action?id=${typ2.id}" target="_blank">${typ2.
            title}</a>
     </td></tr>
 </c:forEach>
 <%--迭代获取综合新闻结束 --%>
  <tr><td height="33"colspan="2">
```

```
<div align="right"><span class="p6">
<a href="<%=basePath%>//listNewsnewsAction.action?typ=2"><font color="#
3333FF">&gt;&gt;更多内容</font></a></span></div></td>
                    </tr>
                </table></td>
                <td bgcolor="B7B7B7" width="1"><img src="IMAGES/dian_
04.gif" width="1" height="1"></td>
              </tr>
            </table></td></tr></table></div></body></html>
```

(5) 设置将引用的 mysql-connector-java jar 包在部署项目时同时部署到服务器。

右击 electrone 项目，从弹出的快捷菜单中选择 Properties，出现图 5-28。

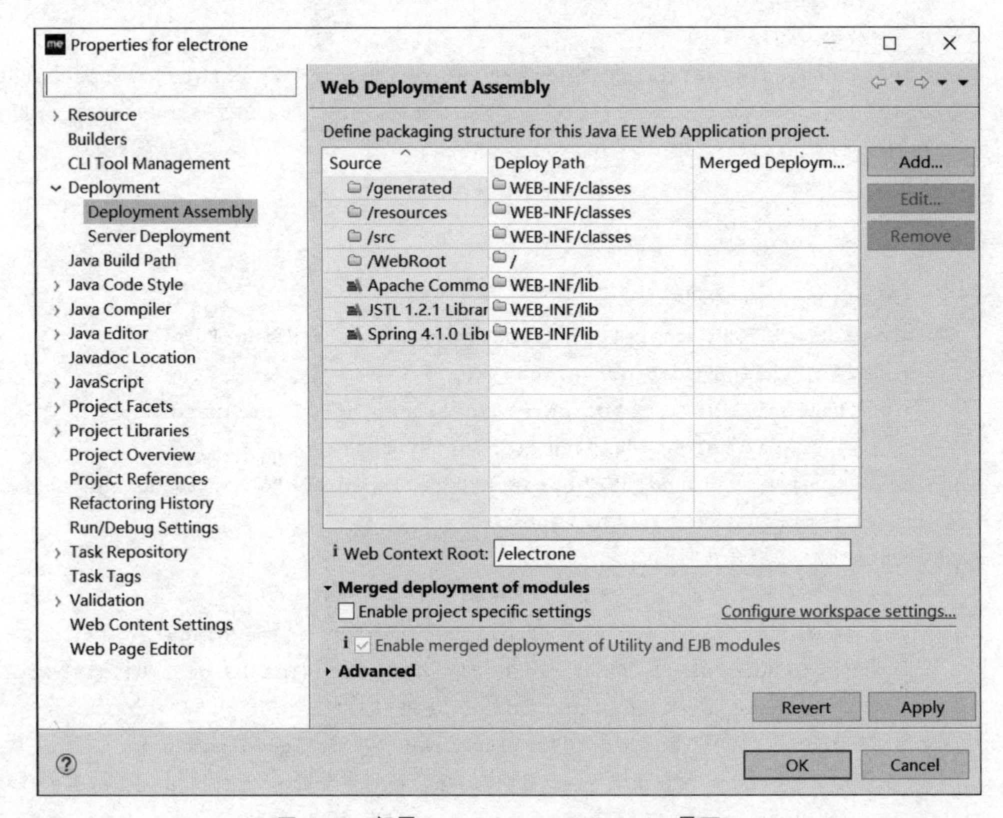

图 5-28　部署 mysql-connector-java jar 界面 1

单击 Add 按钮，出现图 5-29。

选择 Java Build Path Entries，单击 Next 按钮，出现图 5-30。

选择 mysql 驱动包，之后单击 Finish 按钮。

然后单击 OK 按钮。这一步骤很重要，部署项目时，有些被 Build Path 引用或者用户自定义的外部 jar 文件不会被部署到服务器上，需要按照上述步骤手动添加。

(6) 部署项目到 tomcat 中并运行，如图 5-31 和图 5-32 所示。

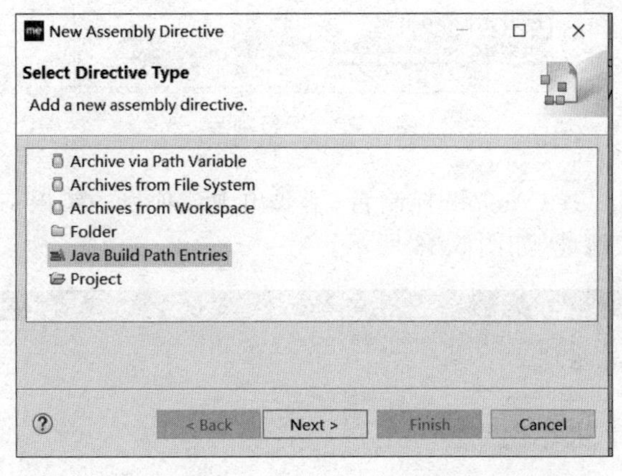

图 5-29 部署 mysql-connector-java jar 界面 2

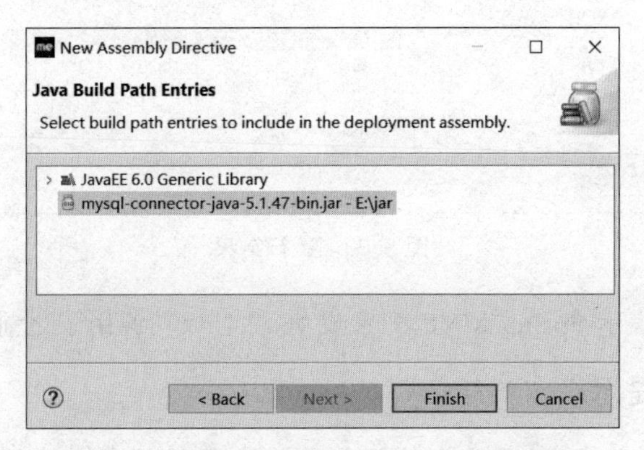

图 5-30 部署 mysql-connector-java jar 界面 3

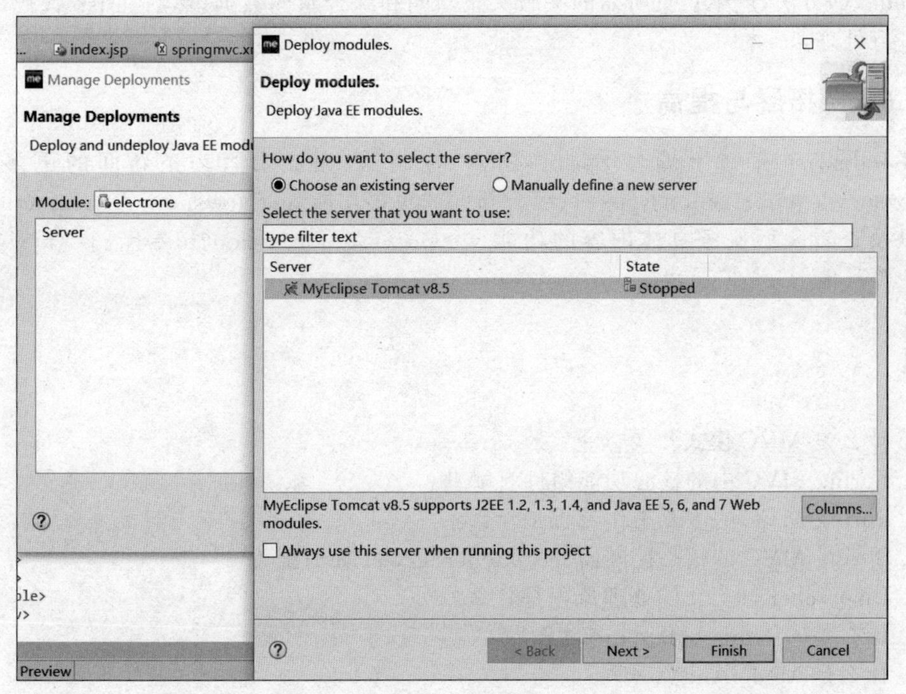

图 5-31 部署项目

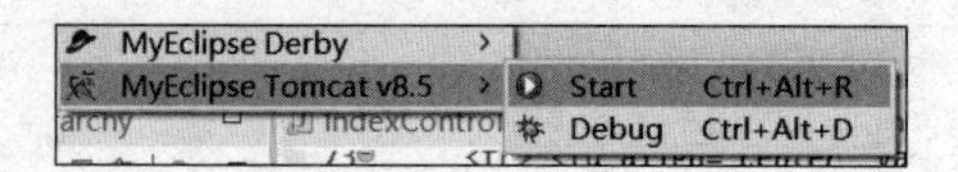

图 5-32 运行项目

单击 Start。注意观察 Console 控制台，不要出现该项目的错误信息。

打开浏览器，输入地址，如图 5-33 所示。

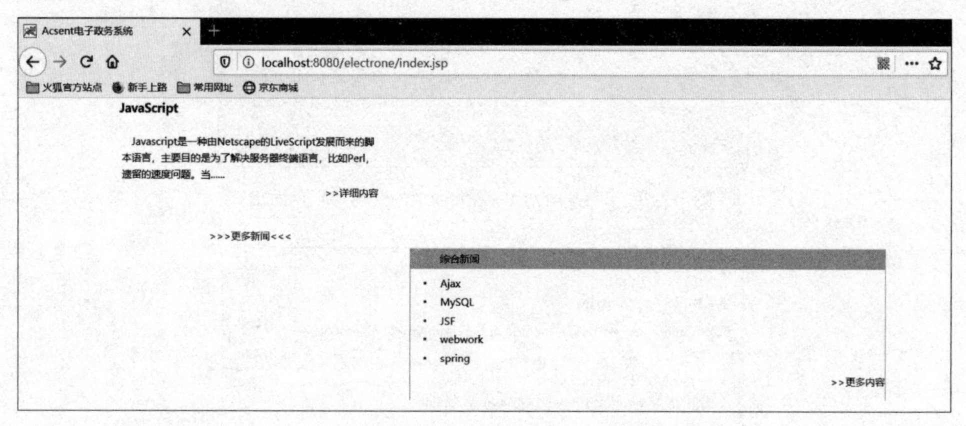

图 5-33 运行结果

输出这些数据，表示 Spring MVC 配置成功，并且成功调用了之前所编写的业务类。

5.6.5 特别提示

index.jsp 中的 jsp:include 标签动态包含调用标签，这里负责执行一次 indexController 中的 index()方法，产生首页所需的头版头条新闻和综合新闻数据，然后在 jsp 代码的后半部分进行处理。

5.6.6 拓展与提高

在 index.jsp 基础上编写 Spring MVC 控制器，完成头版头条新闻的更多功能：listNewsnewsAction.action?typ＝1、综合新闻的更多功能/listNewsnewsAction.action?typ＝2，以及查看头版头条具体内容的功能 viewnewsAction.action?id＝＜％＝news.getId()％＞等其他 eGov 扩展功能。

习题

1. 什么是 MVC 模式？
2. Spring MVC 中的核心功能组件有哪些？
3. Spring MVC 的主要工作流程是什么？
4. Spring MVC 的核心控制器 DispatcherServlet 如何配置？
5. DispatcherServlet 的处理流程是什么？
6. 什么是@RequestMapping 注解？
7. 什么是 ModelAndView？
8. Spring 处理视图的两个接口是什么？

第6章 Spring MVC 高级特性

学习目的与学习要求

学习目的：本章需要掌握Spring MVC 高级特性，包括Spring MVC 表单标签、Spring MVC 数据校验、Spring MVC 拦截器和 Spring MVC 国际化等。

学习要求：熟练掌握 Spring MVC 表单标签、Spring MVC 数据校验、Spring MVC 拦截器和 Spring MVC 国际化等高级特性在 Spring MVC 应用开发中的使用。

本章主要内容

Spring MVC 表单标签、Spring MVC 数据校验、Spring MVC 拦截器和 Spring MVC 国际化。

6.1 Spring MVC 表单标签

Spring MVC 表单标签库中包含可以用在 JSP 页面中渲染 HTML 元素的标签。在 JSP 页面使用 Spring 表单标签库时，必须在 JSP 页面开头处声明 taglib 指令，指令代码如下：

```
<%@ taglib prefix="form" uri="http://www.
springframework.org/tags/form"%>
```

表单标签库中有 form、input、password、hidden、textarea、checkbox、checkboxes、radiobuttton、radiobuttons、select、option、options、errors 等标签，见表 6-1。

表 6-1　表单标签库中的标签

名　称	作　　用
form	渲染表单元素
input	渲染<input type="text"/>元素
password	渲染<input type="password"/>元素
hidden	渲染<input type="hidden"/>元素
textarea	渲染 textarea 元素
checkbox	渲染一个<input type="checkbox"/>元素
checkboxes	渲染多个<input type="checkbox"/>元素
radiobutton	渲染一个<input type="radio"/>元素
radiobuttons	渲染多个<input type="radio"/>元素
select	渲染一个选择元素
option	渲染一个选项元素
options	渲染多个选项元素
errors	在 span 元素中渲染字段错误

1. form 标签

表单标签的语法格式如下：

```
<form:formmodelAttribute="xxx" method="post" action="xxx">
...
</form:form>
```

表单标签除了具有 HTML 表单元素属性以外，还具有 acceptCharset、commandName、cssClass、cssStyle、htmlEscape 和 modelAttribute 等属性。

- acceptCharset：定义服务器接受的字符编码列表。
- commandName：暴露表单对象的模型属性名称，默认为 command。
- cssClass：定义应用到 form 元素的 CSS 类。
- cssStyle：定义应用到 form 元素的 CSS 样式。
- htmlEscape：true 或 false，表示是否进行 HTML 转义。
- modelAttribute：暴露 form backing object 的模型属性名称，默认为 command。

其中，commandName 和 modelAttribute 属性的功能基本一致，属性值绑定一个 JavaBean 对象。假设控制器类 UserController 的方法 inputUser()是返回 userAdd.jsp 的请求处理方法，则它的代码如下：

```
@RequestMapping(value="/input")
public StringinputUser(Model model) {
...
    model.addAttribute("user", new User());
```

```
    return "userAdd";
}
```

userAdd.jsp 的表单标签代码如下：

```
<form:formmodelAttribute="user" method="post" action="user/save">
    ...
</form:form>
```

注意：在 inputUser()方法中，如果没有 Model 属性 user，userAdd.jsp 页面就会抛出异常，因为表单标签无法找到在其 modelAttribute 属性中指定的 form 对象。

2. input 标签

input 标签的语法格式如下：

```
<form:input path="xxx"/>
```

该标签除了有 cssClass、cssStyle、htmlEscape 属性以外，还有一个最重要的属性——path。path 属性将文本框输入值绑定到 form backing object 的一个属性。示例代码如下：

```
<form:formmodelAttribute="user" method="post" action="user/save">
    <form:input path="userName"/>
</form:form>
```

上述代码将输入值绑定到 user 对象的 userName 属性。

3. password 标签

password 标签的语法格式如下：

```
<form:password path="xxx"/>
```

该标签与 input 标签的用法完全一致，只是不显示密码的输入内容。

4. hidden 标签

hidden 标签的语法格式如下：

```
<form:hidden path="xxx"/>
```

该标签与 input 标签的用法基本一致，只不过它不可显示，不支持 cssClass 和 cssStyle 属性。

5. textarea 标签

textarea 基本上就是一个支持多行输入的 input 元素，语法格式如下：

```
<form:textarea path="xxx"/>
```

该标签与 input 标签的用法完全一致，这里不再赘述。

6. checkbox 标签

checkbox 标签的语法格式如下：

```
<form:checkbox path="xxx" value="xxx"/>
```

多个 path 相同的 checkbox 标签，它们是一个选项组，允许多选，选项值绑定到一个数组属性。示例代码如下：

```
<form:checkbox path="friends" value="张三"/>张三
<
form:checkbox path="friends" value="李四"/>李四
<
form:checkbox path="friends" value="王五"/>王五
<
form:checkbox path="friends" value="赵六"/>赵六
```

上述示例代码中，复选框的值绑定到一个字符串数组属性 friends(String[] friends)。该标签的其他用法与 input 标签基本一致，这里不再赘述。

7. checkboxes 标签

checkboxes 标签渲染多个复选框，是一个选项组，等价于多个 path 相同的 checkbox 标签。它有 3 个非常重要的属性，即 items、itemLabel 和 itemValue。

- items：用于生成 input 元素的 Collection、Map 或 Array。
- itemLabel：items 属性中指定的集合对象的属性，为每个 input 元素提供 label。
- itemValue：items 属性中指定的集合对象的属性，为每个 input 元素提供 value。

checkboxes 标签的语法格式如下：

```
<form:checkboxes items="xxx" path="xxx"/>
```

示例代码如下：

```
<form:checkboxes items="${hobbys}" path="hobby"/>
```

上述示例代码将 model 属性 hobbys 的内容(集合元素)渲染为复选框。在 itemLabel 和 itemValue 省略的情况下，如果集合是数组，则复选框的 label 和 value 相同；如果是 Map 集合，复选框的 label 是 Map 的值(value)，复选框的 value 是 Map 的关键字(key)。

8. radiobutton 标签

radiobutton 标签的语法格式如下：

```
<form:radiobutton path="xxx" value="xxx"/>
```

多个 path 相同的 radiobutton 标签，它们是一个选项组，只允许单选。

9. radiobuttons 标签

radiobuttons 标签渲染多个 radio，是一个选项组，等价于多个 path 相同的 radiobutton 标签。radiobuttons 标签的语法格式如下：

```
<form:radiobuttons items="xxx" path="xxx"/>
```

该标签的 itemLabel 和 itemValue 属性与 checkboxes 标签的 itemLabel 和 itemValue 属性完全一样，但只允许单选。

10. select 标签

select 标签的选项可能来自其属性 items 指定的集合,或者来自一个嵌套的 option 标签或 options 标签。其语法格式如下:

```
<form:select path="xxx" items="xxx"/>
```

或

```
<form:select path="xxx" items="xxx">
<option value="xxx">xxx</option>
</form:select>
```

或

```
<form:select path="xxx">
<form:options items="xxx"/>
</form:select>
```

该标签的 itemLabel 和 itemValue 属性与 checkboxes 标签的 itemLabel 和 itemValue 属性完全一样。

11. option 标签

这个标签生成一个 HTML 的 option。根据绑定的值,它会恰当地设置 selected 属性。

```
<tr>
    <td>House:</td>
    <td>
        <form:select path="house">
            <form:option value="Gryffindor"/>
            <form:option value="Hufflepuff"/>
            <form:option value="Ravenclaw"/>
        </form:select>
    </td>
</tr>
```

12. options 标签

这个标签生成多个 HTML 的 option 标签的列表。根据绑定的值,它会恰当地设置 selected 属性。

```
<tr>
    <td>Country:</td>
    <td>
        <form:select path="country">
            <form:option value="-" label="--Please Select"/>
            <form:options items="${countryList}" itemValue="code" itemLabel=
              "name"/>
        </form:select>
```

```
    </td>
</tr>
```

13. errors 标签

errors 标签渲染一个或者多个 span 元素，每个 span 元素包含一个错误消息。它可用于显示一个特定的错误消息，也可以显示所有的错误消息。其语法格式如下：

```
<form:errors path="*"/>
```

或

```
<form:errors path="xxx"/>
```

其中，“*”表示显示所有错误消息；“xxx”表示显示由“xxx”指定的特定的错误消息。

接下来看一个使用标签的实例开发(Spring MVC_taglib_demo)。具体步骤如下：

(1) 新建一个 Web 项目，项目名称为 Spring MVC_taglib_demo。

(2) 添加 Spring MVC 相关 jar 包，右击 Spring MVC_taglib_demo 项目，从弹出的快捷菜单中选择 Configure Facets...→Install Spring Facet。

(3) 建立项目目录。

在 src 下创建包 com.ascent.controller，并存放在 controller 控制类中。

创建包 com.ascent.po，并存放在实体类中。

(4) 修改 web.xml，添加以下内容：

```
<?xml version="1.0" encoding="UTF-8"?>
<web-app xmlns:xsi="http://www.w3.org/2001/XMLSchema-instance"
xmlns="http://xmlns.jcp.org/xml/ns/javaee"
xsi:schemaLocation="http://xmlns.jcp.org/xml/ns/javaee
http://xmlns.jcp.org/xml/ns/javaee/web-app_3_1.xsd" id="WebApp_ID" version=
"3.1">
  <display-name>Spring MVC_validation_demo</display-name>
  <welcome-file-list>
    <welcome-file>index.jsp</welcome-file>
  </welcome-file-list>
  <listener>
<listener-class>org.springframework.web.context.ContextLoaderListener
</listener-class>
  </listener>
  <context-param>
    <param-name>contextConfigLocation</param-name>
    <param-value>classpath:applicationContext.xml</param-value>
  </context-param>
  <servlet>
    <servlet-name>Spring MVC</servlet-name>
    <servlet-class>
        org.springframework.web.servlet.DispatcherServlet
    </servlet-class>
```

```
    <load-on-startup>1</load-on-startup>
  </servlet>
  <servlet-mapping>
    <servlet-name>Spring MVC</servlet-name>
    <url-pattern>/</url-pattern>
  </servlet-mapping>
    <!--避免中文乱码 -->
      <filter>
      <filter-name>characterEncodingFilter</filter-name>
<filter-class>org.springframework.web.filter.CharacterEncodingFilter
</filter-class>
        <init-param>
        <param-name>encoding</param-name>
        <param-value>UTF-8</param-value>
    </init-param>
    <init-param>
        <param-name>forceEncoding</param-name>
        <param-value>true</param-value>
    </init-param>
  </filter>
  <filter-mapping>
    <filter-name>characterEncodingFilter</filter-name>
    <url-pattern>/*</url-pattern>
  </filter-mapping>

</web-app>
```

</web-app>这里将DispatcherServlet命名为Spring MVC,并且让它在Web项目一启动就加载。接下来需要在WEB-INF目录下创建一个Spring MVC -servlet.xml的Spring配置文件,此文件名的命名规则为:在servlet-name名称后面加-servlet.xml。

(5) 添加Spring MVC-servlet.xml配置文件,内容如下:

```
<?xml version="1.0" encoding="UTF-8"?>
<beans xmlns="http://www.springframework.org/schema/beans"
    xmlns:xsi="http://www.w3.org/2001/XMLSchema-instance"
    xmlns:mvc="http://www.springframework.org/schema/mvc"
    xmlns:p="http://www.springframework.org/schema/p"
    xmlns:context="http://www.springframework.org/schema/context"
    xsi:schemaLocation="
        http://www.springframework.org/schema/beans
        http://www.springframework.org/schema/beans/spring-beans.xsd
        http://www.springframework.org/schema/context
        http://www.springframework.org/schema/context/spring-context.xsd
        http://www.springframework.org/schema/mvc
        http://www.springframework.org/schema/mvc/spring-mvc.xsd">
    <!--使用扫描机制扫描控制器类 -->
```

```
    <context:component-scan base-package="com.ascent" />
<!--配置视图解析器 -->
<bean class="org.springframework.web.servlet.view.InternalResourceViewResolver">
    <property name="prefix" value="/WEB-INF/views/" />
    <property name="suffix" value=".jsp" />
</bean>
</beans>
```

（6）在 com.ascent po 包中编写 User 类。

```
package com.ascent.po;

public class User {
    private Integer id;
    private String username;
    private String password;
    public Integer getId() {
        return id;
    }
    public void setId(Integer id) {
        this.id=id;
    }
    public String getUsername() {
        return username;
    }
    public void setUsername(String username) {
        this.username=username;
    }
    public String getPassword() {
        return password;
    }
    public void setPassword(String password) {
        this.password=password;
    }
}
```

（7）在 com.ascent controller 包中编写 UserController 类。

```
package com.ascent.controller;

import com.ascent.po.User;
import org.springframework.stereotype.Controller;
import org.springframework.web.bind.annotation.ModelAttribute;
import org.springframework.web.bind.annotation.RequestMapping;
import org.springframework.web.bind.annotation.RequestMethod;
import org.springframework.web.servlet.ModelAndView;
import org.springframework.ui.ModelMap;
```

```
@Controller
public class UserController {
   @RequestMapping(value="/user", method=RequestMethod.GET)
   publicModelAndView user() {
      return newModelAndView("user", "command", new User());
   }
   @RequestMapping(value="/addUser", method=RequestMethod.POST)
   public String addUser(@ModelAttribute("SpringWeb")User user,
   ModelMap model) {
      model.addAttribute("username", user.getUsername());
      model.addAttribute("password", user.getPassword());
      model.addAttribute("id", user.getId());
      return "result";
   }
}
```

(8) 编写 JSP 文件。

根据 Spring MVC-servlet.xml,在/WEB-INF/views/下编写 user.jsp 和 result.jsp。

```
user.jsp 如下:
<%@taglib uri="http://www.springframework.org/tags/form" prefix="form"%>
<html>
<head>
    <title>Spring MVC Form Handling</title>
</head>
<body>

<h2>User Information</h2>
<form:form method="POST" action="/Spring MVC_taglib_demo/addUser">
    <table>
      <tr>
        <td><form:label path="id">ID</form:label></td>
        <td><form:input path="id" /></td>
      </tr>
      <tr>
        <td><form:label path="username">User Name</form:label></td>
        <td><form:input path="username" /></td>
      </tr>
      <tr>
        <td><form:label path="password">Password</form:label></td>
        <td><form:password path="password" /></td>
      </tr>
      <tr>
        <td colspan="2">
            <input type="submit" value="Submit"/>
        </td>
      </tr>
```

```
</table>
</form:form>
</body>
</html>
```

result.jsp 如下：

```
<%@taglib uri="http://www.springframework.org/tags/form" prefix="form"%>
<html>
<head>
    <title>Spring MVC Form Tag Handling</title>
</head>
<body>

<h2>Submitted User Information</h2>
    <table>
      <tr>
        <td>ID</td>
        <td>${id}</td>
      </tr>
      <tr>
        <td>User Name</td>
        <td>${username}</td>
      </tr>
      <tr>
        <td>Password</td>
        <td>${password}</td>
      </tr>
  </table>
</body>
</html>
```

(9) 部署和运行项目。

选中 Spring MVC_taglib_demo 项目，部署在 Tomcat 8 服务器上，之后启动 Tomcat，在浏览器中输入 http://localhost:8080/Spring MVC_taglib_demo/user，进入 user 页面，如图 6-1 所示。

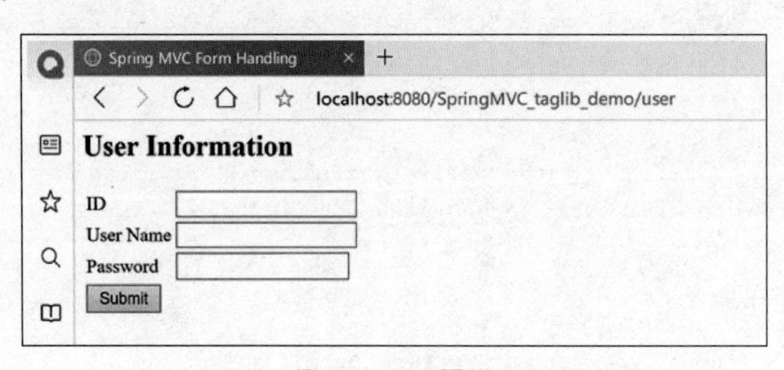

图 6-1　user 页面

在 ID 处输入 1,在 User Name 处输入 Lixin,在 Password 处输入 123456,单击 Submit 按钮,进入 result 页面,如图 6-2 所示。

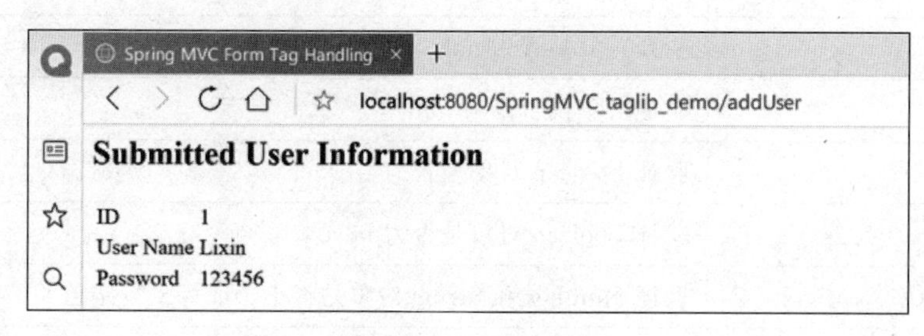

图 6-2 result 页面

6.2 Spring MVC 数据校验

在进行 Web 开发的时候,需要对用户输入的数据进行验证,以防数据不合法。数据验证是所有 Web 应用必须处理的问题。数据验证分为客户端验证和服务器端验证。客户端验证主要是过滤正常用户的一些误操作,一般通过 JavaScript 代码完成。服务器端验证是阻止非法数据的最后防线,通常在应用中编程实现。

1. 客户端验证

在大多数情况下,使用 JavaScript 进行客户端验证的步骤如下。

(1) 编写验证函数。

(2) 在提交表单的事件中调用验证函数。

(3) 根据验证函数判断是否进行表单提交。

客户端验证可以过滤用户的误操作,是第一道防线。但仅有客户端验证是不够的,攻击者还可以绕过客户端验证直接进行非法输入,这样可能会引起系统异常,为了确保数据的合法性,防止用户通过非正常手段提交错误信息,必须加上服务器端验证。

2. 服务器端验证

服务器端验证对系统的安全性、完整性、健壮性起到了至关重要的作用。在 Spring MVC 框架中可以利用 Spring 自带的验证框架验证数据,也可以利用 JSR 303 实现数据验证。

Spring MVC 支持的数据校验是 JSR 303 的标准,通过在需要进行校验的对象的属性上打上@NotNull @Max 等注解,在 Controller()方法中的 bean 对象前加上 @Valid 注解进行验证。JSR 303 提供了很多 annotation 接口,这里使用 hibernate-validator 进行校验,该方法实现了 JSR 303 验证框架支持注解风格的验证。Spring MVC 的控制器可以将所有的错误提示保存在集合里(BindingResult),我们拿到这个集合后将里面的错误信息提取出来交给前端页面展示。

常用的校验注解见表 6-2。

接下来看一个数据验证的实例开发(Spring MVC_validation_demo)。具体步骤如下:

表 6-2 常用的校验注解

注 解	功 能
@Null	验证对象是否为 null
@NotNull	验证对象是否不为 null
@AssertTrue	验证 Boolean 对象是否为 true
@AssertTrue	验证 Boolean 对象是否为 false
@Max(value)	验证 Number 和 String 对象是否小于或等于指定值
@Min(value)	验证 Number 和 String 对象是否大于或等于指定值
@DecimalMax(value)	验证注解的元素值是否小于或等于@DecimalMax 指定的 value 值
@DecimalMin(value)	验证注解的元素值是否大于或等于@DecimalMin 指定的 value 值
@Digits(integer,fraction)	验证字符串是否符合指定格式的数字，integer 指定整数精度，fraction 指定小数精度
@Size(min,max)	验证对象长度是否在给定的范围内
@Past	验证 Date 和 Calendar 对象是否在当前时间之前
@Future	验证 Date 和 Calendar 对象是否在当前时间之后
@Pattern	验证 String 对象是否符合正则表达式的规则
@NotBlank	检查字符串是否为 Null，被 Trim 的长度是否大于 0，只对字符串，且会去掉前后空格
@URL	验证是否是合法的 URL
@Email	验证是否是合法的邮箱
@CreditCardNumber	验证是否是合法的信用卡号
@Length(min,max)	验证字符串的长度是否在指定范围内
@NotEmpty	检查元素是否为 Null 或 Empty
@Range(min,max,message)	验证属性值是否在合适的范围内

(1) 新建一个 Web 项目，项目名称为 Spring MVC_validation_demo。

(2) 添加 Spring MVC 相关 jar 包，右击 Spring MVC_validation_demo 项目，从弹出的快捷菜单中选择 Configure Facets...→Install Spring Facet。

另外，因为采用的是 hibernate validator，所以还需要导入 hibernate validator 相关 jar 包。

读者可以从 https://sourceforge.net/projects/hibernate/files/hibernate-validator/下载，这里用的是 hibernate-validator-6.0.18.Final，在解压文件之后，导入其中的 classmate-1.3.4.jar、hibernate-validator-6.0.18.Final.jar、hibernate-validator-annotation-processor-6.0.18.Final.jar、hibernate-validator-cdi-6.0.18.Final.jar、jboss-logging-3.3.2.Final.jar 和 validation-api-2.0.1.Final.jar 等 jar 包，如图 6-3 所示。

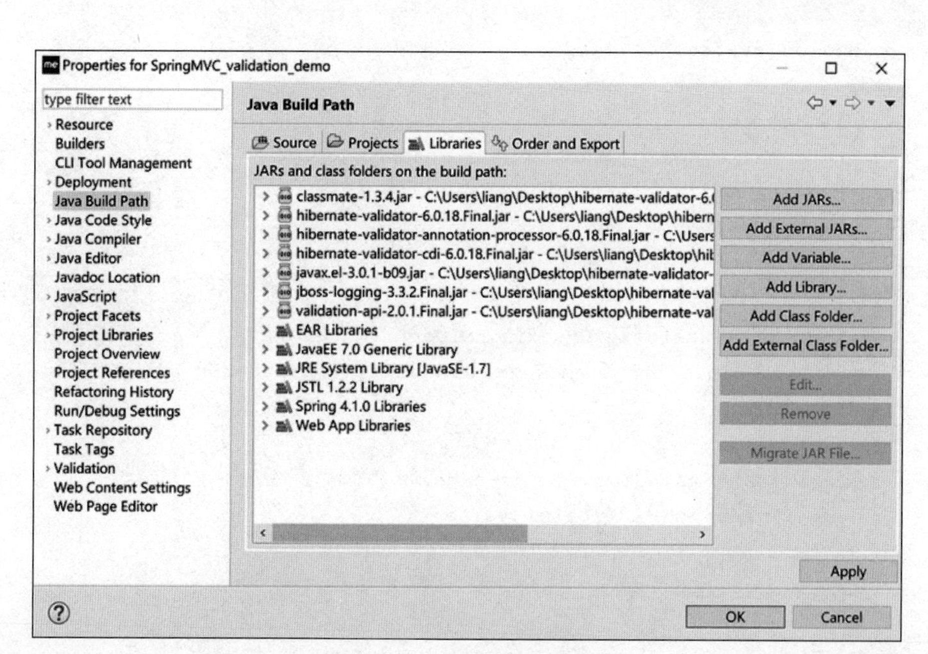

图 6-3　导入相关 jar 包

(3) 建立项目目录。

在 src 下创建包 com.ascent.controller,并存放在 controller 控制类中。

创建包 com.ascent.po,并存放在实体类中。

(4) 修改 web.xml,添入以下内容:

```
<?xml version="1.0" encoding="UTF-8"?>
<web-app xmlns:xsi="http://www.w3.org/2001/XMLSchema-instance"
xmlns="http://xmlns.jcp.org/xml/ns/javaee"
xsi:schemaLocation="http://xmlns.jcp.org/xml/ns/javaee
http://xmlns.jcp.org/xml/ns/javaee/web-app_3_1.xsd" id="WebApp_ID" version=
"3.1">
  <display-name>Spring MVC_login_demo</display-name>

  <listener>
<listener-class>org.springframework.web.context.ContextLoaderListener
  </listener-class>
  </listener>
  <context-param>
    <param-name>contextConfigLocation</param-name>
    <param-value>classpath:applicationContext.xml</param-value>
  </context-param>
  <servlet>
    <servlet-name>Spring MVC</servlet-name>
    <servlet-class>
        org.springframework.web.servlet.DispatcherServlet
    </servlet-class>
```

```
    <load-on-startup>1</load-on-startup>
  </servlet>
  <servlet-mapping>
    <servlet-name>Spring MVC</servlet-name>
    <url-pattern>/</url-pattern>
  </servlet-mapping>

  <filter>
    <filter-name>CharacterEncodingFilter</filter-name>
<filter-class>org.springframework.web.filter.CharacterEncodingFilter
  </filter-class>
    <init-param>
      <param-name>encoding</param-name>
      <param-value>UTF-8</param-value>
    </init-param>
  </filter>
  <filter-mapping>
    <filter-name>CharacterEncodingFilter</filter-name>
    <url-pattern>/*</url-pattern>
  </filter-mapping>
</web-app>
```

这里将 DispatcherServlet 命名为 Spring MVC，并且让它在 Web 项目一启动就加载。接下来需要在 WEB-INF 目录下创建一个 Spring MVC -servlet.xml 的 Spring 配置文件，此文件名的命名规则为：在 servlet-name 名称后面加-servlet.xml。

(5) 添加 Spring MVC-servlet.xml 配置文件，内容如下：

```
<?xml version="1.0" encoding="UTF-8"?>
<beans xmlns="http://www.springframework.org/schema/beans"
    xmlns:xsi="http://www.w3.org/2001/XMLSchema-instance"
xmlns:p="http://www.springframework.org/schema/p"
    xmlns:mvc="http://www.springframework.org/schema/mvc"
xmlns:context="http://www.springframework.org/schema/context"
    xmlns:aop="http://www.springframework.org/schema/aop"
    xsi:schemaLocation="http://www.springframework.org/schema/beans
        http://www.springframework.org/schema/beans/spring-beans-4.2.xsd
        http://www.springframework.org/schema/mvc
        http://www.springframework.org/schema/mvc/spring-mvc-4.2.xsd
        http://www.springframework.org/schema/context
        http://www.springframework.org/schema/context/spring-context-4.2.xsd
        http://www.springframework.org/schema/aop
        http://www.springframework.org/schema/aop/spring-aop-4.2.xsd">

<context:component-scan base-package="com.ascent" />

    <bean id="viewResolver" class="org.springframework.web.servlet.view.
```

```
        InternalResourceViewResolver">
        <property name="prefix" value="/WEB-INF/views/" />
        <property name="suffix" value=".jsp" />
    </bean>
      <!--配置验证器 -->
        <bean id="myvalidator" class="org.springframework.validation.
            beanvalidation.LocalValidatorFactoryBean">
            <property name="providerClass"
value="org.hibernate.validator.HibernateValidator"></property>
    </bean>

    <mvc:annotation-driven validator="myvalidator"/>
</beans>
```

注意：这里添加了 HibernateValidator 的配置，并在 mvc 注解驱动中设置了 validator 属性。

```
<bean id="myvalidator"
class="org.springframework.validation.beanvalidation.LocalValidatorFactoryBean">
    <property name="providerClass"
value="org.hibernate.validator.HibernateValidator"></property>
    </bean>
```

(6) 在 com.ascent po 包中编写 UserInfo 类。

```
package com.ascent.po;

import javax.validation.constraints.Max;
import javax.validation.constraints.Min;
import javax.validation.constraints.NotNull;
import javax.validation.constraints.Pattern;
import javax.validation.constraints.Size;
import org.hibernate.validator.constraints.Length;

public class UserInfo{

    //名字不能为空
    //必须在 8 个字符以上,16 个字符以下
     @NotNull(message="用户名不能为空")
     @Length(min=8,max=16,message="用户名不能少于 8 位,且不能大于 16 位")
    private String name;

    //必须在 0~120
    @Min(value=0,message="年龄最小值为{value}")
    @Max(value=120,message="年龄最大值为{value}")
    private Integer age;
```

```
    //手机号码必须不能为空,必须以 1 开头 第二位为 3,4,5,6,7,8 中的某个数,最后 9 位任何
      数字都可以
    @NotNull(message="手机号码不能为空")
    @Pattern(regexp="^1[3,4,5,6,7,8]\\d{9}$",message="手机号码不正确")
    private String phone;

    public StringgetName() {
        return name;
    }
    public voidsetName(String name) {
        this.name=name;
    }

    public Integer getAge() {
        return age;
    }
    public void setAge(Integer age) {
        this.age=age;
    }

    public String getPhone() {
        return phone;
    }
    public void setPhone(String phone) {
        this.phone=phone;
    }

}
```

在所需校验的对象对应的实体类 UserInfo 的属性上加相应的校验注解。

例如：

```
@NotNull 对象不为空
@Length 验证字符串的长度必须在指定范围内
@Min(value=)验证 Number 对象是否大于或等于指定的值
@Max(value=)验证 Number 对象是否小于或等于指定的值
@Pattern 验证 String 对象是否符合正则表达式的规则
```

(7) 在 com.ascent controller 包中编写 ValidateController 类。

```
package com.ascent.controller;

import javax.validation.Valid;

import org.springframework.stereotype.Controller;
import org.springframework.validation.BindingResult;
import org.springframework.validation.FieldError;
```

```
import org.springframework.web.bind.annotation.RequestMapping;
import org.springframework.web.servlet.ModelAndView;

import com.ascent.po.UserInfo;

@Controller
public classValidateController {
    @RequestMapping("/validate.do")
    public ModelAndView doValidation(@Valid UserInfo info,BindingResult br){
        ModelAndView mv=new ModelAndView();
        mv.setViewName("welcome");

        interrorCount=br.getErrorCount();
        if (errorCount>0) {
            FieldError name=br.getFieldError("name");
            FieldError age=br.getFieldError("age");
            FieldError phone=br.getFieldError("phone");

            if (name!=null){
                mv.addObject("namemsg",name.getDefaultMessage());
            }

            if (age!=null){
                mv.addObject("agemsg",age.getDefaultMessage());
            }

            if (phone!=null){
                mv.addObject("phonemsg",phone.getDefaultMessage());
            }
            mv.setViewName("index");
        }

        return mv ;
    }
}
```

在 Spring MVC 的 Controller()方法中需要进行校验的对象前加上@Valid 注解，并在校验对象后加上一个 BindingResult 对象接收校验信息。

(8) 编写 JSP 文件。

根据 Spring MVC-servlet.xml，在/WEB-INF/views/下编写 index.jsp 和 welcome.jsp。

```
index.jsp 如下：
<%@ page language="java" import="java.util.*" pageEncoding="utf-8"%>
<%
String path=request.getContextPath();
String basePath = request. getScheme () +"://" + request. getServerName () +":" +
```

```
request.getServerPort()+path+"/";
%>

<!DOCTYPE HTML PUBLIC "-//W3C//DTD HTML 4.01 Transitional//EN">
<html>
  <head>
    <base href="<%=basePath%>">

    <title>Spring Validation Sample</title>

  </head>

  <body>
    <h1>数据校验</h1>
    <form action="${pageContext.request.contextPath }/validate.do" method=
    "post">
            姓名:<input name="name"/><span>${namemsg }</span><br/><br/>
            年龄:<input name="age" /><span>${agemsg }</span><br/><br/>
            电话:<input name="phone"/><span>${phonemsg }</span><br/><br/>
      <input type="submit" value="注册"/>
    </form>
  </body>
</html>
```

welcome.jsp 如下：

```
<%@ page language="java" import="java.util.*" pageEncoding="utf-8"%>
<%
String path=request.getContextPath();
StringbasePath= request.getScheme()+"://"+request.getServerName()+":"+
request.getServerPort()+path+"/";
%>

<!DOCTYPE HTML PUBLIC "-//W3C//DTD HTML 4.01 Transitional//EN">
<html>
  <head>
    <base href="<%=basePath%>">

    <title>welcome page</title>

    <meta http-equiv="pragma" content="no-cache">
    <meta http-equiv="cache-control" content="no-cache">
    <meta http-equiv="expires" content="0">
    <meta http-equiv="keywords" content="keyword1,keyword2,keyword3">
    <meta http-equiv="description" content="This is my page">
    <!--
```

```
    <linkrel="stylesheet" type="text/css" href="styles.css">
    -->
  </head>

  <body>
    欢迎！ <br>
  </body>
</html>
```

(9) 部署和运行项目。

选中 Spring MVC_validation_demo 项目，部署在 Tomcat 8 服务器上，之后启动 Tomcat，在浏览器中输入 http://localhost:8080/Spring MVC_validation_demo/validate.do 进入验证页面，如图 6-4 所示。

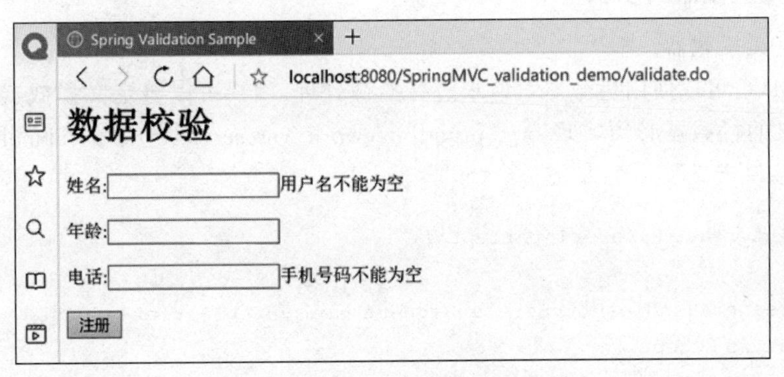

图 6-4 验证页面

如果“输入的用户名不少于 8 位且不大于 16 位，年龄在 0～120，手机号码不为空，且以 1 开头，第二位为 3，4，5，6，7，8 中的某个数，最后 9 位数字任意”，则进入成功页面，如图 6-5 所示。

图 6-5 成功页面

否则将继续停留在原页面。

6.3 Spring MVC 拦截器

1. 拦截器概述

Spring MVC 的拦截器(Interceptor)与 Java Servlet 的过滤器(Filter)类似，主要用于拦截用户的请求并进行相应的处理，通常应用在权限验证、记录请求信息的日志、判断用户是否登录等功能上。

开发者可以自己定义一些拦截器实现特定的功能。

(1) 拦截器链(Interceptor Chain)：将拦截器按一定的顺序连接成一条链。在访问被拦截的方法或者字段时，拦截器链中的拦截器就会按照之前定义的顺序被调用。

(2) 过滤器与拦截器的区别见表 6-3。

表 6-3 过滤器与拦截器的区别

过 滤 器	拦 截 器
servlet 规范中的一部分，任何 Java Web 工程都可以使用	拦截器是 Spring MVC 框架自己的，只有使用了 Spring MVC 框架的工程才能使用
在 url-pattern 中配置/＊之后，可以对所有要访问的资源进行拦截	拦截器只会拦截访问的控制器方法，如果访问的是 jsp/html/css/image/js，则不会进行拦截

2. 自定义拦截器的步骤

(1) 自定义拦截器。

Spring MVC 可以使用拦截器对请求进行拦截处理，用户可以自定义拦截器实现特定的功能。自定义的拦截器必须实现 org.springframework.web.servlet 包下的 HandlerInterceptor 接口。这个接口定义了 3 个方法：

```
public interface HandlerInterceptor {

    booleanpreHandle(HttpServletRequest var1, HttpServletResponse var2, Object
var3) throws Exception;

    voidpostHandle (HttpServletRequest var1, HttpServletResponse var2, Object
var3, ModelAndView var4) throws Exception;

    voidafterCompletion (HttpServletRequest var1, HttpServletResponse var2,
Object var3, Exception var4) throws Exception;
}
```

这 3 个方法为各种类型的前处理和后处理需求提供了足够的灵活性。

① preHandle：在访问到达 Controller 之前执行，如果需要对请求做预处理，可以选择在该方法中完成。

返回值为 true：继续执行后面的拦截器或者 Controller。

返回值为 false：不再执行后面的拦截器和 Controller，并调用返回 true 的拦截器的 afterCompletion()方法。

② postHandle：在执行完 Controller()方法之后，渲染视图之前执行，如果需要对响应相关的数据进行处理，可以选择在该方法中完成。

③ afterCompletion：在调用完 Controller 接口，渲染 View 页面后调用。返回 true 的拦截器都会调用该拦截器的 afterCompletion()方法，顺序相反。

(2) 配置拦截器。

拦截器在 Spring 的配置文件中进行配置后才可以生效。拦截器可以通过 interceptors

属性配置，在配置拦截器时，根节点是<mvc:interceptors>，用于配置拦截器链，即任何一个 Spring MVC 项目，都可以有若干个拦截器，从而形成拦截器链，如果某个请求符合多个拦截器的拦截配置，则会依次被各拦截器进行处理。任何一个拦截都会导致后续不再将请求交给 Controller 处理。

在<mvc:interceptor>中，<bean>节点用于指定拦截器类；<mvc:mapping>节点用于配置拦截的请求路径，每个拦截器可以配置多个该节点，并且支持使用星号 * 作为通配符，如<mvc:mapping path="/user/ * "/>。为了避免拦截范围过大，可以通过<mvc:exclude-mapping />配置排除在外的请求路径，可以理解为白名单，该节点也可以配置多个。

例如：

```
<!--配置拦截器 -->
<mvc:interceptors>
    <mvc:interceptor>
        <mvc:mapping path="/* * "/>
        --用于指定拦截的 URL,可以配置多个
        <mvc:exclude-mapping path="/xxx"/>
        --用于指定排除的 RUL,可以配置多个
        <bean id="handlerInterceptorOne"
              class="com.ascent.myInterceptor.HanderInterceptorOne"/>
    </mvc:interceptor>
</mvc:interceptors>
```

接下来看一个拦截器的实例开发(Spring MVC_Interceptor_demo)，具体步骤如下。

(1) 新建一个 Web 项目，项目名称为 Spring MVC_Interceptor_demo。

(2) 添加 Spring MVC 相关 jar 包，右击 Spring MVC_Interceptor_demo 项目，从弹出的快捷菜单中选择 Configure Facets...→Install Spring Facet。

(3) 建立项目目录。

在 src 下创建包 com.ascent.controller，并存放在 controller 控制类中。

创建包 com.ascent.po，并存放在实体类中。

创建包 com.ascent.interceptor，并存放在 interceptor 拦截器中。

(4) 修改 web.xml，添入以下内容：

```
<?xml version="1.0" encoding="UTF-8"?>
<web-app xmlns:xsi="http://www.w3.org/2001/XMLSchema-instance"
xmlns="http://xmlns.jcp.org/xml/ns/javaee"
xsi:schemaLocation="http://xmlns.jcp.org/xml/ns/javaee
http://xmlns.jcp.org/xml/ns/javaee/web-app_3_1.xsd" id="WebApp_ID" version=
"3.1">
  <display-name>Spring MVC_validation_demo</display-name>
  <welcome-file-list>
    <welcome-file>index.jsp</welcome-file>
  </welcome-file-list>
  <listener>
```

```
<listener-class>org.springframework.web.context.ContextLoaderListener
</listener-class>
  </listener>
  <context-param>
    <param-name>contextConfigLocation</param-name>
    <param-value>classpath:applicationContext.xml</param-value>
  </context-param>
  <servlet>
    <servlet-name>Spring MVC</servlet-name>
    <servlet-class>
        org.springframework.web.servlet.DispatcherServlet
    </servlet-class>
    <load-on-startup>1</load-on-startup>
  </servlet>
  <servlet-mapping>
    <servlet-name>Spring MVC</servlet-name>
    <url-pattern>/</url-pattern>
  </servlet-mapping>
    <!--避免中文乱码 -->
      <filter>
        <filter-name>characterEncodingFilter</filter-name>
<filter-class>org.springframework.web.filter.CharacterEncodingFilter
</filter-class>
        <init-param>
            <param-name>encoding</param-name>
            <param-value>UTF-8</param-value>
        </init-param>
        <init-param>
            <param-name>forceEncoding</param-name>
            <param-value>true</param-value>
        </init-param>
    </filter>
    <filter-mapping>
        <filter-name>characterEncodingFilter</filter-name>
        <url-pattern>/*</url-pattern>
    </filter-mapping>

</web-app>
```

</web-app>这里将DispatcherServlet命名为Spring MVC,并且让它在Web项目一启动就加载。接下来需要在WEB-INF目录下创建一个Spring MVC -servlet.xml的Spring配置文件,此文件名的命名规则为:在servlet-name名称后面加-servlet.xml。

(5)添加Spring MVC-servlet.xml配置文件,内容如下:

```
<?xml version="1.0" encoding="UTF-8"?>
<beans xmlns="http://www.springframework.org/schema/beans"
```

```
    xmlns:xsi="http://www.w3.org/2001/XMLSchema-instance"
    xmlns:mvc="http://www.springframework.org/schema/mvc"
    xmlns:p="http://www.springframework.org/schema/p"
    xmlns:context="http://www.springframework.org/schema/context"
    xsi:schemaLocation="
        http://www.springframework.org/schema/beans
        http://www.springframework.org/schema/beans/spring-beans.xsd
        http://www.springframework.org/schema/context
        http://www.springframework.org/schema/context/spring-context.xsd
        http://www.springframework.org/schema/mvc
        http://www.springframework.org/schema/mvc/spring-mvc.xsd">
    <!--使用扫描机制扫描控制器类 -->
    <context:component-scan base-package="com.ascent" />
    <!--配置视图解析器 -->
    <bean class="org.springframework.web.servlet.view.
InternalResourceViewResolver">
        <property name="prefix" value="/WEB-INF/views/" />
        <property name="suffix" value=".jsp" />
    </bean>
    <!--配置拦截器 -->
    <mvc:interceptors>
        <mvc:interceptor>
            <!--配置拦截器作用的路径 -->
            <mvc:mapping path="/**" />
            <bean class="com.ascent.interceptor.LoginInterceptor" />
        </mvc:interceptor>
    </mvc:interceptors>
</beans>
```

注意：这里配置了拦截器的内容。

```
<mvc:interceptors>
    <mvc:interceptor>
        <!--配置拦截器作用的路径 -->
        <mvc:mapping path="/**" />
        <bean class="com.ascent.interceptor.LoginInterceptor" />
    </mvc:interceptor>
    </mvc:interceptors>
```

(6) 在 com.ascent po 包中编写 User 类。

```
package com.ascent.po;

public class User {
    private Integer id;
    private String username;
    private String password;
```

```
    public Integer getId() {
        return id;
    }
    public void setId(Integer id) {
        this.id=id;
    }
    public String getUsername() {
        return username;
    }
    public void setUsername(String username) {
        this.username=username;
    }
    public String getPassword() {
        return password;
    }
    public void setPassword(String password) {
        this.password=password;
    }
}
```

(7) 在 com.ascent controller 包中编写 UserController 类。

```
package com.ascent.controller;

import javax.servlet.http.HttpSession;
import org.springframework.stereotype.Controller;
import org.springframework.ui.Model;
import org.springframework.web.bind.annotation.RequestMapping;
import com.ascent.po.User;
@Controller
public classUserController {
    /**
     * 进入登录页面
     * /
    @RequestMapping("/toLogin")
    public String toLogin() {
        return "login";
    }
    /**
     * 处理登录
     * /
    @RequestMapping("/login")
    public String login(User user, Model model,HttpSession session) {

        if("Lixin".equals(user.getUsername())
                && "123456".equals(user.getPassword())) {
```

```
            session.setAttribute("user", user);
            //登录过了,重定向到主页面
            return "redirect:main";
        }
        model.addAttribute("msg", "用户名或密码错误,请重新登录!");
        return "login";
    }
    /**
     * 跳转到主页面
     */
    @RequestMapping("/main")
    public String toMain() {
        return "main";
    }
    /**
     * 退出登录
     */
    @RequestMapping("/logout")
    public String logout(HttpSession session) {
        //清除 session
        session.invalidate();
        return "login";
    }
}
```

(8) 在 com.ascent.interceptor 包中编写 LoginInterceptor 类。

```
package com.ascent.interceptor;

import javax.servlet.http.HttpServletRequest;
import javax.servlet.http.HttpServletResponse;
import javax.servlet.http.HttpSession;
import org.springframework.web.servlet.HandlerInterceptor;
import org.springframework.web.servlet.ModelAndView;

public class LoginInterceptor implements HandlerInterceptor {
    @Override
    public boolean preHandle(HttpServletRequest request,
            HttpServletResponse response, Object handler) throws Exception {
        String url=request.getRequestURI();
        //登录页面或登录请求,不拦截
        if (url.indexOf("/toLogin") >=0 || url.indexOf("/login") >=0) {
            return true;
        }
        //获取 session
        HttpSession session=request.getSession();
```

```
        Object obj=session.getAttribute("user");
        if (obj !=null)
            return true;
        //没有登录且不是登录页面,转发到登录页面,并给出提示错误信息
        request.setAttribute("msg", "还没有完成登录,请先登录!");
        request.getRequestDispatcher("/WEB-INF/views/login.jsp").forward
            (request, response);
        return false;
    }
    @Override
    public void afterCompletion(HttpServletRequest arg0,
            HttpServletResponse arg1, Object arg2, Exception arg3)
            throws Exception{

    }
    @Override
    public void postHandle(HttpServletRequest arg0, HttpServletResponse arg1,
            Object arg2,ModelAndView arg3) throws Exception {

    }
}
```

(9) 编写 JSP 文件。

根据 Spring MVC-servlet.xml,在/WEB-INF/views/下编写 login.jsp 和 main.jsp。

```
login.jsp 如下:
<%@ page language="java" contentType="text/html; charset=UTF-8"
    pageEncoding="UTF-8"%>
<!DOCTYPE html PUBLIC "-//W3C//DTD HTML 4.01 Transitional//EN"
"http://www.w3.org/TR/html4/loose.dtd">
<html>
<head>
<meta http-equiv="Content-Type" content="text/html; charset=UTF-8">
<title>登录页面</title>
</head>
<body>
    ${msg }
    <form action="${pageContext.request.contextPath }/login" method="post">
        用户名: <input type="text" name="username" /><br>
        密码: <input type="password" name="password" /><br>
        <input type="submit" value="登录" />
    </form>
</body>
</html>

main.jsp 如下:
```

```
<%@ page language="java" contentType="text/html; charset=UTF-8"
pageEncoding="UTF-8"%>
<!DOCTYPE html PUBLIC "-//W3C//DTD HTML 4.01 Transitional//EN"
"http://www.w3.org/TR/html4/loose.dtd">
<html>
<head>
<meta http-equiv="Content-Type" content="text/html; charset=UTF-8">
<title>主页面</title>
</head>
<body>
        欢迎,当前用户：${user.username }<br />
    <a href="${pageContext.request.contextPath }/logout">退出</a>
</body>
</html>
```

(10) 部署和运行项目。

选中 Spring MVC_ Interceptor _demo 项目,部署在 Tomcat 8 服务器上,之后启动 Tomcat,在浏览器中输入 http://localhost:8080/Spring MVC_Interceptor_demo/main,希望进入 main 页面,但因为还没有完成登录,所以被拦截,转向登录页面,如图 6-6 所示。

图 6-6　被拦截后转向登录页面

在用户名处输入 Lixin,在密码处输入 123456,单击“登录”按钮,成功进入 main 页面,如图 6-7 所示。

图 6-7　成功进入 main 页面

6.4　Spring MVC 国际化

1. Spring MVC 国际化简介

几年之前,应用程序开发者能够考虑到仅支持他们本国的只使用一种语言(或者有时是两种)和通常只有一种数量表现方式(如日期、数字、货币值)的应用。然而,基于 Web 技术的应用程序的爆炸性增长,以及将这些应用程序部署在 Internet 或其他被广泛访问的网络之上,已经在很多情况下使得国家的边界淡化到不可见。这种情况转变成为一种对于应用程序支持国际化(internationalization，经常被称作 i18n，因为 18 是字母 i 和字母 n 之间的字母个数)和本地化的需求。国际化是商业系统中不可或缺的一部分，所以无论读者学习的是什么 Web 框架，它都是必须掌握的技能。

Spring MVC 的国际化建立在 Java 国际化的基础之上，它也是通过提供不同国家/语言环境的消息资源，然后通过 Resource Bundle 加载指定 Locale 对应的资源文件，再取得该资源文件中指定 key 对应的消息。这个过程与 Java 程序的国际化完全相同，只是 Spring MVC 框架对 Java 程序国际化进行了进一步封装，从而简化了应用程序的国际化。

2. Spring MVC 国际化的知识

(1) LocaleResolver 接口：确定语言区域。

① accept-langage：基于浏览器的语言区域选择，为默认方式，不需要配置。

② SessionLocaleResolver：基于会话的语言区域选择，需要配置(常用)。

③ CookieLocaleResolver：基于 Cookie 的语言区域选择，需要配置。

(2) MessageSource 接口：告诉系统国际资源文件的存储位置。

可以使用 org.springframework.context.support.ResourceBundleMessageSource 类。

(3) LocaleChangeInterceptor 拦截器：国际化的拦截器，当语言区域发生改变时，该拦截器将进行拦截，根据传递的参数改变应用的语言环境。需要在 Spring MVC 的配置文件中进行注册。

(4) 输出国际化。Spring MVC 输出国际化消息有两种方式：

① 在视图页面上输出国际化消息，需要使用 Spring MVC 的标签库中的 message 标签。一般情况采用这种方法。

② 在 Controller 的处理方法中输出国际化消息，需要使用 org.springframework.web.servlet.support Requestcontext 的 getMessage()方法完成。

3. Spring MVC 国际化步骤

1) 配置国际化相关的类和拦截器

```
<!--基于 SessionLocaleResolver 的语言区域选择器 -->
<bean id="localeResolver" class="org.springframework.web.servlet.i18n.
SessionLocaleResolver"/>
<!--注册 MessageSource,明确资源文件的位置 -->
<bean id="messageSource" class="org.springframework.context.support.
   ResourceBundleMessageSource">
   <property name="basename" value="messages"></property>
  </bean>

  <!--配置拦截器 -->
  <mvc:interceptors>
  <bean class="org.springframework.web.servlet.i18n.LocaleChangeInterceptor">
    <property name="paramName" value="lang"></property>
  </bean>
</mvc:interceptors>
```

2）根据配置文件中资源文件的位置，创建相关资源文件

```
messages_zh_CN.properties
messages_en_US.properties
```

资源文件的内容通常采用“关键字=值”的形式，根据关键字检索值显示在页面上。一个资源包中的所有资源文件的关键字必须相同，值则为相应国家的文字。资源文件中采用的是 properties 格式文件，文件中的所有字符都必须是 ASCII 字码，属性文件是不能保存中文的，对于像中文这样的非 ACSII 字符，必须先进行编码。

Java 提供了一个 native2ascii 工具，用于对中文字符进行编码处理。jdk 的安装目录下会有一个 bin 目录，其中就有 native2ascii.exe 命令，用来将文本类文件（如 *.txt，*.ini，*.properties，*.java 等）编码转为 Unicode 编码。Unicode（统一码、万国码、单一码）是一种在计算机上使用的字符编码，它为每种语言中的每个字符设定了统一并且唯一的二进制编码，以满足跨语言、跨平台进行文本转换、处理的要求。

native2ascii 命令的格式如下：

```
native2ascii -[options][inputfile[outputfile]]
```

说明：

-[options]：表示命令开关，有两个选项可供选择。

-reverse：将 Unicode 编码转为本地编码或者指定编码，在不指定编码的情况下，将转为本地编码。

-encoding encoding_name：转换为指定编码，encoding_name 为编码名称。

```
[inputfile[outputfile]]
```

inputfile：表示输入文件全名。

outputfile：输出文件名。如果缺少此参数，将输出到控制台。

3）在页面中使用 message 标签输出国际化信息

在 Spring MVC 中显示本地化消息最容易的方法是使用 Spring 的 message 标签。为了使用这个标签，要在使用该标签的所有 JSP 页面最前面声明一个 taglib 指令：

```
<%@taglib prefix="spring" uri="http://www.springframework.org/tags" %>
```

message 标签的属性见表 6-4，所有这些属性都是可选的。

表 6-4 message 标签的属性

属 性	描 述
arguments	该标签的参数写成一个有界的字符串、一个对象数组或者单个对象
argumentSeparator	用来分割该标签参数的字符
code	获取消息的 key
htmlEscape	接受 true 或者 false，表示被渲染文本都应该进行 HTML 转义
javaScriptEscape	接受 true 或者 false，表示被渲染文本都应该进行 JavaScript 转义

续表

属　性	描　述
message	MessageSourceResolvable 参数
scope	保存 var 属性中定义的变量的范围
text	如果 code 属性不存在，或者指定码无法获取消息，所显示的默认文本
var	用于保存消息的有界变量

例如：

```
<spring:message code="language"/>:
  <a href="?lang=zh_CN">
    <spring:message code="language.cn"/>
  </a>  ------
  <a href="?lang=en_US">
    <spring:message code="language.en"/>
  </a>
  <br><br>
  <div align="center">
    <h2>
      <spring:message code="userlogin"/>
    </h2>
    <hr><br>
      <spring:message code="username"/>:
    <input type="text">
    <br><br>
    <spring:message code="password"/>
  <input type="password">
  <br><br>
  <input type="submit" value="<spring:message code="submit"/>">
      
  <input type="reset" value="<spring:message code="reset"/>">
</div>
```

接下来看一个国际化的实例开发(Spring MVC_i8n_demo)，具体步骤如下。

(1) 新建一个 Web 项目，项目名称为 Spring MVC_i8n_demo。

(2) 添加 Spring MVC 相关 jar 包，右击 Spring MVC_i8n_demo 项目，从弹出的快捷菜单中选择 Configure Facets...→Install Spring Facet。

(3) 建立项目目录。

在 src 下创建包 com.ascent.controller，并存放在 controller 控制类中。

(4) 修改 web.xml，添加以下内容：

```
<?xml version="1.0" encoding="UTF-8"?>
<web-app xmlns:xsi="http://www.w3.org/2001/XMLSchema-instance"
xmlns="http://xmlns.jcp.org/xml/ns/javaee"
```

```
 xsi:schemaLocation="http://xmlns.jcp.org/xml/ns/javaee
 http://xmlns.jcp.org/xml/ns/javaee/web-app_3_1.xsd" id="WebApp_ID" version=
"3.1">
   <display-name>Spring MVC_validation_demo</display-name>
   <welcome-file-list>
     <welcome-file>index.jsp</welcome-file>
   </welcome-file-list>
   <listener>
 <listener-class>org.springframework.web.context.ContextLoaderListener
 </listener-class>
   </listener>
   <context-param>
     <param-name>contextConfigLocation</param-name>
     <param-value>classpath:applicationContext.xml</param-value>
   </context-param>
   <servlet>
     <servlet-name>Spring MVC</servlet-name>
     <servlet-class>
         org.springframework.web.servlet.DispatcherServlet
     </servlet-class>
     <load-on-startup>1</load-on-startup>
   </servlet>
   <servlet-mapping>
     <servlet-name>Spring MVC</servlet-name>
     <url-pattern>/</url-pattern>
   </servlet-mapping>
     <!--避免中文乱码 -->
       <filter>
         <filter-name>characterEncodingFilter</filter-name>
     <filter-class>org.springframework.web.filter.CharacterEncodingFilter
         </filter-class>
         <init-param>
           <param-name>encoding</param-name>
           <param-value>UTF-8</param-value>
         </init-param>
         <init-param>
           <param-name>forceEncoding</param-name>
           <param-value>true</param-value>
         </init-param>
     </filter>
     <filter-mapping>
         <filter-name>characterEncodingFilter</filter-name>
         <url-pattern>/*</url-pattern>
     </filter-mapping>
```

```
</web-app>
```

</web-app>这里将 DispatcherServlet 命名为 Spring MVC,并且让它在 Web 项目一启动就加载。接下来需要在 WEB-INF 目录下创建一个 Spring MVC -servlet.xml 的 Spring 配置文件,此文件名的命名规则为:在 servlet-name 名称后面加-servlet.xml。

(5)添加 Spring MVC-servlet.xml 配置文件,内容如下:

```
<?xml version="1.0" encoding="UTF-8"?>
<beans xmlns="http://www.springframework.org/schema/beans"
    xmlns:xsi="http://www.w3.org/2001/XMLSchema-instance"
    xmlns:mvc="http://www.springframework.org/schema/mvc"
    xmlns:p="http://www.springframework.org/schema/p"
    xmlns:context="http://www.springframework.org/schema/context"
    xsi:schemaLocation="
        http://www.springframework.org/schema/beans
        http://www.springframework.org/schema/beans/spring-beans.xsd
        http://www.springframework.org/schema/context
        http://www.springframework.org/schema/context/spring-context.xsd
        http://www.springframework.org/schema/mvc
        http://www.springframework.org/schema/mvc/spring-mvc.xsd">
    <!--使用扫描机制扫描包 -->
    <context:component-scan base-package="com.ascent" />
    <!--配置视图解析器 -->
    <bean
        class="org.springframework.web.servlet.view.InternalResourceViewResolver">
        <property name="prefix" value="/WEB-INF/views/" />
        <property name="suffix" value=".jsp" />
    </bean>
    <!--国际化操作拦截器 -->
    <mvc:interceptors>
        <bean class="org.springframework.web.servlet.i18n.
            LocaleChangeInterceptor" />
    </mvc:interceptors>
    <!--存储区域设置信息 -->
    <bean id="localeResolver"
        class="org.springframework.web.servlet.i18n.SessionLocaleResolver">
        <property name="defaultLocale" value="zh_CN"></property>
    </bean>
    <!--加载国际化资源文件信息 -->
    <bean id="messageSource" class="org.springframework.context.support.
        ReloadableResourceBundleMessageSource">
    <!--<property name="basename" value="classpath:messages" />-->
    <property name="basename" value="/WEB-INF/resource/messages" />
</bean>
</beans>
```

注意：这里配置了国际化操作拦截器、存储区域设置信息和加载国际化资源文件信息。

（6）WEB-INF /resource 中编写属性文件 messages_en_US.properties 和 messages_zh_CN.properties。

messages_en_US.properties 内容如下：

```
first=This is first page
second=This is second page
language.en=English
language.cn=Chinese
```

messages_zh_CN.properties 内容如下：

```
first=\u8FD9\u662F\u7B2C\u4E00\u9875
second=\u8FD9\u662F\u7B2C\u4E8C\u9875
language.cn=\u4E2D\u6587
language.en=\u82F1\u6587
```

注意：messages_zh_CN.properties 是经过 native2ascii 工具处理后的结果。

（7）在 com.ascent controller 包中编写 MyController 和 I18NTestController 类。

```
//MyController.java
package com.ascent.controller;

import org.springframework.stereotype.Controller;
import org.springframework.web.bind.annotation.RequestMapping;
@Controller
@RequestMapping("/i18n_sample")
public classMyController {
    @RequestMapping("/firstPage")
    public String firstPage() {
        return "first";
    }
    @RequestMapping("/secondPage")
    public String secondPage() {
        return "second";
    }
}

//I18NTestController.java
package com.ascent.controller;

import java.util.Locale;
import org.springframework.stereotype.Controller;
import org.springframework.web.bind.annotation.RequestMapping;
@Controller
public class I18NTestController {
```

```
    @RequestMapping("/i18nTest")
    /**
     * locale 接收请求参数 locale 值,并存储到 session 中
     */
    public String firstPage(Locale locale) {
        return "first";
    }
}
```

(8) 编写 JSP 文件。

根据 Spring MVC-servlet.xml,在/WEB-INF/views/下编写 first.jsp 和 second.jsp。

```
first.jsp 如下:
<%@ page language="java" contentType="text/html; charset=UTF-8"
    pageEncoding="UTF-8"%>
<%@taglib prefix="spring" uri="http://www.springframework.org/tags"%>
<!DOCTYPE html PUBLIC "-//W3C//DTD HTML 4.01 Transitional//EN"
"http://www.w3.org/TR/html4/loose.dtd">
<html>
<head>
<meta http-equiv="Content-Type" content="text/html; charset=UTF-8">
<title>I18N Demo</title>
</head>
<body>
    <a href="${pageContext.request.contextPath }/i18nTest?locale=zh_CN">
        <spring:message code="language.cn" /></a>--
    <a href="${pageContext.request.contextPath }/i18nTest?locale=en_US">
        <spring:message code="language.en" /></a>
    <br>
    <br>
    <spring:message code="first" />
    <br>
    <br>
    <a href="${pageContext.request.contextPath }/i18n_sample/secondPage">
    <spring:message code="second" /></a>
</body>
</html>
```

```
second.jsp 如下:
<%@ page language="java" contentType="text/html; charset=UTF-8"
    pageEncoding="UTF-8"%>
<%@taglib prefix="spring" uri="http://www.springframework.org/tags"%>
<!DOCTYPE html PUBLIC "-//W3C//DTD HTML 4.01 Transitional//EN"
"http://www.w3.org/TR/html4/loose.dtd">
<html>
```

```
<head>
<meta http-equiv="Content-Type" content="text/html; charset=UTF-8">
<title>I18N Demo</title>
</head>
<body>
    <spring:message code="second"/><br><br>
    <a href="${pageContext.request.contextPath }/i18n_sample/firstPage">
        <spring:message code="first" />
    </a>
</body>
</html>
```

(9) 部署和运行项目。

选中 Spring MVC_i18n_demo 项目,部署在 Tomcat 8 服务器上,之后启动 Tomcat,在浏览器中输入 http://localhost:8080/Spring MVC_i18n_demo/i18n_sample/firstPage 进入首页(这里是中文页面),如图 6-8 所示。

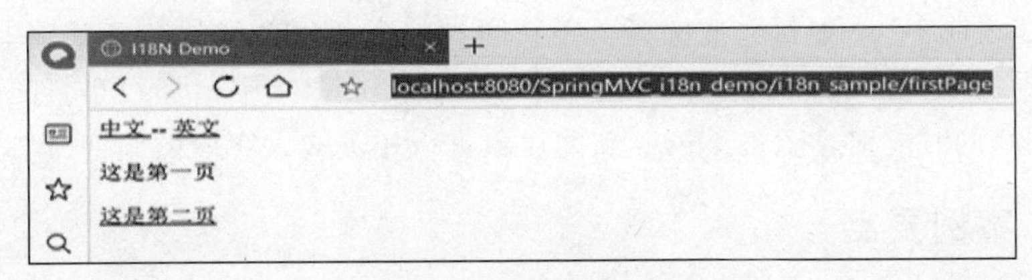

图 6-8 中文首页页面

单击“这是第二页”链接,进入第二页,如图 6-9 所示。

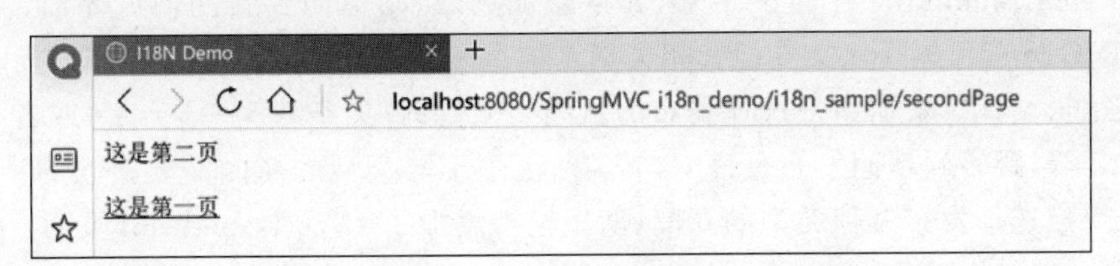

图 6-9 中文第二页页面

单击“这是第一页”链接,重新回到第一页,之后单击“英文”链接,看到英文首页,如图 6-10 所示。

图 6-10 英文首页页面

单击 This is second page 链接,进入英文第二页,如图 6-11 所示。

图 6-11　英文第二页页面

6.5　项目案例

6.5.1　学习目标

本章学习了 Spring MVC 的标签、国际化以及表单验证和拦截器，通过实现 eGov 的发布新闻的功能，完成数据校验，提示信息国际化。

6.5.2　案例描述

本案例在之前 eGov 项目完成头版头条新闻和综合新闻查询的功能基础上，依然借助 JdbcTemplate 完成发布新闻的业务功能。在 IndexController 中增加发布新闻的控制器方法 addNews()，同时要对输入的 News 新闻对象进行相关输入字段的验证。

6.5.3　案例要点

在发布新闻过程中，使用注解风格的方法针对 News 新闻类添加校验规则，在 IndexController 中使用@Validate 标签告知 Spring MVC 对输入的 News 对象进行规则校验，使用 ResultBinding 保存验证结果，如果验证失败，则退回到新闻输入界面，同时使用 Spirng MVC 标签和国际化文件设置相关提示信息。

页面逻辑设计是根据用户选择发布的新闻类型{头版头条新闻 typ=1} 或者 {综合新闻 typ=2} 等，跳转到新闻发布页面 issue.jsp，由于不同类型的新闻需要输入的字段不同，所以需要用户选择发布哪种类型的新闻，所以模拟增加了 selectIssue.html 页面。

编写 IndexController.java 的 addNews()方法，发布新闻，同时使用@Validate 标签针对输入的 News 对象属性数据进行校验。

编写 message_zh_CN.properties，针对每个被校验的字段编写错误提示中文信息。

修改 Spring MVC.xml，增加国际化资源配置文件信息。

注意：使用 Spring MVC 标签库时，要求 web.xml 中要注册 ContextLoaderListener，所以 Spring MVC Form 表单标签库依赖于 Spring 框架。

6.5.4　案例实施

(1) 修改 web.xml，添加如下指令：

```
<listener>
    <listener-class>org.springframework.web.context.ContextLoaderListener</
listener-class>
</listener>
```

```
<context-param>
       <param-name>contextConfigLocation</param-name>
       <param-value></param-value><!—暂时不配置 Spring  -->
</context-param>
```

(2) 在 WebRoot 下新建 jsp 文件夹，新建 selectIssue.jsp。

```
<%@ page language="java" import="java.util.* "pageEncoding="UTF-8"%>
<%
String path=request.getContextPath();
StringbasePath=request.getScheme()+"://"+request.getServerName()+":"+
             request.getServerPort()+path+"/";
%>
<!DOCTYPE HTML PUBLIC "-//W3C//DTD HTML 4.01 Transitional//EN">
<html>
  <head>
    <base href="<%=basePath%>">
    <title>选择发布新闻种类</title>
<meta http-equiv="pragma" content="no-cache">
  </head>
    <body>
    <p><a href="<%=basePath%>/jsp/issue.jsp?typ=1">发布头版头条新闻</a></p>
    <p><a href="<%=basePath%>/jsp/issue.jsp?typ=2">发布综合新闻</a></p>
    <!--可以增加其他新闻种类,如通知 -->
  </body>
</html>
```

(3) 在 jsp 目录下新建 issue.jsp。

```
<%@ page language="java" import="java.util.*,com.ascent.po.*"pageEncoding=
"utf-8"%>

<%@taglib uri="http://java.sun.com/jsp/jstl/core" prefix="c"%>
<%@taglib prefix="sf" uri="http://www.springframework.org/tags/form" %>

<%
String path=request.getContextPath();
StringbasePath= request. getScheme ( ) +"://" + request. getServerName ( ) +":" +
        request.getServerPort()+path+"/";
%>

<!DOCTYPE HTML PUBLIC "-//W3C//DTD HTML 4.01 Transitional//EN">
<html>
  <head>
    <title>Acsent 电子政务系统</title>

    <meta http-equiv="pragma" content="no-cache">
```

```
    <meta http-equiv="cache-control" content="no-cache">
    <meta http-equiv="expires" content="0">
    <meta http-equiv="keywords" content="keyword1,keyword2,keyword3">
    <meta http-equiv="description" content="This is my page">
    <link rel="stylesheet" href="<%=basePath%>/IMAGES/001.css" type="text/
    css">
    <script type="text/javascript">
      <!--使用 JavaScript 提交表单-->
        function sub(){
            form0.action="<%=basePath%>/addNewsnewsAction.action";
            form0.method="post";
            form0.submit();
        }
    </script>

  </head>

  <body bgcolor="#FFFFFF" text="#000000"topmargin="0">
    <%
        String type=request.getParameter("typ");
        pageContext.setAttribute("typ",type);
    %>

<table width="780" border="0"cellspacing="0" cellpadding="0" align="center">
  <tr>
    <td width="11" valign="top" background="<%=basePath%>/IMAGES/118.gif">
            </td>
    <td valign="top">
      <div align="center">
        <!--Spring MVC Form 标签要求 modelAttribute 属性必须经过实例化-->

        <%
        News news=(News)request.getAttribute("news");
        if(null==news)
        request.setAttribute("news", new News());
        %>
    <!--使用 Spring MVC 表单标签库  -->
        <sf:form name="form0" action="/addNewsnewsAction.action" method=
            "post" modelAttribute="news">
          <table width="700" border="0"cellspacing="1" cellpadding="0"
              align="center" bordercolor="78B5C9" bgcolor="78B5C9">
            <tr bgcolor="78B5C9">
              <td height="40"colspan="2">
                <div align="center">
                  <strong>
```

```
            <%
                if("1".equals(type)){
            %>
                    头版头条发布
            <%
                } else {
            %>
                    综合新闻发布
            <%
                }
            %>
          </strong>
        </div>
      </td>
    </tr>
    <tr bgcolor="E5F1F5">
      <td bgcolor="E5F1F5" height="30" width="105">
        <div align="center">标      题: </div>
      </td>
      <td bgcolor="E5F1F5" height="22" width="592">
        <sf:input path="title" size="50"/><sf:errors path="title">
            </sf:errors>

      </td>
    </tr>
    <tr bgcolor="E5F1F5">
      <td bgcolor="E5F1F5" height="137" width="105">
        <div align="center">正      文: </div>
      </td>
      <td bgcolor="E5F1F5" height="137" width="592">
        <sf:textarea path="content" cols="70" rows="8"/>
        <sf:errors path="content"></sf:errors>
      </td>
    </tr>
    <%
        if(!"1".equals(type)){
    %>
          <tr bgcolor="E5F1F5">
            <td bgcolor="E5F1F5" height="30" width="105">
              <div align="center">是否跨栏: </div>
            </td>
            <td bgcolor="E5F1F5" height="30" width="592">
              <sf:radiobutton path="crossstatus" value="0" label="是"
                  checked="true"/>
                 
```

```
                    <sf:radiobutton path="crossstatus" value="1" label="不
                    是"/>
                  </td>
                </tr>
            <%
                }
            %>
            <tr bgcolor="E5F1F5">
              <td bgcolor="E5F1F5" height="20"colspan="2">  </td>
            </tr>
          </table>
          <table width="700" border="0"cellspacing="0" cellpadding="0"
              align="center">
            <tr>
              <td bgcolor="78B5C9"valign="middle">
                <div align="center"><br>
                  <table width="555" border="0"cellspacing="0" cellpadding="0"
                      align="center">
                    <tr>
                      <td width="89">
                        <div align="center">
                              <!--JavaScript 控制提交表单 -->
<a href="javascript:login()"><img src="<%=basePath%>/IMAGES/74.gif" width=
"44" height="18" border="0"></a>

                        </div>
                      </td>
                      <td width="93">
                        <div align="center"></div>
                      </td>
                      <td width="93">
                        <div align="center"></div>
                      </td>
                      <td width="106">
                        <div align="center"></div>
                      </td>
                      <td width="174">
                        <div align="center"></div>
                      </td>
                    </tr>
                  </table>
                  <br>
                </div>
              </td>
            </tr>
```

```
            </table>
            <sf:hidden path="type" value="${pageScope.type}%>"/>
          </sf:form>
        </div>
      </td>
      </tr>
      </table>
</body>
</html>
```

运行页面链接测试：头版头条新闻页面，如图 6-12 所示。

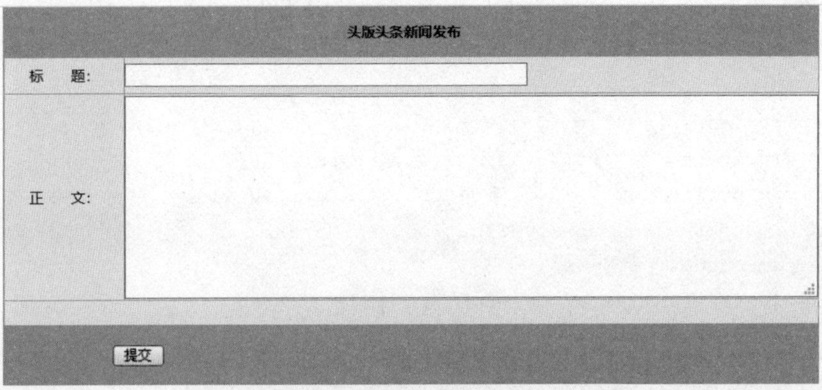

图 6-12 头版头条新闻页面

综合新闻页面如图 6-13 所示。

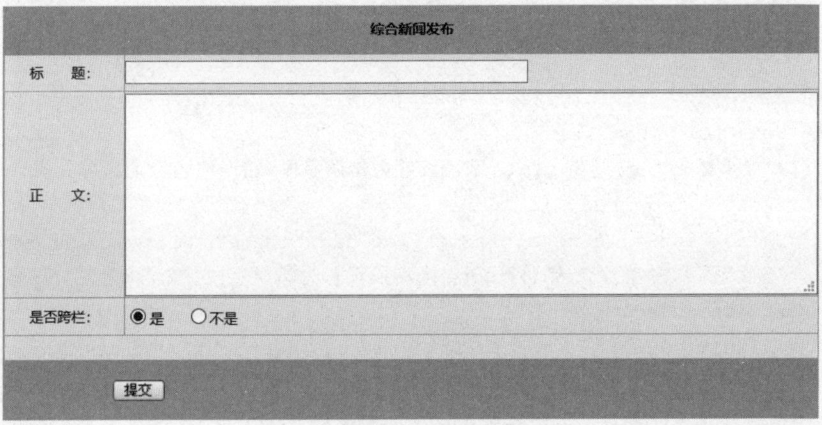

图 6-13 综合新闻页面

(4) 在 electrone 项目中添加 validator 框架。

右击 electrone 项目，从弹出的快捷菜单中选择 Build Path→Add Libraries，出现图 6-14。

选择 MyEclipse Library，之后选择 Validation-1.0 Library，单击 Finish 按钮。

(5) 编写 NewsValidator，并编写新闻数据验证规则以及错误提示字段（内容写在国际

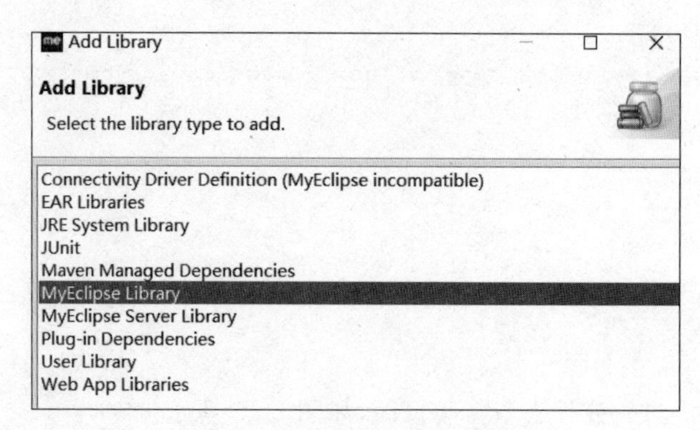

图 6-14　添加 validator 框架

资源文件中)。

```
package com.ascent.mvc.validator;
import org.springframework.validation.Errors;
import org.springframework.validation.ValidationUtils;
import org.springframework.validation.Validator;
import com.ascent.po.News;
public classNewsValidator implements Validator {

    @Override
    public boolean supports(Class<?>news) {

        return news.equals(News.class);
    }

    @Override
    public void validate(Object newsObject, Errors errors) {
        News news=(News)newsObject;
        //如果标题为空,则将 title 字段设置为错误字段,错误内容对应国际化文件中的
        //title.required 字段
        ValidationUtils.rejectIfEmpty(errors,"title", "title.required");
        //如果新闻内容为空,则将 content 字段设置为错误字段,错误内容对应国际化文件中的
        //content.required 字段
        ValidationUtils.rejectIfEmpty(errors, "content", "content.required");
        //如果新闻内容中全部由空格组成,则将 content 字段设置为错误字段,错误内容对应
        //国际化文件中的 content.invalide 字段
        if(news.getContent().trim().length()==0)
            errors.reject("content", "content.invalide");
        }
}
```

(6) 编写国际资源化文件 message_zh_CN.properties,并配置国际化。

在 WEB-INF 目录下新建 message 目录,并新建 message_zh_CN.properties 文件,如图 6-15 所示。

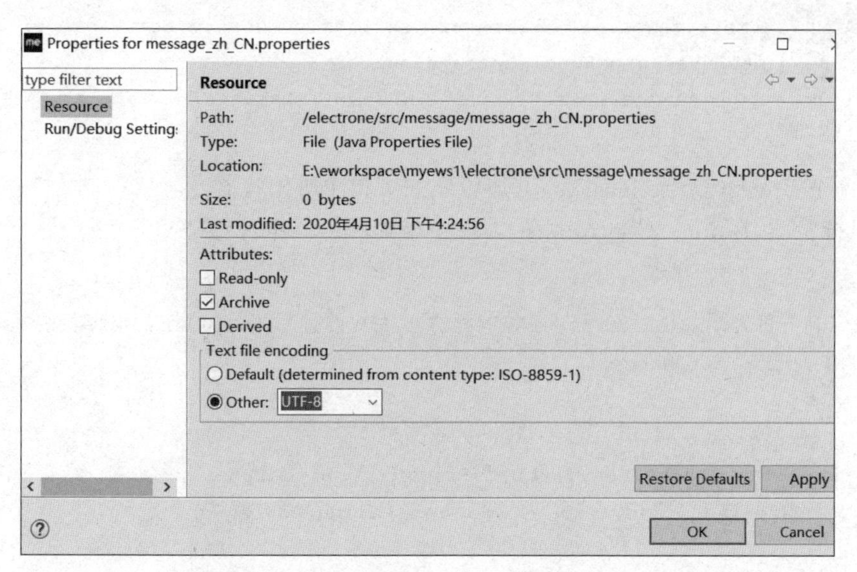

图 6-15 新建 message_zh_CN.properties 文件

注意：该文件的编码方式为 UTF-8，需要 MyEclipse 安装 PropertiesEditor 插件，如图 6-16 所示。

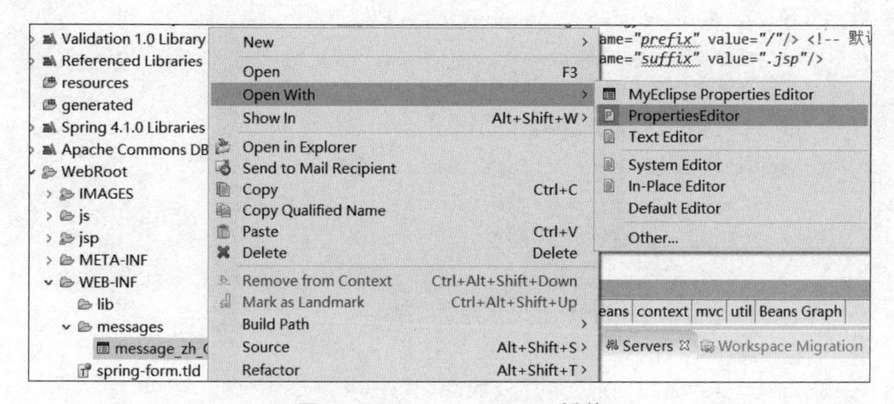

图 6-16 PropertiesEditor 插件

message_zh_CN.properties 内容如下：

```
title.required=缺少新闻标题
content.required=缺少新闻内容
content.invalide=新闻内容非法
```

在 EditPlus 或者其他文本编辑器中查看应该是 Unicode 编码，如下：

```
title.required=\u7f3a\u5c11\u65b0\u95fb\u6807\u9898
content.required=\u7f3a\u5c11\u65b0\u95fb\u5185\u5bb9
content.invalide=\u65b0\u95fb\u5185\u5bb9\u975e\u6cd5
```

修改 Spring MVC.properties，增加以下内容：

```
<!--国际资源化文件 message_zh_CN.properties 配置 -->
```

```
    <bean id="messageSource" class="org.springframework.context.support.
        ReloadableResourceBundleMessageSource">
      <property name="basename" value="/WEB-INF/messages/message_zh_CN " />
  </bean>
```

(7) 修改 IndexController.java，增加验证方法和注册校验器。

注册校验器，在 IndexController()方法中增加 initBinder()方法。

```
/**
    * 所有进入 IndexController 控制的请求，都先经过 initBinder 的校验
    * @param dataBinder
    * /
  @InitBinder
  public void initBinder(WebDataBinder dataBinder) {
      dataBinder.setValidator(new NewsValidator());
  }
  //注意，使用的是@Validate 注解，而不是@Valid
  @RequestMapping(value="addNewsnewsAction")
  public ModelAndView addNews(@Validate News news,Errors errors) {

      ModelAndView mv=new ModelAndView();
      if(errors.hasErrors()) {
      //追加了查询字符串 typ=新闻类型 ID
      //如 jsp/issue.jsp?typ=1
      //jsp 中可以使用 request.getParameter 获取
      mv.addObject("typ",news.getType());
      mv.setViewName("jsp/issue");
              }else {
      //调用相关 Service，将数据保存到数据库中
      //保存成功，回到管理界面 selectIssue.jsp
      mv.setViewName("jsp/selectIssue");
  }
  return mv;
}
```

(8) 测试 Spring MVC 自定义校验器和国际资源化文件配置。

部署项目，启动 Tomcat，如图 6-17 所示。

localhost:8080/electrone/jsp/selectIssue.jsp

火狐官方站点　新手上路　常用网址　京东商城

发布头版头条新闻

发布综合新闻

图 6-17　测试页面

(9) 单击“发布综合新闻”链接,出现图 6-18 所示。

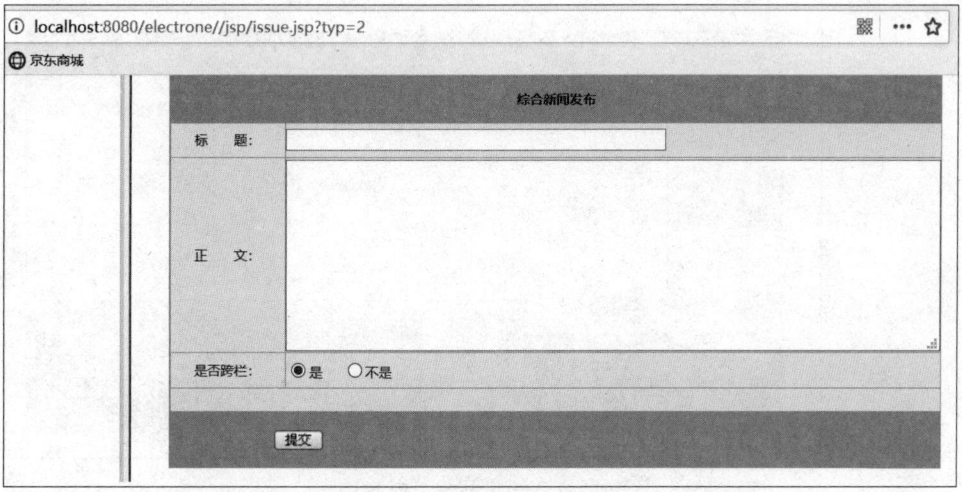

图 6-18 发布综合新闻页面

(10) 单击“提交”按钮,出现图 6-19。

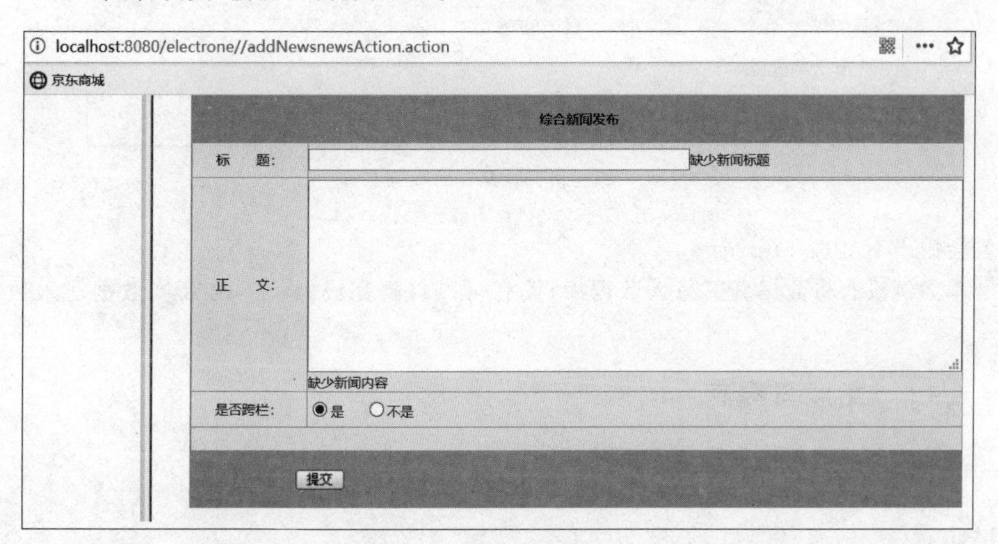

图 6-19 提交综合新闻页面

(11) 填写相关内容,如图 6-20 所示。

(12) 再次单击“提交”按钮,如图 6-21 所示。

6.5.5 特别提示

本案例中的国际资源化文件一定是 Unicode 编码,使用 PropertiesEditor 插件或者 native2ascii 命令将含有中文字符的 properties 转换为 Unicode 编码的文件。

在本案例的 issue.jsp 中,由于是通过浏览器直接读取 jsp 文件,没有经过 Spring MVC 控制器进行视图跳转,所以 Spring MVC form 标签中使用的 news 对象必须通过传统 jsp 代码<%request.setAttribute("news",new News());%>生成 news 实例,否则 sf:form

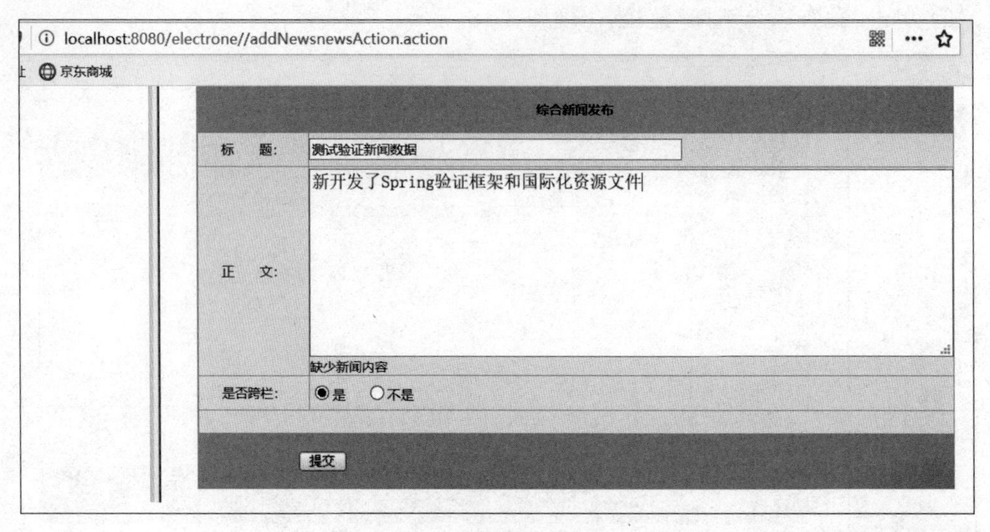

图 6-20　填写相关内容

图 6-21　再次提交综合新闻页面

标签会抛出 NullPointer 异常。

本案例没有将数据真正写入数据库,读者可以自己完成,将 news 新闻数据写入数据库中。

6.5.6　拓展与提高

完成新闻修改的控制器验证功能。

习题

1. Spring MVC 表单标签主要包括哪些?
2. 数据验证分为哪两种验证?
3. Spring MVC 数据验证常用的校验注解有哪些?
4. 过滤器与拦截器的区别是什么?
5. 分别描述 HandlerInterceptor 接口中定义的 3 个方法。
6. Spring MVC 国际化步骤是什么?

第7章 MyBatis基础

学习目的与学习要求

学习目的：了解 MyBatis 框架概述，掌握 MyBatis 组件和流程，熟悉 MyBatis 基本原理。

学习要求：搭建 MyBatis 开发环境，熟悉 MyBatis 核心配置及基本操作。

本章主要内容

本章主要讲解 MyBatis 框架的概述及核心流程、MyBatis 的对象/关系数据库映射的原理。

介绍完 Spring MVC 之后，下面讨论数据持久层的处理。实际项目都需要数据的支持，而 MyBatis 就是目前最好的数据持久层框架之一。

MyBatis 本是 Apache 的一个开源项目 iBATIS，2010 年这个项目由 Apache Software Foundation 迁移到 Google Code，并且改名为 MyBatis，2013 年 11 月迁移到 Github。

iBATIS 一词来源于 internet 和 abatis 的组合，是一个基于 Java 的持久层框架。iBATIS 提供的持久层框架包括 SQL Maps 和 Data Access Objects(DAOs)。

7.1 MyBatis 概述

在今日的企业环境中，一起使用面向对象的软件和关系数据库可能相当麻烦，并且很浪费时间。MyBatis 是一款优秀的持久层框架，它支持定制化 SQL、存储过程以及高级映射。MyBatis 避免了几乎所有的 JDBC 代码和手动设置参数以及获取的结果集。MyBatis 可以使用简单的 XML 或注解配置和映射原生类型、接口和 Java 的 POJO(Plain Old Java Objects，普通老式 Java 对象)为数据库中的记录。

每个基于 MyBatis 的应用都以一个 SqlSessionFactory 的实例为核心。SqlSessionFactory 的实例可以通过 SqlSessionFactoryBuilder 获得。而 SqlSessionFactoryBuilder 则可以从 XML 配置文件或一个预先定制的 Configuration 的实例构建出 SqlSessionFactory 的实例。

从 XML 文件中构建 SqlSessionFactory 的实例非常简单,建议使用类路径下的资源文件进行配置。但是,也可以使用任意的输入流(InputStream)实例,包括字符串形式的文件路径或者 file://的 URL 形式的文件路径来配置。MyBatis 包含一个名叫 Resources 的工具类,它包含一些实用方法,可使从 classpath 或其他位置加载资源文件更加容易。

构建了 SqlSessionFactory,就可以从中获得 SqlSession 的实例了。SqlSession 包含了面向数据库执行 SQL 命令所需的所有方法。可以通过 SqlSession 实例直接执行已映射的 SQL 语句。

除了使用 SqlSession 执行 SQL 语句之外,更流行的做法是使用正确描述每个语句的参数和返回值的接口,通常称之为 Mapper 类。Mapper 类可以使用 XML 文件配置每个接口方法的 SQL 语句,或者使用注解方式配置。这样做不仅可以执行更清晰和类型安全的代码,而且还不用担心易错的字符串值以及进行强制类型转换。

7.2 MyBatis 组件和流程

MyBatis 框架总体上可分为三层,如图 7-1 所示。

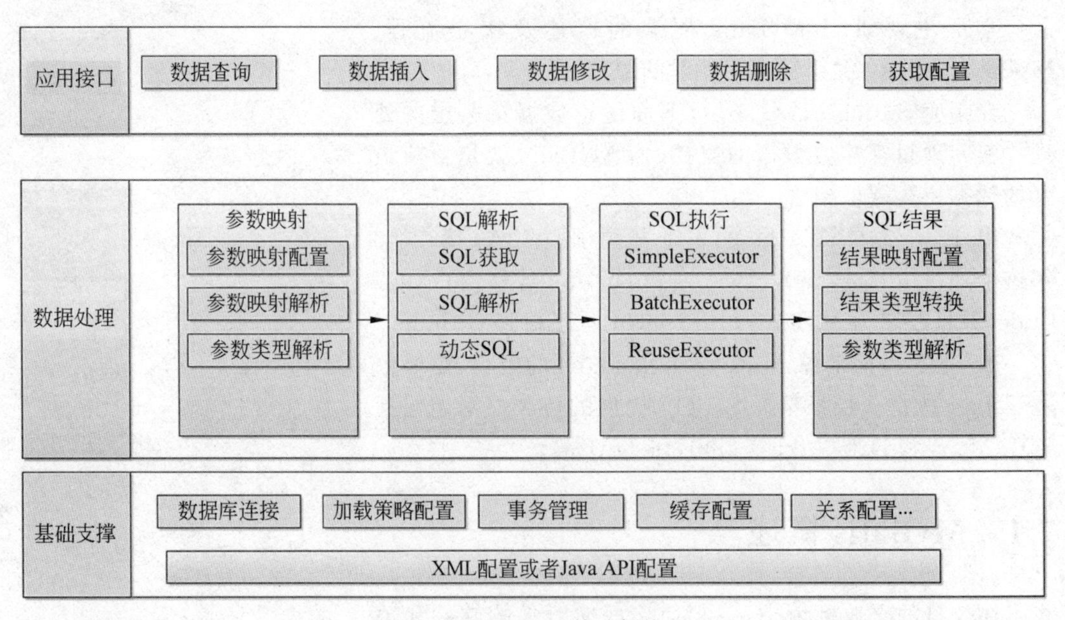

图 7-1 MyBatis 基础架构

(1) 应用接口层:提供给外部应用语言使用的接口 API,如 Java Interface,开发人员通过这些 API 接口操作数据库。接口层接收到调用请求会调用 SQL 层完成具体的数据处理。

(2) 数据处理层:负责具体的 SQL 查找、SQL 解析、SQL 执行和执行结果映射处理等。它的主要目的是根据调用的请求完成一次数据库操作,SQL 层依赖于基础支撑层。

（3）基础支撑层：负责最基础的功能支撑，包括连接管理、事务管理、配置加载和缓存处理，这些都是共用的东西，将它们抽取出来作为最基础的组件，为上层的数据处理层提供最基础的支撑。

MyBatis 工作架构如图 7-2 所示。

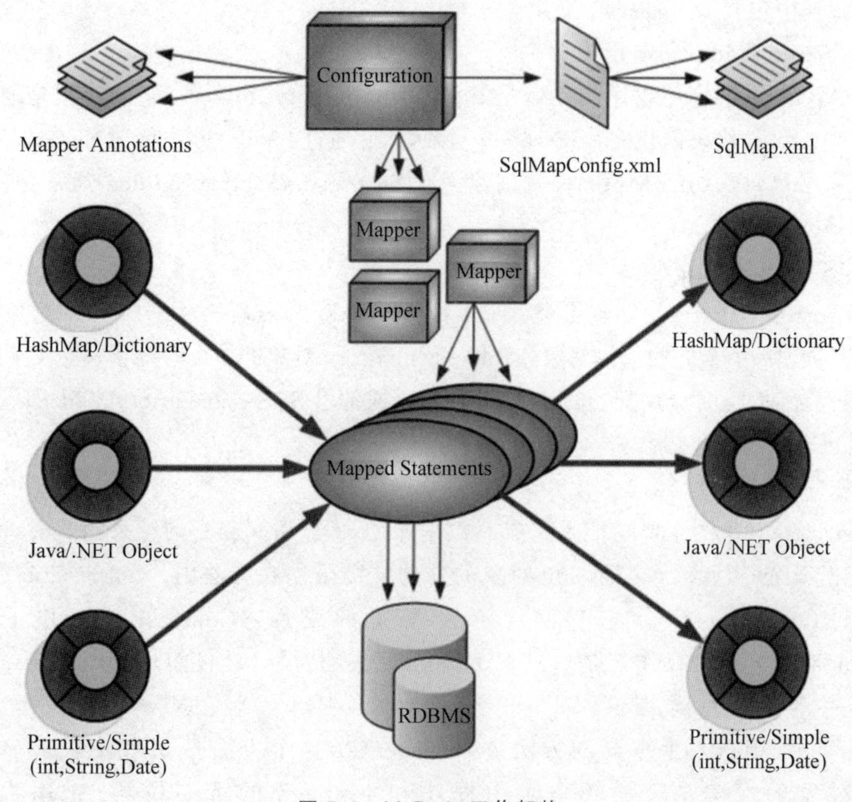

图 7-2　MyBatis 工作架构

MyBatis 工作时的主要流程如下：

（1）MyBatis 最初的初始化是从 sqlMapConfig.xml 配置文件，解析构造成 Configuration 对象。

（2）加载 XML 配置或者解析 Java 注解，将 SQL 语句的具体配置信息加载成为一个个独立的 MappedStatement 对象（包括传入参数映射配置、执行的 SQL 语句、结果映射配置等），存储在 JVM 内存中。

（3）进行 SQL 语句的解析：当 Java API 接口层接收到 Maper()方法的调用请求时，会接收到传入 SQL 的 ID（每个 SQL 配置都有唯一的 ID，这个 ID 在 XML 中和 Java 接口方法中的方法名称一致）和传入对象（可以是 Map、JavaBean 或者基本数据类型，也可以使用注解标识参数），MyBatis 会根据 SQL 的 ID 找到对应的 MappedStatement，然后根据传入参数对象对 MappedStatement 进行解析，解析后可以得到最终要执行的 SQL 语句和参数。

（4）执行 SQL 操作：将解析后得到的 SQL 语句和所需注入的参数传递到数据库中进

行执行，然后得到数据库操作的结果。

(5) SQL 语句执行结果映射：将操作数据库的结果按照映射的配置进行转换，可以转换成 HashMap、JavaBean 或者基本数据类型，并将最终结果返回 Java API 接口调用层。

MyBatis 中包含如下的核心类：

- SqlSessionFactoryBuilder

每个 MyBatis 的应用程序的入口是 SqlSessionFactoryBuilder。它的作用是通过 XML 配置文件或者 Java 代码创建 Configuration 对象，然后通过 build()方法创建 SqlSessionFactory 对象。没有必要每次访问 MyBatis 就创建一次 SqlSessionFactoryBuilder，通常是创建一个全局的单体对象。

- SqlSessionFactory

使用 SqlSessionFactory 创建 SqlSession 对象。和 SqlSessionFactoryBuilder 对象一样，通常的做法是创建一个全局的单体对象。SqlSessionFactory 对象的一个必要属性是 Configuration 对象，它是保存 MyBatis 全局配置的一个配置对象，通常由 SqlSessionFactoryBuilder 从 XML 配置文件创建。

- SqlSession

SqlSession 对象的主要功能是完成对数据库的 SQL 访问和查询结果的映射，由于该对象不是线程安全的，所以 SqlSession 对象的作用域需限制在方法内。SqlSession 的默认实现类是 DefaultSqlSession，它有两个必须配置的属性：Configuration 和 Executor。SqlSession 对数据库的操作都是通过 Executor 完成的。默认创建 DefaultSqlSession 时开启一级缓存，创建执行器、赋值。

SqlSession 中有一个重要的方法 getMapper()，这个方法是用来获取 Java 接口层 Mapper 对象的。该方法是联系应用程序和 MyBatis 的重要桥梁，应用程序访问 getMapper 时，MyBatis 会根据传入的接口类型和对应的 XML 配置文件生成一个代理对象，这个代理对象就叫 Mapper 对象。应用程序获得 Mapper 对象后访问 MyBatis 的 SqlSession 对象，这样就达到了插入 MyBatis 流程的目的。

- Executor

Executor 对象在创建 Configuration 对象的时候创建，并且缓存在 Configuration 对象里。Executor 对象的主要功能是调用 StatementHandler 访问数据库，并将查询结果存入缓存中(如果配置了缓存)。

- StatementHandler

StatementHandler 是真正访问数据库，执行 SQL 语句的对象，如果配置了返回结果，则调用 ResultSetHandler 处理查询结果。

- ResultSetHandler

处理封装 SQL 语句的查询结果。

先看一个简单的实例 MyBatis_basic_demo，之后再做相关理论的讲解。

(1) 使用 MySQL 建立 test 数据库，建立 usr 表并插入数据：

```
USE `test`;

/* Table structure for table `usr` */

DROP TABLE IF EXISTS `usr`;

CREATE TABLE `usr` (
  `id` int NOT NULL AUTO_INCREMENT,
  `username` varchar(255) DEFAULT NULL,
  `password` varchar(255) DEFAULT NULL,
  PRIMARY KEY (`id`)
) ENGINE=InnoDB AUTO_INCREMENT=4 DEFAULT CHARSET=latin1;

/* Data for the table `usr` */

insert  into `usr`(`id`,`username`,`password`) values (1,'Lixin','123456'),(2,
'admin','123456'),(3,'Linda','123456');
```

（2）新建 Java 项目，如图 7-3 所示。

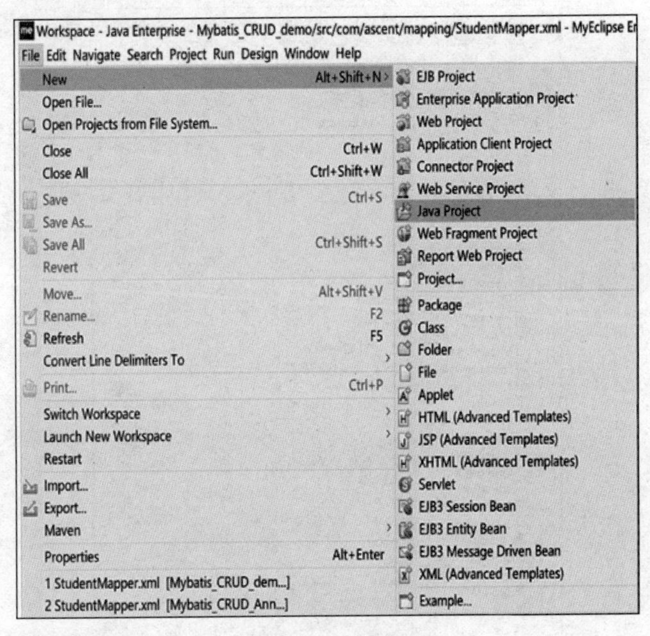

图 7-3 新建项目

在 Project name 中输入 MyBatis_basic_demo，如图 7-4 所示。

单击 Next 按钮，采用默认配置，之后单击 Finish 按钮，完成项目的创建。

右击 MyBatis_basic_demo 项目，从弹出的快捷菜单中选择 Properties，如图 7-5 所示。

从弹出的项目属性对话框中选择 Java Build Path→Libraries 命令，最后单击 Add External JARs 按钮，如图 7-6 所示。

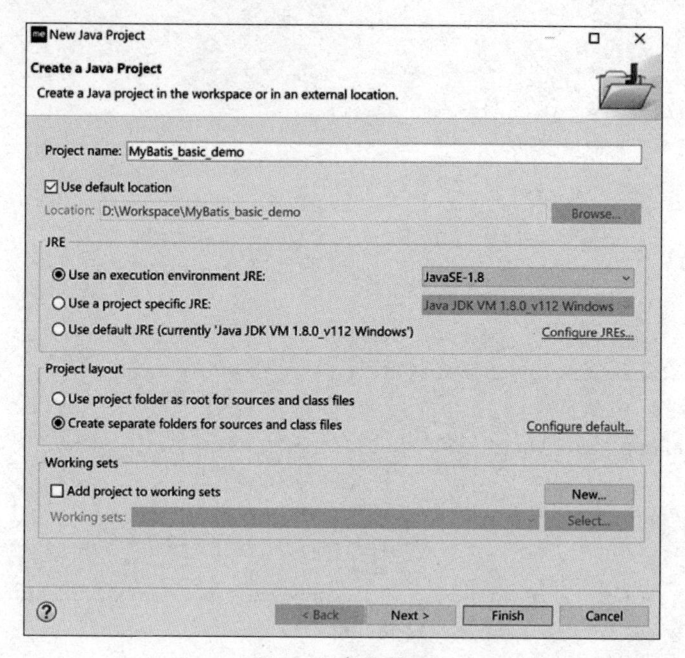

图 7-4　命名项目

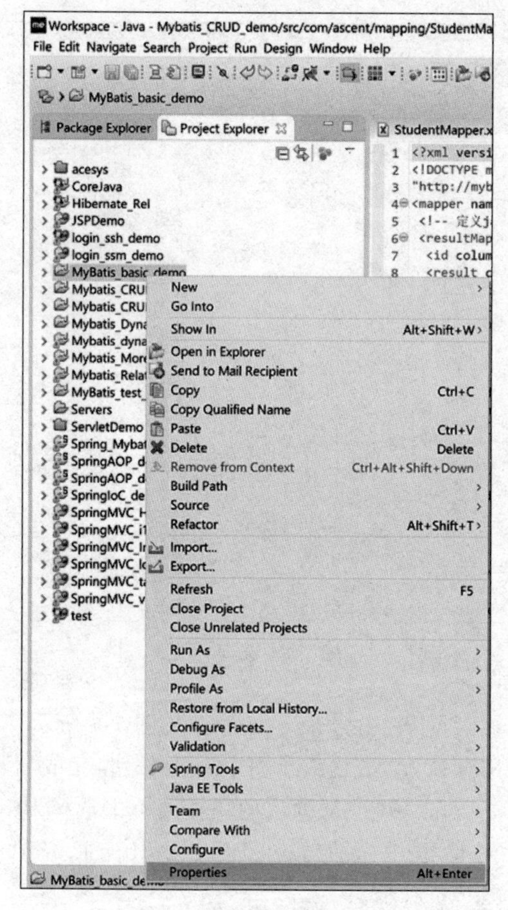

图 7-5　选择 Properties

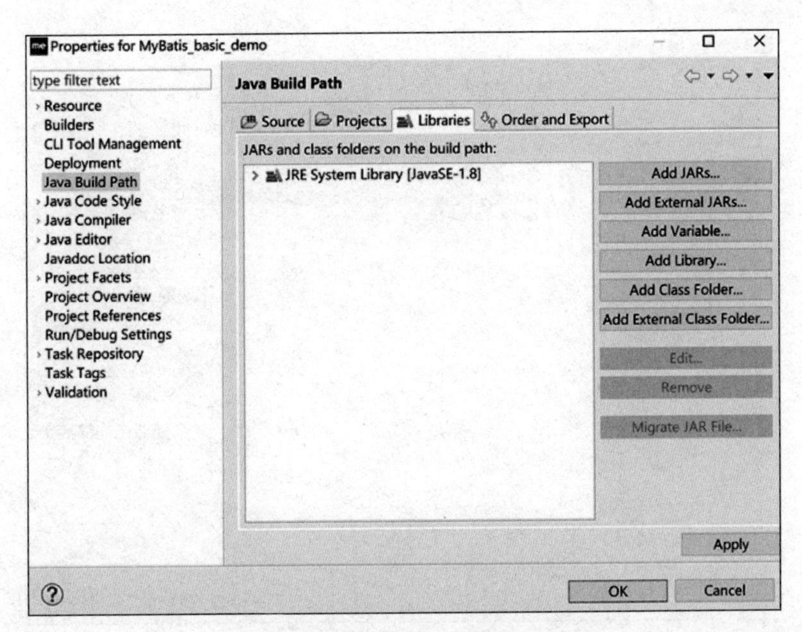

图 7-6 添加 jar 包界面 1

选中 mybatis-3.4.2.jar 和 mysql-connector-java-5.1.6-bin.jar(MySQL 的 Java 驱动包),如图 7-7 所示。

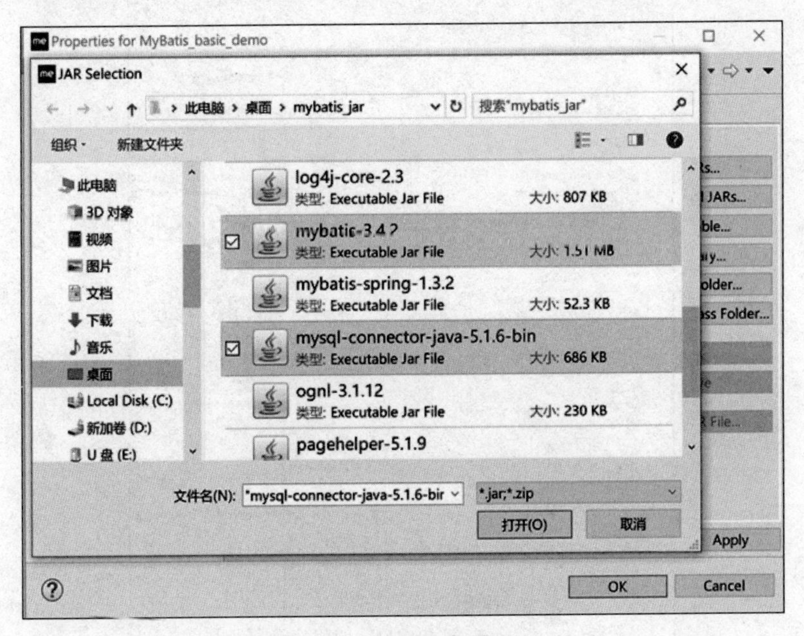

图 7-7 添加 jar 包界面 2

注意:每个用户存放 jar 包的位置可能不同,要修改成用户自己的路径。

之后单击 Apply 按钮,并单击 OK 按钮,结果如图 7-8 所示。

(3) 建立项目包结构。

右击 src,从弹出的快捷菜单中选择 New→Package,如图 7-9 所示。

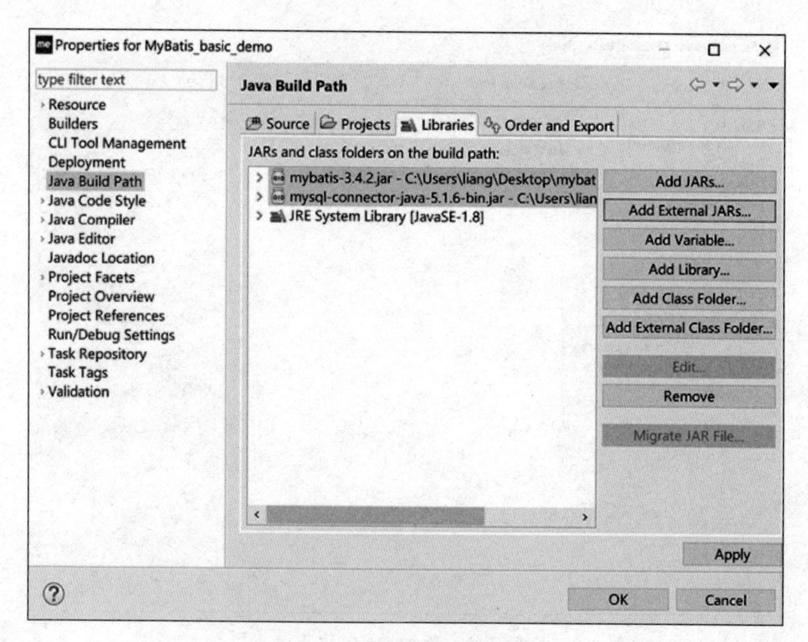

图 7-8　添加 jar 包界面 3

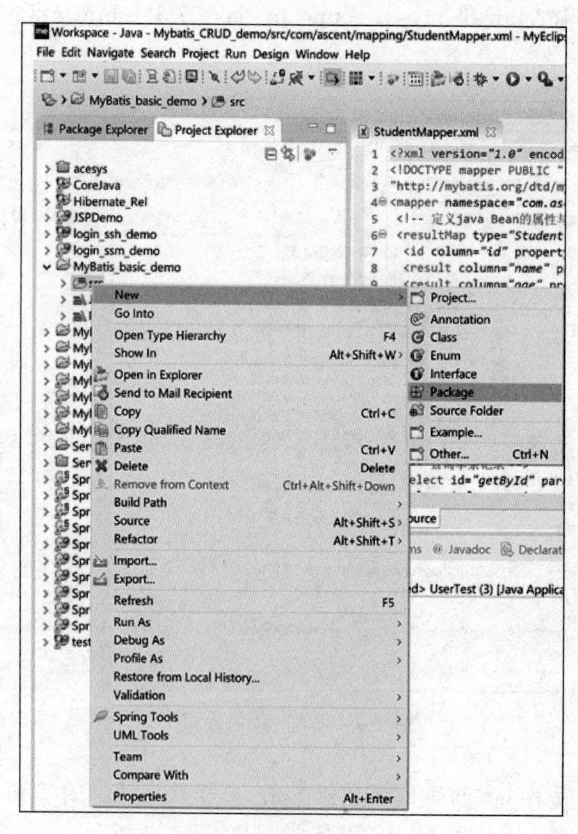

图 7-9　新建 Package

在 Name 输入框中输入 com.ascent.mybatis.mapper，单击 Finish 按钮，如图 7-10 所示。

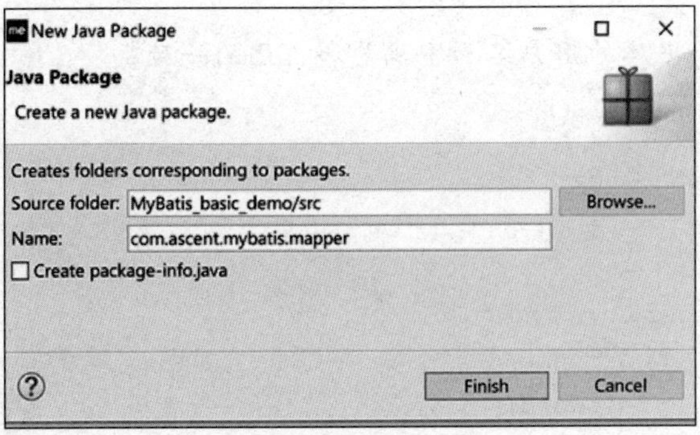

图 7-10　命名 Package

类似操作，建立 com.ascent.mybatis.mapping、com.ascent.mybatis.po 和 com.ascent.mybatis.test 包。

```
src/com/ascent/mybatis/mapper     存放 Java API 接口文件 UserMapper.java
src/com/ascent/mybatis/mapping    存放 SQL XML 映射文件 UserMapper.xml
src/com/ascent/po                 存放 Java POJO 类,映射 User 实体
src/com/ascent/test               存放测试运行类 UserTest
```

另外，在 src 目录中存放 mybatis-config.xml 文件。

(4) 在 src 目录下新建 mybatis-config.xml 文件。

```
<?xml version="1.0" encoding="UTF-8" ?>
<!DOCTYPE configuration
PUBLIC "-//mybatis.org//DTD Config 3.0//EN"
"http://mybatis.org/dtd/mybatis-3-config.dtd">
<configuration>
   <typeAliases>
      <package name="com.ascent.mybatis.po"/>
   </typeAliases>
   <environments default="development">
      <environment id="development">
      <transactionManager type="JDBC"/>
         <dataSource type="POOLED">
         <property name="driver" value="com.mysql.jdbc.Driver"/>
           <property name="url" value="jdbc:mysql://localhost:3306/test?
           characterEncoding=UTF-8"/>
         <property name="username" value="root"/>
         <property name="password" value="root"/>
      </dataSource>
   </environment>
</environments>
<mappers>
```

```
    <mapper resource="com/ascent/mybatis/mapping/UserMapper.xml"/>
</mappers>
</configuration>
```

注意：MySQL 数据库的用户名和密码要和自己的一致。

(5) 在 po 目录下新建 User 类。

```
package com.ascent.mybatis.po;

public class User {
    private Integer id;
    private String username;
    private String password;
    public Integer getId() {
        return id;
    }
    public void setId(Integer id) {
        this.id=id;
    }
    public String getUsername() {
        return username;
    }
    public void setUsername(String username) {
        this.username=username;
    }
    public String getPassword() {
        return password;
    }
    public void setPassword(String password) {
        this.password=password;
    }
}
```

(6) 在 mapper 目录下新建 Java 接口 UserMapper。

```
package com.ascent.mybatis.mapper;
import com.ascent.mybatis.po.*;
import java.util.List;

public interface UserMapper {

    public List<User>listUser();

}
```

(7) 在 mapping 目录下新建 UserMapper.xml。

```
<?xml version="1.0" encoding="UTF-8"?>
```

```
<!DOCTYPE mapper
    PUBLIC "-//mybatis.org//DTD Mapper 3.0//EN"
    "http://mybatis.org/dtd/mybatis-3-mapper.dtd">

    <mapper namespace="com.ascent.mybatis.mapper.UserMapper">
    <select id="listUser" resultType="User">
        select * from usr
    </select>
</mapper>
```

(8) 在 test 目录下建立 UserTest 测试类,含 main()方法。

```
package com.ascent.mybatis.test;

import java.io.IOException;
import java.io.InputStream;
import java.util.List;

import org.apache.ibatis.io.Resources;
import org.apache.ibatis.session.SqlSession;
import org.apache.ibatis.session.SqlSessionFactory;
import org.apache.ibatis.session.SqlSessionFactoryBuilder;

import com.ascent.mybatis.mapper.UserMapper;
import com.ascent.mybatis.po.User;;

public class UserTest {

    public static void main(String[]args) throws IOException {
        String resource="mybatis-config.xml";
        InputStream inputStream=Resources.getResourceAsStream(resource);
        SqlSessionFactory sqlSessionFactory=new SqlSessionFactoryBuilder().
            build(inputStream);
        SqlSession session=sqlSessionFactory.openSession();
        UserMapper mapper=session.getMapper(UserMapper.class);
        List<User>cs=mapper.listUser();
        for (User c : cs){
            System.out.println(c.getUsername());
        }

    }
}
```

(9) 右击 UserTest,从弹出的快捷菜单中选择 Run As->Java Application,运行这个类,可看到如下结果:

```
Lixin
```

```
admin
Linda
```

接下来详细介绍相关知识点。

7.3 MyBatis 原理

每个 MyBatis 都应该以一个 SqlSessionFactory 实例为中心。一个 SqlSessionFactory 实例可以使用 SqlSessionFactoryBuilder 创造。从配置类中创造的定制 SqlSessionFactoryBuilder 实例,可以使用 XML 配置文件生成一个 SqlSessionFactory 实例。

7.3.1 从 XML 中创造 SqlSessionFactory

从 XML 文件中创造 SqlSessionFactory 实例非常简单。推荐使用一个类路径资源进行配置,也可以使用一个 Reader 实例,甚至使用 URL 路径。

MyBatis 有一个 Resources 通用类,类中有许多方法可以简单地从类路径和其他地址中加载资源。

```
String resource="com/ascent/config/Configuration.xml";
Reader reader=Resources.getResourceAsReader(resource);
SqlSessionFactory sqlMapper=new SqlSessionFactoryBuilder().build(reader);
```

XML 文件包含了许多 MyBatis 的核心设置,包括一个获取数据库连接(Connection)实例的数据源(DataSource)和一个决定事务作用域和操作的 TransactionManager。全部 XML 配置文件的内容将在以后提到,先给出一个简单的样子。

```
<?xml version="1.0" encoding="UTF-8" ?>
<!DOCTYPE configuration
PUBLIC "-//mybatis.org//DTD Config 3.0//EN"
"http://mybatis.org/dtd/mybatis-3-config.dtd">
<configuration>
  <environments default="development">
    <environment id="development">
      <transactionManager type="JDBC"/>
      <dataSource type="POOLED">
        <property name="driver" value="${driver}"/>
        <property name="url" value="${url}"/>
        <property name="username" value="${username}"/>
        <property name="password" value="${password}"/>
      </dataSource>
    </environment>
  </environments>
<mappers>
<mapper resource="org/mybatis/example/BlogMapper.xml"/>
    </mappers>
  </configuration>
```

XML 配置文件中还有其他许多内容,上面的例子只给出最重要的部分。

注意:这个 XML 的标头,需要一个 DTD 验证文档。environment 项中包含了事务管理和连接池的环境配置。mappers 项中包含了一系列 SQL 语句映射定义的 XML 文件。

7.3.2 不使用 XML 文件新建 SqlSessionFactory

如果更想直接使用 Java 语言,而不是 XML 生成这些配置,或者想使用自己的配置生成器,MyBatis 提供了一个完整的配置类完成 XML 文件一样的配置。

```
DataSource dataSource=BlogDataSourceFactory.getBlogDataSource();
TransactionFactory transactionFactory=new JdbcTransactionFactory();
Environment environment=new Environment("development",transactionFactory,
        dataSource);
Configuration configuration=new Configuration(environment);
configuration.addMapper(UsrMapper.class);
SqlSessionFactory sqlSessionFactory=new SqlSessionFactoryBuilder().build
        (configuration);
```

这个配置里加载了一个映射类。映射类是包含了 SQL 映射注解的 Java 类,可以用来取代 XML。然而,由于 Java 注解的一些限制和 MyBatis 映射的复杂性,一些高级的映射还是要用 XML 配置,如嵌套映射等。由于这个原因,MyBatis 会自动查找和加载已经存在的 XML。例如上面的代码,UsrMapper.xml 将会被类路径中的 UsrMapper.class 加载。

7.3.3 通过 SqlSessionFactory 获取 SqlSession

假设有一个 SqlSessionFactory,就可以获取一个 SqlSession 实例,SqlSession 包含了针对数据库执行语句的每一个方法。可以直接使用 SqlSession 执行已经映射的每个 SQL 语句。例如:

```
SqlSession session=sqlMapper.openSession();
try{
  Usr usr=(Usr) session.select("com.ascent.mybatis.mapper.UsrMapper.
      selectUsr", 21);
} finally {
  session.close();
}
```

上述步骤对于使用 MyBatis 的上一个版本(即 iBATIS 2)的用户来说比较熟悉。现在有一个更加清晰的方式。使用一个有正确参数和返回值的接口,就可以更加清晰和安全地编写代码,从而避免出错。像这样:

```
SqlSession session=sqlSessionFactory.openSession();
try{
  UsrMapper usrMapper=session.getMapper(UsrMapper.class);
  Usr usr=usrMapper.selectUsr(21);
} finally {
  session.close();
}
```

7.3.4 SQL 映射语句简介

对于上面所说的,你可能很好奇 SqlSession 或 Mapper 类具体是怎么执行的。为了概括执行过程,现在给出两个例子。在上面的例子中,语句已经由 XML 或注解所定义。先看一下 XML,以前 MyBatis 提供的所有特性都是基于 XML 的映射语句实现的。

XML 的映射配置文档有了许多改进,以后将会变得越来越简单、清晰。下面这个基于 XML 的映射语句可以完成上面的 SqlSession 调用。

```
<?xml version="1.0" encoding="UTF-8" ?>
<!DOCTYPE mapper
PUBLIC "-//mybatis.org//DTD Mapper 3.0//EN"
"http://mybatis.org/dtd/mybatis-3-mapper.dtd">
<mapper namespace="com.ascent.mybatis.mapper.UsrMapper">
  <select id="selectUsr" parameterType="int" resultType="com.ascent.po.
    Usr">
    select * fromUsr where id=#{id}
  </select>
</mapper>
```

虽然这个简单的例子有点生涩,但是却非常简约。可以定义多个文件,也可以在一个 XML 文件里定义任意个映射语句,这样可以省去 XML 标头。文件的其他部分都是自身的描述。它定义了一个 com.ascent.mybatis.mapper.UsrMapper 命名空间,在这个空间里又定义了一个 selectUsr 语句。也可以使用 com.ascent.mybatis.mapper.UsrMapper.selectUsr 全名称调用。可以按如下方式调用上面这个文件:

```
Usr usr=(Usr) session.select("com.ascent.mybatis.mapper.UsrMapper.selectUsr",
21);
```

这和调用一个普通的 Java 类非常相似。这个名字可以直接映射为一个与命名空间相同名称的 Mapper 类,语句名对应类的方法名,参数和返回值也相对应。

还可以用下列语句简单地针对 Mapper 接口进行调用,代码如下:

```
UsrMapper usrMapper=session.getMapper(UsrMapper.class);
Usr usr=mapper.selectUsr(21);
```

这种方式有许多优点。①它不依赖字符串,可以减少出错;②如果你的 IDE 有代码自动完成功能,可以很快导航到你的 SQL 语句(因为已经转化为方法名);③不再需要设定返回值类型,因为接口限定了返回值和参数。

还有一个关于 Mapper 类的技巧。它们的映射语句完全不需要使用 XML 配置,可以使用 Java 注解方式来取代。例如,上面的 XML 语句可以替换为

```
packagecom.ascent.mybatis.mapper;
public interface UsrMapper {
  @Select("SELECT * FROMusr WHERE id=#{id}")
  Usr selectUsr(int id);
}
```

注解是非常简单明了的，但是 Java 注解有局限性，在语句比较复杂的情况下容易混乱。所以，如果语句比较复杂，最好使用 XML 映射语句。

7.3.5 MyBatis 对象的作用域与生命周期

理解作用域和生命周期类非常重要，如果使用不当，会造成各种各样的问题。

1. SqlSessionFactoryBuilder

这个类可以被初始化、使用和丢弃，如果已经创建好一个 SqlSessionFactory，就不用再保留它。因此，SqlSessionFactoryBuilder 的作用域最好是方法体内，比如说定义一个方法变量。你可以重复使用 SqlSessionFactoryBuilder 生成多个 SqlSessionFactory 实例，但是最好不要强行保留，因为 XML 的解析资源要用来做其他更重要的事。

2. SqlSessionFactory

一旦创建，SqlSessionFactory 就会在整个应用过程中始终存在。所以，没有理由销毁和再创建它，一个应用运行中也不建议多次创建 SqlSessionFactory。如果真那样做，会显得很笨拙。因此，SqlSessionFactory 的作用域最好是 Application，可以有多种方法实现。最简单的方法是单例模式或者静态单例模式。然而，这不是广泛赞成和好用的，使用 Google Guice 或 Spring 进行依赖反射更好。这些框架允许生成管理器管理 SqlSessionFactory 的单例生命周期。

3. SqlSession

每个线程都有自己的 SqlSession 实例，SqlSession 实例不能被共享，也不是线程安全的，因此最好使用 Request 作用域或者方法体作用域。不要使用类的静态变量引用一个 SqlSession 实例，甚至不要使用类的一个实例变更来引用。不要在一个被管理域中引用 SqlSession，比如说在 Servlet 中的 HttpSession 中。如果正在使用 Web 框架，应该让 SqlSession 跟随 HTTP 请求的相似作用域。也就是说，在收到一个 HTTP 请求后，打开 SqlSession，等返回一个回应后立刻关掉这个 SqlSession。关闭 SqlSession 非常重要。必须确保 SqlSession 在 finally()方法体中正常关闭。可以使用下面的标准方式关闭：

```
SqlSession session=sqlSessionFactory.openSession();
try{
  //do work
} finally {
  session.close();
}
```

使用这种模式贯穿所有代码，以确保所有数据库资源都被完全关闭。

4. Mapper 实例

Mapper 是一种你创建的用于绑定映射语句的接口。Mapper 接口的实例是用 SqlSession 获得的。同样，从技术上来说，最广泛的 Mapper 实例作用域像 SqlSession 一样，使用请求作用域。确切地说，在方法被使用的时候调用 Mapper 实例，保持 Mapper 在方法体作用域内，之后就自动销毁掉。下面演示了如何操作：

```
SqlSession session=sqlSessionFactory.openSession();
try{
  BlogMapper mapper=session.getMapper(BlogMapper.class);
  //do work
} finally {
  session.close();
}
```

7.3.6 XML 配置文件

XML 配置文件包含一些设置和属性，层次如下：

```
configuration
          |---properties
          |---settings
          |---typeAliases
          |---typeHandlers
          |---objectFactory
          |---plugins
          |---environments
          |---|---environment
          |---|---|---transactionManager
          |---|---|__ dataSource
          |__mappers
```

XML 配置文件的根元素是 configuration，其他是 configuration 元素的子元素。

1. 属性（properties）

属性元素可以用来配置直观的、可代替的属性，或者是属性项的子项。例如：

```
<properties resource="config/db.properties">
<property name="username" value="root"/>
<property name="password" value="123456"/>
</properties>
```

通过动态配置，这些属性可以被属性文件的值所替换。例如：

```
<dataSource type="POOLED">
  <property name="driver" value="${jdbc.driver}"/>
  <property name="url" value="${jdbc.url}"/>
  <property name="username" value="${jdbc.username}"/>
  <property name="password" value="${jdbc.password}"/>
</dataSource>
```

例子中的 username、password、driver 和 url 属性将被 db.properties 文件中的值所替换，这为配置提供了多种选择。

属性值也可以设入 SqlSessionBuilder.build()方法中，例如：

```
SqlSessionFactory factory=sqlSessionFactoryBuilder.build(reader, props);
//... or ...
SqlSessionFactory factory = sqlSessionFactoryBuilder. build ( reader, environment,
props);
```

如果一个属性项在多个地方出现，那么 MyBatis 将按以下顺序加载：

- 属性文件中的属性项首先被读取。
- 在类路径或 URL 资源中读取的属性项以第二顺序加载，并且可以覆盖第一顺序加载的值。
- 在方法体中给定的参数值最后加载，但是可以覆盖上述两种加载的值。

也就是说，最高级别的属性值是方法体中设定的参数值，接下来是类路径和 URL，最后才是属性文件。

2. 设置(settings)

这是 MyBatis 修改操作运行过程细节的重要步骤。表 7-1 描述了这些设置项、含义和默认值。

表 7-1　设置项、含义和默认值

设　置　项	描述	允许值	默认值
cacheEnabled	对在此配置文件下的所有 cache 进行全局性开/关设置	true/false	true
lazyLoadingEnabled	全局性设置惰性加载。如果设为 false，则所有相关联的对象都会被初始化加载	true/false	true
aggressiveLazyLoading	当设置为 true 的时候，惰性加载的对象可以被任何惰性属性全部加载，否则，每个属性都按需加载	true/false	true
multipleResultSetsEnabled	是否允许单条语句返回多个数据集(取决于驱动需求)	true/false	true
useColumnLabel	使用列标签代替列名称。不同的驱动器有不同的做法。参考驱动器文档，或者用这两个不同的选项进行测试	true/false	true
useGeneratedKeys	允许 JDBC 生成主键。需要驱动器支持。如果设为 true，这个设置将强制使用被生成的主键，有一些驱动器不兼容，不过仍然可以执行	true/false	false
autoMappingBehavior	指定 MyBatis 是否并且如何自动映射数据表字段与对象的属性。PARTIAL 将只自动映射简单的、没有嵌套的结果。FULL 将自动映射所有复杂的结果	NONE、PARTIAL、FULL	PARTIAL

续表

设 置 项	描述	允许值	默认值
defaultExecutorType	配置和设定执行器,SIMPLE 执行器执行其他语句。REUSE 执行器可以重复使用 prepared statements 语句,BATCH 执行器可以重复执行语句和批量更新	SIMPLE、REUSE、BATCH	SIMPLE
defaultStatementTimeout	设置一个时限,以决定驱动器等待数据库回应多长时间为超时	正整数	Not Set(null)

下面列出关于设置的完整例子:

```
<settings>
  <setting name="cacheEnabled" value="true"/>
  <setting name="lazyLoadingEnabled" value="true"/>
  <setting name="multipleResultSetsEnabled" value="true"/>
  <setting name="useColumnLabel" value="true"/>
  <setting name="useGeneratedKeys" value="false"/>
  <setting name="enhancementEnabled" value="false"/>
  <setting name="defaultExecutorType" value="SIMPLE"/>
  <setting name="defaultStatementTimeout" value="25000"/>
</settings>
```

3. 类型别名(typeAliases)

类型别名是 Java 类型的简称,可以用来简化冗长的 Java 类名。例如:

```
<typeAliases>
  <typeAlias alias="Author" type="domain.blog.Author"/>
  <typeAlias alias="Blog" type="domain.blog.Blog"/>
  <typeAlias alias="Comment" type="domain.blog.Comment"/>
  <typeAlias alias="Post" type="domain.blog.Post"/>
  <typeAlias alias="Section" type="domain.blog.Section"/>
  <typeAlias alias="Tag" type="domain.blog.Tag"/>
</typeAliases>
```

使用这个配置,Blog 就能在任何地方代替 domain.blog.Blog 使用。

还有一些通用 Java 类型建立的别名,它们是大小写敏感的。

注意:Java 基本类型的别名使用_(下画线)命名,见表 7-2。

4. 类型处理器(typeHandlers)

当 MyBatis 对 PreparedStatement 设入一个参数或者是从 ResultSet 返回一个值的时候,类型处理器被用来将值转换为相匹配的 Java 类型。表 7-3 描述了默认的类型处理器。

表 7-2　Java 基本类型的别名

别　名	映射的类型	别　名	映射的类型
_byte	byte	double	double
_long	long	float	float
_short	short	boolean	boolean
_int	int	date	date
_integer	int	decimal	BigDecimal
_double	double	bigdecimal	BigDecimal
_float	float	object	object
_boolean	boolean	map	map
string	string	hashmap	HashMap
byte	byte	list	list
long	long	arraylist	ArrayList
short	short	collection	collection
int	integer	iterator	iterator
integer	integer		

表 7-3　默认的类型处理器

类型处理器	Java 类型	JDBC 类型
BooleanTypeHandler	Boolean,boolean	任何兼容的布尔值
ByteTypeHandler	Byte,byte	任何兼容的数字或字节类型
ShortTypeHandler	Short,short	任何兼容的数字或短整型
IntegerTypeHandler	Integer,int	任何兼容的数字和整型
LongTypeHandler	Long,long	任何兼容的数字或长整型
FloatTypeHandler	Float,float	任何兼容的数字或单精度浮点型
DoubleTypeHandler	Double,double	任何兼容的数字或双精度浮点型
BigDecimalTypeHandler	BigDecimal	任何兼容的数字或十进制小数类型
StringTypeHandler	String	CHAR 和 VARCHAR 类型
ClobTypeHandler	String	CLOB 和 LONGVARCHAR 类型
NStringTypeHandler	String	NVARCHAR 和 NCHAR 类型
NClobTypeHandler	String	NCLOB 类型
ByteArrayTypeHandler	byte[]	任何兼容的字节流类型
BlobTypeHandler	byte[]	BLOB 和 LONGVARBINARY 类型
DateTypeHandler	Date(java.util)	TIMESTAMP 类型

续表

类型处理器	Java 类型	JDBC 类型
DateOnlyTypeHandler	Date(java.util)	DATE 类型
TimeOnlyTypeHandler	Date(java.util)	TIME 类型
SqlTimestampTypeHandler	Timestamp(java.sql)	TIMESTAMP 类型
SqlDateTypeHandler	Date(java.sql)	DATE 类型
SqlTimeTypeHandler	Time(java.sql)	TIME 类型
ObjectTypeHandler	Any	其他或未指定类型
EnumTypeHandler	Enumeration 类型	VARCHAR-任何兼容的字符串类型,作为代码存储(而不是索引)

可以重写(override)类型句柄或者创建自己的方式处理不支持或者是非标准的类型。只需要简单地实现 org.mybatis.type 包里的 TypeHandler,并且映射新的类型句柄类到一个 Java 类型,再选定一个 JDBC 类型。例如:

```
public class ExampleTypeHandler implements TypeHandler {
  public void setParameter ( PreparedStatement ps, int i, Object parameter,
JdbcType jdbcType)
throws SQLException {
    ps.setString(i, (String) parameter);
  }
  public Object getResult(ResultSet rs, String columnName) throws SQLException {
    return rs.getString(columnName);
  }
  public ObjectgetResult ( CallableStatement cs, int columnIndex ) throws
SQLException {
    return cs.getString(columnIndex);
  }
}
//MapperConfig.xml
<typeHandlers>
  < typeHandler javaType = "String" jdbcType = "VARCHAR" handler = "org.mybatis.
example.ExampleTypeHandler"/>
</typeHandlers>
```

使用像这样的类型处理器,将会覆盖现有的处理 Java String 属性与 Varchar 和返回值的类型句柄。注意,MyBatis 无法审查数据库的元数据,从而决定类型,所以必须指定参数是一个 Varchar 类型,并且将结果映射到正确的类型处理器上。这么做主要是由于 MyBatis 没有执行语句之类,无法得知数据的类型。

5. 对象工厂(ObjectFactory)

MyBatis 每次为结果对象创建一个新实例,都会用到 ObjectFactory。默认的

ObjectFactory 与使用目标类的构造函数创建一个实例毫无区别。如果有已经映射的参数,也可以使用带参数的构造函数。如果重写 ObjectFactory 的默认操作,就可以创建自己的对象工厂。例如:

```
public class ExampleObjectFactory extends DefaultObjectFactory{
  public Object create(Class type){
    return super.create(type);
  }
  public Object create(Class type, List<Class>constructorArgTypes, List<Object
>constructorArgs) {
    return super.create(type, constructorArgTypes, constructorArg
  }
  public void setProperties(Properties properties) {
    super.setProperties(properties);
  }
}
//MapperConfig.xml
<objectFactory type="org.mybatis.example.ExampleObjectFactory">
  <property name="someProperty" value="100"/>
</objectFactory>
```

ObjectFactory 接口非常简单,只包含两个方法:一个是默认构造函数;一个是带参数的构造函数。最后,setProperties()方法也可以使用 ObjectFactory 配置。可以在 ObjectFactory 实例化后通过 setProperties()方法在对象工厂中定义属性。

6. 插件(plugins)

MyBatis 允许在映射语句执行过程的某点上拦截调用。默认 MyBatis 允许插件拦截以下调用:

- Executor (update, query, flushStatements, commit, rollback, getTransaction, close, isClosed)
- ParameterHandler(getParameterObject, setParameters)
- ResultSetHandler(handleResultSets, handleOutputParameters)
- StatementHandler(prepare, parameterize, batch, update, query)

这些类的细节在每个方法签名中均可以找到,源代码在 MyBatis 每次发布时都可以下载。如果要做的事不仅是调用,还有重写方法,就需要了解要重写的方法的动作。

使用插件是非常简单而又有用的。只简单地实现这个 Interceptor 接口,确定好要拦截的标识即可。

```
//ExamplePlugin.java
@ Intercepts ({@ Signature (type = Executor. class, method =" update", args =
{MappedStatement.class,Object.class})})
public classExamplePlugin implements Interceptor {
  public Object intercept(Invocation invocation) throwsThrowable {
    return invocation.proceed();
```

```
    }
    public Object plugin(Object target) {
      return Plugin.wrap(target, this);
    }
    public void setProperties(Properties properties) {
    }
}
//MapperConfig.xml
<plugins>
  <plugin interceptor="org.mybatis.example.ExamplePlugin">
    <property name="someProperty" value="100"/>
  </plugin>
</plugins>
```

上面这个插件可以在执行器上拦截所有 update()方法的调用。

7. 环境(environments)

MyBatis 可以配置多个环境。这可以帮助 SQL 映射多种数据库等。比如说,想为开发、测试、发布产品配置不用的环境,或者想为多个数据库产品共享相同的模式,或者想使用相同的 SQL 映射等。

需要记住一个重要的事情:虽然可以配置多重环境,但只能选择一个对应的 SqlSessionFactory 实例。所以,如果想连接两个数据库,需要使用 SqlSessionFactory 创建两个实例,每个数据库一个。如果要连 3 个数据库,就创建 3 个实例,以此类推。记住:一个 SqlSessionFactory 实例对应一个数据库。想指定生成哪个环境,只要简单地把它作为一个可选参数代入 SqlSessionFactoryBuilder 即可。下面两种方式都可以:

```
SqlSessionFactory factory=sqlSessionFactoryBuilder.build(reader, environment);
SqlSessionFactory factory = sqlSessionFactoryBuilder. build ( reader, environment,
properties);
```

如果环境变量省略了,就会载入默认的环境变量。像这样:

```
SqlSessionFactory factory=sqlSessionFactoryBuilder.build(reader);
SqlSessionFactory factory=sqlSessionFactoryBuilder.build(reader,properties);
```

环境元素用来定义这些环境是如何被配置的。

```
<environments default="development">
  <environment id="development">
    <transactionManager type="JDBC">
      <property name="" value=""/>
    </transactionManager>
    <dataSource type="POOLED">
      <property name="driver" value="${jdbc.driver}"/>
      <property name="url" value="${jdbc.url}"/>
      <property name="username" value="${jdbc.username}"/>
      <property name="password" value="${jdbc.password}"/>
```

```
    </dataSource>
  </environment>
</environments>
```

注意这些关键段：

- 设定一个默认环境 ID。
- 这个环境 ID 对每个环境都起作用。
- 配置事务管理器。
- 配置数据源。

默认的环境和环境 ID 是对自己起作用的，可以随意起名字，只要它们不重复就可以。

1）事务管理器

MyBatis 有两个事务管理类型：

(1) JDBC。这个类型直接全部使用 JDBC 的提交和回滚功能。它依靠使用连接的数据源管理事务的作用域。

(2) MANAGED。这个类型什么也不做，它从不提交、回滚和关闭连接，而是让容器管理事务的全部生命周期(如 Spring 或者 JavaEE 服务器)，它们都不需要任何属性。既然它们是类型别名，直接把类名称或者类型别名指向 TransactionFactory 接口实现类就可以了。

```
public interface TransactionFactory {
  void setProperties(Properties props);
  Transaction newTransaction(Connection conn, boolean autoCommit);
}
```

实例化后，在 XML 中已经被配置的任何属性都可以代入 setProperties()方法中。你的实现也需要创建一个事务实现，它的接口是非常简单的：

```
public interface Transaction{
  Connection getConnection();
  void commit() throws SQLException;
  void rollback() throws SQLException;
  void close() throwsSQLException;
}
```

通过使用这两个接口，可以定制 MyBatis 如何处理事务。

2）数据源

数据源元素用来配置使用 JDBC 数据源。

大部分的 MyBatis 应用都像上面例子中那样配置数据源，但是这并不是必需的。只有使用了惰性加载，才必须使用数据源。

数据源类型有 3 种：UNPOOLED、POOLED、JNDI。

(1) UNPOOLED。这个数据源实现只是在每次请求的时候简单地打开和关闭一个连接。虽然这有点慢，但作为一些不需要性能和立即响应的简单应用来说，不失为一种好选择。不同的数据库在性能方面也有所不同，所以相对于连接池来说倒是不重要，这个配置会比较适合。

UNPOOLED 数据源有几个属性：

- driver：指定 JDBC 驱动器的 Java 类，而不是数据类。
- url：连接数据库实例的 URL 路径。
- username：登录数据库的用户名。
- password：登录数据库的密码。
- defaultTransactionsolationLevel：指定连接的默认事务隔离层。

另外，也可以为数据驱动器设置属性。只简单地取 driver.开头就行了，例如：

```
driver.encoding=UTF8
```

这就会把属性为 encoding，值为 UTF-8，通过 DriverManager.getConnection(url, driverProperties) 方法传递给数据库驱动器。

(2) POOLED。这个数据源缓存 JDBC 连接对象，用于避免每次都要连接和生成连接实例而需要的验证时间。对于并发 Web 应用，这种方式非常流行，因为它有最快的响应时间。在 UNPOOLED 的属性之上，POOLED 数据还有其他许多配置属性。

- poolMaximumActiveConnections：特定时间里可同时使用的连接数。
- poolMaximumIdleConnections：特定时间里闲置的连接数。
- poolMaximumCheckoutTime：在连接池强行返回前，一个连接可以进行检出的总计时间。
- poolTimeToWait：这是一个底层的设置，给连接一个机会打印 log 状态，并尝试重新连接，免得长时间等待。
- poolPingQuery：Ping Query 是发送给数据库的 Ping 信息，测试数据库连接是否良好和是否准备好了接受请求。默认值是“NO PING QUERY SET”，让大部分数据库都不使用 Ping，返回一个友好的错误信息。
- poolPingEnabled：设置 PingQuery 是否可用。如果可用，可以使用一个最简单的 SQL 语句测试一下，默认是 false。
- poolPingConnectionsNotUsedFor：配置 poolPingQuery 多长时间可以用。通常匹配数据库连接的超时设置，避免无谓的 ping。默认值这 0，表示随时允许 ping，当然是在 poolPingEnabled 设为 true 的前提下。

(3) JNDI。这个数据源实现是为了和 Spring 或应用服务一起使用，可以在外部，也可以在内部配置这个数据源，然后在 JNDI 上下文中引用它。这个数据源配置只需要两个属性：

- initial_context：这个属性被用于上下文从 InitialContext 中(如 initialContext.lookup(initial_context))查找。这个属性是可选的，如果被省略，InitialContext 将会直接查找 data_source 属性。
- data_source：这是数据源实例能搜索到的上下文路径。它会直接查找 initial_context 搜索返回的值，如果 initial_context 没有值，则直接使用 InitialContext 查找。像数据源的其他配置一样，可以使用 env.属性直接设给 InitialContext，如 env.encoding＝UTF8，这样就可以把值为 UTF8 的属性直接代入 InitialContext 实例化的构造器。

8. 映射器(Mappers)

现在 MyBatis 的基本元素已经通过以上步骤配置好了,可以开始定义映射 SQL 语句了。首先需要告诉 MyBatis 去哪寻找映射文件。可以使用类路径中的资源引用,或者使用字符,输入确切的 URL 引用。例如:

```
<mappers>
  <mapper resource="com/ascent/mybatis/mapper/UsrMapper.xml"/>
  <mapper resource="com/ascent/mybatis/mapper/BlogMapper.xml"/>
  <mapper resource="com/ascent/mybatis/mapper/PostMapper.xml"/>
</mappers>
//Using url fully qualified paths
<mappers>
  <mapper url="file:///var/sqlmaps/UsrMapper.xml"/>
  <mapper url="file:///var/sqlmaps/BlogMapper.xml"/>
  <mapper url="file:///var/sqlmaps/PostMapper.xml"/>
</mappers>
```

7.3.7 XML 映射文件

MyBatis 真正强大的地方是映射语句。就所有功能来说,SQL 映射 XML 文件相对来说比较简单。如果比较了映射文件与 JDBC 代码,就会发现 XML 映射文件节省了 95%的代码。MyBatis 为 SQL 而建,却又给你极大的空间。SQL 映射 XML 文件有一些基本的元素:

- select:映射 SELECT 语句。
- insert:映射 INSERT 语句。
- update:映射 UPDATE 语句。
- delete:映射 DELETE 语句。
- SQL:一个可以被其他语句复用的 SQL 块。
- resultMap:这是最复杂、最强大的一个元素,它描述如何从结果集中加载对象。
- cache:配置给定模式的缓存。
- cache-ref:从别的模式中引用一个缓存。

1. select

select 语句可能是 MyBatis 中使用最多的元素,通常会把数据存储在数据库中,然后读取,所以大部分的应用中查询远远多于修改。对于每个 insert、update 或者 delete,都会伴有众多 select。这是 MyBatis 基础原则之一,也是为什么要把更多的焦点和努力都放在查询和结果映射上的原因。一个 select 元素非常简单,例如:

```
<select id="selectPerson" parameterType="int" resultType="hashmap">
   SELECT * FROM PERSON WHERE ID=#{id}
</select>
```

这条语句就叫作'selectPerson',有一个 int(或是 Integer)参数,并返回一个 HashMap

结果集。

注意：参数的标识是：

```
#{id}
```

这告诉 MyBatis 生成一个 PreparedStatement 的参数。对于 JDBC，这样的参数在 SQL 中用？标识，传递给一个 PreparedStatement 语句。通常像这样：

```
//Similar JDBC code, NOT MyBatis…
StringselectPerson="SELECT * FROM PERSON WHERE ID=?";
PreparedStatement ps=conn.prepareStatement(selectPerson);
ps.setInt(1,id);
```

当然，如果只用 JDBC 单独解析这个结果集并映射到对象上，则需要非常多的代码，而这些 MyBatis 都已经做好了。select 语句有众多的属性可用于配置语句的执行细节，例如：

```
<select
  id="selectPerson"
  parameterType="int"
  resultType="hashmap"
  resultMap="personResultMap"
  flushCache="false"
  useCache="true"
  timeout="10000"
  fetchSize="256"
  statementType="PREPARED"
  resultSetType="FORWARD_ONLY"
>
```

select 属性见表 7-4。

表 7-4　select 属性

属　性	描　　述
id	在命名空间中唯一的标识符可以用来引用这条语句
parameterType	将会传入这条语句的参数类的完全限定名或别名。这个属性是可选的，因为 MyBatis 可以通过类型处理器(TypeHandler)推断出具体传入语句的参数，默认值为未设置(unset)
resultType	从这条语句中返回的期望类型的类的完全限定名或别名。注意：如果返回的是集合，就应该设置为集合包含的类型，而不是集合本身。可以使用 resultType 或 resultMap，但不能同时使用
resultMap	外部 resultMap 的命名引用。结果集的映射是 MyBatis 最强大的特性，如果对其理解透彻，许多复杂映射的情形都能迎刃而解。可以使用 resultMap 或 resultType，但不能同时使用

续表

属　性	描　　述
flushCache	将其设置为 true 后，只要语句被调用，就会导致本地缓存和二级缓存被清空，默认值为 false
useCache	其设置为 true 后，会导致本条语句的结果被二级缓存缓存起来，默认值：对 select 元素为 true
timeout	这个设置是在抛出异常之前，驱动程序等待数据库返回请求结果的秒数。默认值为未设置(unset)(依赖驱动)
fetchSize	这是一个给驱动的提示，尝试让驱动程序每次批量返回的结果行数和这个设置值相等。默认值为未设置(unset)(依赖驱动)
statementType	STATEMENT、PREPARED 或 CALLABLE 中的一个。这会让 MyBatis 分别使用 Statement、PreparedStatement 或 CallableStatement，默认值为 PREPARED
resultSetType	FORWARD_ONLY、SCROLL_SENSITIVE、SCROLL_INSENSITIVE 或 DEFAULT(等价于 unset)中的一个，默认值为 unset(依赖驱动)

2. insert、update、delete

语句 insert、update、delete 都非常相似，例如：

```
<insert
id="insertAuthor"
parameterType="domain.blog.Author"
flushCache="true"
statementType="PREPARED"
keyProperty=""
useGeneratedKeys=""
timeout="20000">
<update
id="insertAuthor"
parameterType="domain.blog.Author"
flushCache="true"
statementType="PREPARED"
timeout="20000">
<delete
id="insertAuthor"
parameterType="domain.blog.Author"
flushCache="true"
statementType="PREPARED"
timeout="20000">
```

insert、update、delete 属性见表 7-5。

表 7-5　insert、update、delete 属性

属　性	描　述
id	命名空间中的唯一标识符，可用来代表这条语句
parameterType	将要传入语句的参数的完全限定类名或别名。这个属性是可选的，因为 MyBatis 可以通过类型处理器推断出具体传入语句的参数，默认值为未设置(unset)
flushCache	将其设置为 true 后，只要语句被调用，都会导致本地缓存和二级缓存被清空，默认值为 true(对于 insert、update 和 delete 语句)
timeout	这个设置是在抛出异常之前，驱动程序等待数据库返回请求结果的秒数。默认值为未设置(unset)(依赖驱动)
statementType	STATEMENT、PREPARED 或 CALLABLE 中的一个。这会让 MyBatis 分别使用 Statement、PreparedStatement 或 CallableStatement，默认值为 PREPARED
useGeneratedKeys	(仅对 insert 和 update 有用)这会令 MyBatis 使用 JDBC 的 getGeneratedKeys () 方法取出由数据库内部生成的主键(如 MySQL 和 SQL Server 这样的关系数据库管理系统的自动递增字段)，默认值为 false
keyProperty	(仅对 insert 和 update 有用)唯一标记一个属性，MyBatis 会通过 getGeneratedKeys 的返回值或者通过 insert 语句的 selectKey 子元素设置它的键值，默认值为未设置(unset)。如果希望得到多个生成的列，也可以是逗号分隔的属性名称列表

下面是一些 insert、update、delete 语句的示例：

```
<insert id="insertAuthor" parameterType="domain.blog.Author">
  insert into Author (id,username,password,email,bio)
  values (#{id},#{username},#{password},#{email},#{bio})
</insert>
<update id="updateAuthor" parameterType="domain.blog.Author">
  update Author set
  username=#{username},
  password=#{password},
  email=#{email},
  bio=#{bio}
  where id=#{id}
</update>
<delete id="deleteAuthor" parameterType="int">
  delete from Author where id=#{id}
</delete>
```

insert 语句有更多的属性拥有子元素，可以使用多种方式生成主键处理。首先，如果使用的数据库支持自动生成主键，就可以设置 useGeneratedKeys＝"true"，然后把 keyProperty 设成对应的列即可。例如，上面的 Author 使用 auto-generated 为 id 列生成主键，语句可以修改如下：

```
<insert id="insertAuthor" parameterType="domain.blog.Author"
```

```
  useGeneratedKeys="true" keyProperty="id">
  insert into Author (username,password,email,bio)
  values (#{username},#{password},#{email},#{bio})
</insert>
```

MyBatis 还有另外一种方式处理在数据库不支持自动生成主键的情况生成主键,或者是在 JDBC 驱动不能自动生成主键的情况下生成主键。

下面简单示例一下如何生成随机的 ID。也许不需要这么做,这里仅演示一下 MyBatis 的复杂功能。

```
<insert id="insertAuthor" parameterType="domain.blog.Author">
  <selectKey keyProperty="id" resultType="int" order="BEFORE">
    select CAST(RANDOM() * 1000000 as INTEGER) a from SYSIBM.SYSDUMMY1
  </selectKey>
  insert into Author
  (id,username, password, email,bio, favourite_section)
  values
  (#{id}, #{username}, #{password}, #{email}, #{bio},
  #{favouriteSection,jdbcType=VARCHAR}
  )
</insert>
```

在上面的例子中,selectKey 语句将会首先运行,Author 表的 id 列将会被设置,然后再调用这个 insert 语句,这相当于自动在数据库里生成主键,而不需要写复杂的代码。

selectKey 属性见表 7-6。

表 7-6 selectKey 属性

属 性	描 述
keyProperty	它是 selectKey 语句结果应该被设置的目标属性。如果希望得到多个生成的列,也可以是逗号分隔的属性名称列表
resultType	结果的类型。MyBatis 通常可以推断出来,但是,为了更加精确,写上也不会有问题。MyBatis 允许将任何的简单类型用作主键的类型,包括字符串。如果希望作用于多个生成的列,则可以使用一个包含期望属性的 Object 或一个 Map
order	这可以被设置为 BEFORE 或 AFTER。如果设置为 BEFORE,它会首先生成主键,然后设置 keyProperty,最后执行插入语句。如果设置为 AFTER,那么首先执行插入语句,然后执行 selectKey 中的语句(这和 Oracle 数据库的行为相似,在插入语句内部可能有嵌入索引调用)
statementType	与前面相同,MyBatis 支持 STATEMENT、PREPARED 和 CALLABLE 语句的映射类型,分别代表 Statement、PreparedStatement 和 CallableStatement 类型

3. SQL

这个元素用来定义一个可以复用的 SQL 语句段,供其他语句调用。例如:

```
<sql id="userColumns">id,username,password</sql>
```

这个语句块可以包含到别的语句中,例如:

```
<select id="selectUsers" parameterType="int" resultType="hashmap">
select <include refid="userColumns"/>
from some_table
where id=#{id}
</select>
```

4. resultMap

1) 简单结果映射

resultMap 是 MyBatis 中最重要、最强大的元素,比使用 JDBC 调用结果集省掉 90%的代码,也可以做许多 JDBC 不支持的事情。实际上,写一个等同于交互映射这样的复杂语句,可能需要上千行代码。resultMap 的目的就是实现简单的语句,而不需要多余的结果映射。对于更多复杂的语句,除了只要一些绝对必需的语句描述关系外,不需要其他语句。你可能已经看到这样的简单映射语句,其中没有一个 resultMap,例如:

```
<select id="selectUsers" parameterType="int" resultType="hashmap">
  select id,username, hashedPassword
  from some_table
  where id=#{id}
</select>
```

像这样的简单语句的所有列结果将会自动映射到以列名为 key 的 HashMap,使用 resultType 指定。虽然这在许多场合下有用,但是 HashMap 却不是非常好的领域模型。你的应用更可能使用 JavaBeans 或者 POJOs 来领域建模。MyBatis 对这两者都支持。

```
package com.someapp.model;
public class User{
  private int id;
  private String username;
  private String hashedPassword;
  public int getId() {
    return id;
  }
  public void setId(int id) {
    this.id=id;
  }
  public String getUsername() {
    return username;
  }
  public void setUsername(String username) {
    this.username=username;
  }
  public String getHashedPassword() {
    return hashedPassword;
  }
```

```
  public void setHashedPassword(String hashedPassword) {
    this.hashedPassword=hashedPassword;
  }
}
```

基于 JavaBeans 规范,上面的类有 3 个属性：id、username 和 hashedPassword。它们对应 select 语句的列名。这样的 JavaBean 可以像 HashMap 一样简单地映射到 ResultSet。

```
<select id="selectUsers" parameterType="int" resultType="com.someapp.model.
User">
  select id,username, hashedPassword
  from some_table
  where id=#{id}
</select>
```

TypeAliases 可以避免很长的类的全路径名。例如：

```
<!—in config XML file →
  <typeAlias type="com.someapp.model.User" alias="User"/>
<!--In SQL Mapping XML file -->
<select id="selectUsers" parameterType="int" resultType="User">
  select id,username, hashedPassword
  from some_table
  where id=#{id}
</select>
```

在这些情况下,MyBatis 自动在后台生成一个 ResultMap 映射列名到 JavaBean 的属性,如果列的名称与属性名确实不符,可以使用标准 SQL 的特性生成别名。例如：

```
<select id="selectUsers" parameterType="int" resultType="User">
  select
    user_id as "id",
    user_name as "userName",
    hashed_password as "hashedPassword"
  from some_table
  where id=#{id}
</select>
```

ResultMap 的优点有许多,其中一个优点是具有 resultMap 扩展功能,可解决列表不配对的问题。

```
<resultMap id="userResultMap" type="User">
  <id property="id" column="user_id" />
  <result property="username" column="username"/>
  <result property="password" column="password"/>
</resultMap>
```

这个语句将会被 resultMap 引用。注意,不能再使用 resultType 属性。

```
<select id="selectUsers" parameterType="int " resultMap="userResultMap">
  select user_id, user_name, hashed_password
  from some_table
  where id=#{id}
</select>
```

2）高级结果映射

我们总是希望只用一条简单的数据库映射，就能完美解决应用中的问题，但是实际上很难做到。高级结果映射就是 MyBatis 为解决这些问题提供的答案。

举一个例子，下面的语句怎么映射？

```
<!--Very Complex Statement -->
<select id="selectBlogDetails" parameterType="int" resultMap=
"detailedBlogResultMap">
  select
    B.id as blog_id,
    B.title as blog_title,
    B.author_id as blog_author_id,
    A.id as author_id,
    A.username as author_username,
    A.password as author_password,
    A.email as author_email,
    A.bio as author_bio,
    A.favourite_section as author_favourite_section,
    P.id as post_id,
    P.blog_id as post_blog_id,
    P.author_id as post_author_id,
    P.created_on as post_created_on,
    P.section as post_section,
    P.subject as post_subject,
    P.draft as draft,
    P.body as post_body,
    C.id as comment_id,
    C.post_id as comment_post_id,
    C.name as comment_name,
    C.comment as comment_text,
    T.id as tag_id,
    T.name as tag_name
  from Blog B
    left outer join Author A on B.author_id=A.id
    left outer join Post P on B.id=P.blog_id
    left outer join Comment C on P.id=C.post_id
    left outer join Post_Tag PT on PT.post_id=P.id
    left outer join Tag T on PT.tag_id=T.id
  where B.id=#{id}
</select>
```

你可能想把它映射到一个智能的对象模型,包括由一个作者写的一个博客,由多项交互,有 0 个或者多个评论和标签。下面会一步步看这个复杂 ResultMap 的完整例子。

```
<!--Very Complex Result Map -->
<resultMap id="detailedBlogResultMap" type="Blog">
  <constructor>
    <idArg column="blog_id" javaType="int"/>
  </constructor>
  <result property="title" column="blog_title"/>
  <association property="author" column="blog_author_id"javaType=" Author">
    <id property="id" column="author_id"/>
    <result property="username" column="author_username"/>
    <result property="password" column="author_password"/>
    <result property="email" column="author_email"/>
    <result property="bio" column="author_bio"/>
    <result property="favouriteSection" column="author_favourite_section"/>
  </association>
  <collection property="posts"ofType="Post">
    <id property="id" column="post_id"/>
    <result property="subject" column="post_subject"/>
    <association property="author" column="post_author_id" javaType="Author"/>
    <collection property="comments" column="post_id"ofType=" Comment">
      <id property="id" column="comment_id"/>
    </collection>
    <collection property="tags" column="post_id"ofType=" Tag" >
      <id property="id" column="tag_id"/>
    </collection>
    <discriminator javaType="int" column="draft">
      <case value="1"resultType="DraftPost"/>
    </discriminator>
  </collection>
</resultMap>
```

这个 resultMap 的元素的子元素比较多,讨论起来比较宽泛。下面从概念上了解一下这个 resultMap 的元素。

- constructor：用于在实例化类时,注入结果到构造方法中。
  - oloidArg：ID 参数;标记出作为 ID 的结果可以帮助提高整体性能。
  - oarg：将被注入构造方法的一个普通结果。
- id：一个 ID 结果;标记出作为 ID 的结果可以帮助提高整体性能。
- result：注入字段或 JavaBean 属性的普通结果。
- association：一个复杂类型的关联;许多结果将包装成这种类型。
  - 嵌套结果映射：关联本身可以是一个 resultMap 元素,或者从别处引用一个 resultMap 元素。
- collection：一个复杂类型的集合。

◆ 嵌套结果映射：集合本身可以是一个 resultMap 元素，或者从别处引用一个 resultMap 元素。

● discriminator：使用结果值决定使用哪个 resultMap 元素。

◆ case：基于某些值的结果映射。

◆ 嵌套结果映射：case 本身可以是一个 resultMap 元素，因此具有相同的结构和元素；或者引用其他的结果映射。

从目前实际的 MyBatis 使用情况看：一步步生成复杂的 resultMap，对理解原理和单元测试非常有用。如果快速生成像上面这样巨大的 resultMap，可能会出错，并且工作起来非常吃力。

(1) id、result。

```
<id property="id" column="post_id"/>
<result property="subject" column="post_subject"/>
```

这些是结果映射最基本的内容。id 和 result 都映射一个单独列的值到简单数据类型(字符串、整型、双精度浮点数、日期等)的单独属性或字段。id 的不同之处是，当和其他对象实例对比的时候，它作为唯一的标识。id、result 映射属性见表 7-7。

表 7-7　id、result 映射属性

属　性	描　述
property	映射到列结果的字段或属性。如果用来匹配的 JavaBean 存在给定名字的属性，那么它将会被使用，否则 MyBatis 将会寻找给定名称的字段。无论是哪种情形，都可以使用通常的点式分隔形式进行复杂属性导航。例如，可以这样映射一些简单的东西：username，或者映射到一些复杂的东西上：address.street.number
column	数据库中的列名，或者是列的别名。一般情况下，这和传递给 resultSet.getString(columnName)方法的参数一样
javaType	一个 Java 类的完全限定名，或一个类型别名。如果映射到一个 JavaBean，MyBatis 通常可以推断类型。然而，如果映射到的是 HashMap，那么应该明确指定 javaType，以保证行为与期望的一致
jdbcType	JDBC 类型。所支持的 JDBC 类型参见这个表格之后的"支持的 JDBC 类型"。只需要在可能执行插入、更新和删除的且允许空值的列上指定 JDBC 类型。这是 JDBC 的要求，而非 MyBatis 的要求。如果直接面向 JDBC 编程，需要对可能存在空值的列指定这个类型
typeHandler	使用这个属性，可以覆盖默认的类型处理器。这个属性值是一个类型处理器实现类的完全限定名，或者是类型别名

支持的 JDBC 类型包括：BIT、FLOAT 、CHAR、TIMESTAMP 、OTHER、UNDEFINED、TINYINT、REAL、VARCHAR、BINARY、BLOB、NVARCHAR、SMALLINT、DOUBLE、LONGVARCHAR、VARBINARY、CLOB、NCHAR、INTEGER、NUMERIC、DATE、LONGVARBINARY、BOOLEAN、NCLOB、BIGINT　DECIMAL、TIME、NULL、CURSOR、ARRAY。

(2) 构造器(constructor)。

```
<constructor>
```

```
  <idArg column="id" javaType="int"/>
  <arg column="username" javaType="String"/>
</constructor>
```

当属性与数据传输对象，和你自己的领域模型进行交互的时候，许多场合要用到不变类。通常，数据表很少(甚至不会)与不变类相匹配。构造器反射允许给实例化后的类设置值，不用通过 public()方法。MyBatis 同样也支持 private 属性和 JavaBeans 的私有属性这么做，但是，有一些用户可能更喜欢使用构造器反射。构造器元素可以做到这一点。

考虑下面的构造器。

```
public class User{
  //…
  public User(int id, Stringusername) {
  //…
  } //…
}
```

为了将值注入构造器中，MyBatis 需要使用它的参数类型标识构造器。Java 没有办法通过参数名称反射，所以，当创建一个构造器元素时，要确定参数是否良好，类型是否指定。

```
<constructor>
  <idArg column="id" javaType="int"/>
  <arg column="username" javaType="String"/>
</constructor>
```

其他的属性和规则与 id 和 result 元素一模一样。

(3) 联合(association)。

```
<association property="author" column="blog_author_id"javaType=" Author">
    <id property="id" column="author_id"/>
    <result property="username" column="author_username"/>
</association>
```

联合元素用来处理“一对一”的关系。例如，一个博客对应一个作者。联合映射和其他结果映射类似，只需要指定目标属性、要返回值的列、属性的 javaType(通常 MyBatis 自己会识别)。如果需要设定 jdbcType，或者想重写返回结果的值，则需要指定 typeHandler。不同的地方是，需要告诉 MyBatis 如何加载一个联合。MyBatis 可以用以下两种方式加载。

- Nested Select：执行一个嵌套的 SQL 语句返回一个期望的复杂类型。
- Nested Results：使用一个嵌套的结果映射处理复杂的结果。

首先查看一下元素的属性，见表 7-8，它不同于普通只有 select 和 resultMap 的结果集映射属性。

联合的嵌套选择见表 7-9。

表 7-8 联合元素属性

属 性	描 述
property	映射到列结果的字段或属性。如果用来匹配的 JavaBean 存在给定名字的属性,那么它将会被使用,否则 MyBatis 将会寻找给定名称的字段。无论是哪一种情形,都可以使用通常的点式分隔形式进行复杂属性导航。例如,可以这样映射一些简单的东西：username,或者映射到一些复杂的东西上：address.street.number
javaType	一个 Java 类的完全限定名,或一个类型别名。如果映射到一个 JavaBean,MyBatis 通常可以推断类型。然而,如果映射到的是 HashMap,那么应该明确指定 javaType 保证行为与期望的一致
jdbcType	JDBC 类型。所支持的 JDBC 类型参见这个表格之前的“支持的 JDBC 类型”。只需要在可能执行插入、更新和删除的且允许空值的列上指定 JDBC 类型。这是 JDBC 的要求,而非 MyBatis 的要求。如果直接面向 JDBC 编程,需要对可能存在空值的列指定这个类型
typeHandler	使用这个属性,可以覆盖默认的类型处理器。这个属性值是一个类型处理器实现类的完全限定名,或者是类型别名

表 7-9 联合的嵌套选择

属 性	描 述
select	用于加载复杂类型属性的映射语句的 ID,它会从 column 属性指定的列中检索数据,作为参数传递给目标 select 语句。注意,在使用复合主键的时候,可以使用 column="{prop1=col1,prop2=col2}"这样的语法指定多个传递给嵌套 select 语句的列名。这会使得 prop1 和 prop2 作为参数对象,被设置为对应嵌套 select 语句的参数

示例：

```
<resultMap id="blogResult" type="Blog">
  <association property="author" column="blog_author_id" javaType="Author"
    select="selectAuthor"/>
</resultMap>
<select id="selectBlog" parameterType="int" resultMap="blogResult">
    SELECT * FROM BLOG WHERE ID=#{id}
</select>
<select id="selectAuthor" parameterType="int" resultType="Author">
    SELECT * FROM AUTHOR WHERE ID=#{id}
</select>
```

就是这个样子,有两个 select 语句,一个用于加载博客,另一个用于加载作者。博客这一项的映射结果中声明了一个 selectAuthor 语句,用来加载它的作者属性。

如果它们的列名和属性名称相匹配,则所有其他的属性都会自动加载。

这个例子比较简单,但是对于大数据集或列表,可能有一点问题。这就是“$N+1$ 选择问题”。概要地说,$N+1$ 选择问题是这样的：

- 执行一个单条语句获取一个列表记录(“+1”)。
- 对每条记录,再执行一个 select 语句加载每条记录的详细信息(“N”)。

这个问题会在成千上万条语句发生，是无法预料的。

针对上面的问题，MyBatis 使用惰性加载这些查询，可以节省资源。然而，如果加载一个列表然后立刻又迭代访问内嵌的数据，再调用所有的惰性加载，那么执行情况就非常糟糕了。鉴于此，有另一个方式，也就是联合的嵌套结果集，见表 7-10。

表 7-10　联合的嵌套结果集

属　性	描　　述
resultMap	结果映射的 ID，可以将此关联的嵌套结果集映射到一个合适的对象树中。它可作为使用嵌套 select 语句的替代方案。它可以将多表连接操作的结果映射成一个单一的 ResultSet。这样的 ResultSet 有部分数据是重复的。为了将结果集正确地映射到嵌套的对象树中，MyBatis 允许"串联"结果映射，以便解决嵌套结果集的问题

下面这个例子演示了如何不执行分离的语句，而是把博客和作者表都连接在一起。像这样：

```
<select id="selectBlog" parameterType="int" resultMap="blogResult">
  select
    B.id as blog_id,
    B.title as blog_title,
    B.author_id as blog_author_id,
    A.id as author_id,
    A.username as author_username,
    A.password as author_password,
    A.email as author_email,
    A.bio as author_bio
  from Blog B left outer join Author A on B.author_id=A.id
  where B.id=#{id}
</select>
```

注意：这个连接结果都被别名为一个唯一且明确的名称。这将使映射变得容易。还可以这样做：

```
<resultMap id="blogResult" type="Blog">
  <id property="blog_id" column="id" />
  <result property="title" column="blog_title"/>
  <association property="author" column="blog_author_id" javaType="Author"
resultMap="authorResult"/>
</resultMap>
<resultMap id="authorResult" type="Author">
  <id property="id" column="author_id"/>
  <result property="username" column="author_username"/>
  <result property="password" column="author_password"/>
  <result property="email" column="author_email"/>
  <result property="bio" column="author_bio"/>
</resultMap>
```

从上面的例子中可以看到，博客的作者放到了 authorResult 的结果映射中，用来加载作者实例。

注意：id 元素是嵌套映射中非常重要的角色。需要指定一个或更多的属性用来标记结果，如果没有这么做，MyBatis 也可能会正常运行，但是会有较大的性能损耗。尽量用数量少的列唯一标识结果，使用主键是一个好的选择。

上面的例子使用一个扩展的 resultMap 元素联合映射，这样可使作者结果映射可重复使用。如果不需要重用它，或者只是简单地想协同定位结果映射到一个描述性的结果映射中，则可以嵌套这个联合结果映射。下面的例子就使用了这样的方式：

```
<resultMap id="blogResult" type="Blog">
  <id property="blog_id" column="id" />
  <result property="title" column="blog_title"/>
  <association property="author" column="blog_author_id" javaType="Author">
    <id property="id" column="author_id"/>
    <result property="username" column="author_username"/>
    <result property="password" column="author_password"/>
    <result property="email" column="author_email"/>
    <result property="bio" column="author_bio"/>
  </association>
</resultMap>
```

联合元素用来处理“一对一”的关系，第 8 章将会详细介绍。在要处理“一对多”关系的时候，就要使用聚集了。

(4) 聚集(collection)。

```
<collection property="posts"ofType="domain.blog.Post">
  <id property="id" column="post_id"/>
  <result property="subject" column="post_subject"/>
  <result property="body" column="post_body"/>
</collection>
```

聚集元素的作用和联合元素的作用几乎一样，下面着重看它们的不同。

继续上面的例子，一个博客只有一个作者，但是一个博客却有许多评论。在一个博客类里，可能会表述成这样：

```
private List<Post>posts;
```

要映射一个嵌套的结果集到一个这样的列表中，可以使用聚集元素。就像联合元素一样，可以使用一个嵌套选择，或者是一个连接的嵌套的结果。

- 集合的嵌套选择。

通过嵌套选择加载博客的评论，如下：

```
<resultMap id="blogResult" type="Blog">
  <collection property="posts"javaType="ArrayList" column="blog_id" ofType=
"Post" select="selectPostsForBlog"/>
</resultMap>
```

```
<select id="selectBlog" parameterType="int" resultMap="blogResult">
  SELECT * FROM BLOG WHERE ID=#{id}
</select>
<select id="selectPostsForBlog" parameterType="int" resultType="Post">
  SELECT * FROM POST WHERE BLOG_ID=#{id}
</select>
```

聚集元素包含一个 ofType 属性。这个属性用来区别 JavaBean(或者字段)的属性类型和集合所包括的类型。所以

```
<collection property="posts"javaType="ArrayList" column="blog_id" ofType="
Post" select="selectPostsForBlog"/>
```

它的意思是:"一组名为 posts 的 ArrayList 集合,它的类型是 Post。"

javaType 属性可以省略,MyBatis 会自动识别,所以通常可以这样简写:

```
< collection property =" posts " column =" blog _ id" ofType =" Post " select ="
selectPostsForBlog"/>
```

- 集合的嵌套结果。

到目前为止,嵌套结果集到一个集合中和联合的工作原理一样,不同的是多了一个 ofType,上面已经提到它的作用,接下来看一下这个 SQL:

```
<select id="selectBlog" parameterType="int" resultMap="blogResult">
  select
    B.id as blog_id,
    B.title as blog_title,
    B.author_id as blog_author_id,
    P.id as post_id,
    P.subject as post_subject,
    P.body as post_body,
  from Blog B
    left outer join Post P on B.id=P.blog_id
  where B.id=#{id}
</select>
```

这里已经连接了博客和评论表,并且把列名标签也进行了明确的处理,以方便映射。现在可以像这样映射一个博客和它的一组评论:

```
<resultMap id="blogResult" type="Blog">
  <id property="id" column="blog_id" />
  <result property="title" column="blog_title"/>
  <collection property="posts"ofType="Post">
    <id property="id" column="post_id"/>
    <result property="subject" column="post_subject"/>
    <result property="body" column="post_body"/>
  </collection>
</resultMap>
```

再次强调这个 id 属性非常重要。同样,如果使用稍长一点的代码而达到可重复使用的目的,可以使用下面这种方式:

```
<resultMap id="blogResult" type="Blog">
  <id property="id" column="blog_id" />
  <result property="title" column="blog_title"/>
  <collection property="posts"ofType="Post" resultMap="blogPostResult"/>
</resultMap>
<resultMap id="blogPostResult" type="Post">
  <id property="id" column="post_id"/>
  <result property="subject" column="post_subject"/>
  <result property="body" column="post_body"/>
</resultMap>
```

注意:这里没有深度、宽度、联合和聚合数目的限制,但是一定要把性能牢记在心。单元测试和性能测试能帮助你调整采取的方式。MyBatis 允许在以后还能修改之前的想法,只修改少量的代码即可。

同样,第 8 章将会详细介绍如何使用聚集(collection)。

(5) 识别器(discriminator)。

```
<discriminatorjavaType="int" column="draft">
  <case value="1"resultType="DraftPost"/>
</discriminator>
```

有时候一条数据库查询返回的结果可能包括各种不同的数据类型。识别器元素就是用来处理这种情况的,它还包括一些继承层次。识别器非常容易理解,它很像 Java 里的 switch 语句。一个识别器定义了指定列和 javaType 属性。列就是 MyBatis 将要取出进行比较的值,JavaType 是需要被用来保证等价测试的合适类型(尽管字符串在很多情形下都有用)。示例如下:

```
<resultMap id="vehicleResult" type="Vehicle">
  <id property="id" column="id" />
  <result property="vin" column="vin"/>
  <result property="year" column="year"/>
  <result property="make" column="make"/>
  <result property="model" column="model"/>
  <result property="color" column="color"/>
  <discriminatorjavaType="int" column="vehicle_type">
    <case value="1"resultMap="carResult"/>
    <case value="2"resultMap="truckResult"/>
    <case value="3"resultMap="vanResult"/>
    <case value="4"resultMap="suvResult"/>
  </discriminator>
</resultMap>
```

上面的实例中,MyBatis 将会从数据集中获取每条记录,并比较 vehicle_type 的值。如果它匹配了识别器的条件，就会使用对应的 resultMap。这种行为有排他性，也就是说，只要匹配到一项，剩余部分都会被忽略，不再进行比较（使用扩展属性除外）。如果没有任何一项能匹配到，MyBatis 就会简单地使用识别器外面定义的 resultMap。因此，如果像下面这样定义一个 carResult：

```
<resultMap id="carResult" type="Car">
  <result property="doorCount" column="door_count" />
</resultMap>
```

那么，仅只是加载了 doorCount 这个属性。这样做是为了保留完全独立的聚集识别器的选项，尽管和父结果映射可能没有什么关系。在刚才的例子里我们当然知道 cars 和 vehicles 的关系，所以也要把其他部分加载进来。下面稍稍改动一下 resultMap。

```
<resultMap id="carResult" type="Car" extends="vehicleResult">
  <result property="doorCount" column="door_count" />
</resultMap>
```

现在，vehicleResult 和 carResult 都将被加载。

可能有人会认为这样的扩展映射有一点单调，所以还有另外一种可选的语法更简明地映射。例如：

```
<resultMap id="vehicleResult" type="Vehicle">
  <id property="id" column="id" />
  <result property="vin" column="vin"/>
  <result property="year" column="year"/>
  <result property="make" column="make"/>
  <result property="model" column="model"/>
  <result property="color" column="color"/>
  <discriminatorjavaType="int" column="vehicle_type">
    <case value="1"resultType="carResult">
      <result property="doorCount" column="door_count" />
    </case>
    <case value="2"resultType="truckResult">
      <result property="boxSize" column="box_size" />
      <result property="extendedCab" column="extended_cab" />
    </case>
    <case value="3"resultType="vanResult">
      <result property="powerSlidingDoor" column="power_sliding_door" />
    </case>
    <case value="4"resultType="suvResult">
      <result property="allWheelDrive" column="all_wheel_drive" />
    </case>
  </discriminator>
</resultMap>
```

记住：对于这么多的结果映射，如果不指定任何结果，那么 MyBatis 会自动将列名与属性相匹配。

5. Cache

1）缓存概述

MyBatis 内置了一个强大的事务性查询缓存机制，它可以非常方便地配置和定制。MyBatis 3 中改进了很多的缓存实现，使得它更加强大，而且易于配置。

默认情况下只启用本地的会话缓存，也就是一级缓存，它仅对一个会话中的数据进行缓存。要启用全局的二级缓存，只需要在 SQL 映射文件中添加一行：

```
<cache/>
```

这个简单语句的效果如下：

- 映射语句文件中的所有 select 语句的结果将会被缓存。
- 映射语句文件中的所有 insert、update 和 delete 语句会刷新缓存。
- 缓存会使用最近最少使用(Least Recently Used,LRU)算法清除不需要的缓存。
- 缓存不会定时刷新(也就是说，没有刷新间隔)。
- 缓存会保存列表或对象(无论查询方法返回哪种)的 1024 个引用。
- 缓存会被视为读/写缓存，这意味着获取到的对象并不是共享的，可以安全地被调用者修改，而不干扰其他调用者或线程所做的潜在修改。

注意：缓存只作用于 cache 标签所在的映射文件中的语句。如果混合使用 Java API 和 XML 映射文件，在共用接口中的语句将不会被默认缓存。通常需要使用 @CacheNamespaceRef 注解指定缓存作用域。

cache 元素的属性可以修改，例如：

```
<cache
  eviction="FIFO"
  flushInterval="60000"
  size="512"
readOnly="true"/>
```

这个更高级的配置创建了一个 FIFO 缓存，每隔 60s 刷新一次，最多可以存储结果对象或列表的 512 个引用，而且返回的对象被认为是只读的，因此，对它们进行修改可能会在不同线程中的调用者之间产生冲突。

可用的清除(eviction)策略有：

- LRU(最近最少使用)：移除最长时间不被使用的对象。
- FIFO(先进先出)：按对象进入缓存的顺序移除它们。
- SOFT(软引用)：基于垃圾回收器状态和软引用规则移除对象。
- WEAK(弱引用)：更积极地基于垃圾收集器状态和弱引用规则移除对象。

默认的清除策略是 LRU。

flushInterval(刷新间隔)属性可以被设置为任意的正整数，设置的值应该是一个以毫秒为单位的合理时间量。默认情况是不设置，也就是没有刷新间隔，缓存仅会在调用语句时刷新。

size(引用数目)属性可以被设置为任意正整数，要注意欲缓存对象的大小和运行环境中可用的内存资源。默认值是 1024。

readOnly(只读)属性可以被设置为 true 或 false。只读的缓存会给所有调用者返回缓存对象的相同实例。因此,这些对象不能被修改。这就提供了可观的性能提升。而可读写的缓存会(通过序列化)返回缓存对象的副本,速度上会慢一些,但是更安全,因此默认值是 false。

注意:二级缓存是事务性的。这意味着,当 SqlSession 完成并提交时,或是完成并回滚,但没有执行 flushCache=true 的 insert/delete/update 语句时,缓存会获得更新。

2) 使用自定义缓存

除了上述使用缓存的方式,也可以通过实现自己的缓存,或为其他第三方缓存方案创建适配器,完全覆盖缓存行为。

```
<cache type="com.domain.something.MyCustomCache"/>
```

这个示例展示了如何使用一个自定义的缓存实现。type 属性指定的类必须实现 org.apache.ibatis.cache.Cache 接口,且提供一个接受 String 参数作为 id 的构造器。这个接口是 MyBatis 框架中许多复杂的接口之一,但是行为却非常简单。

```
public interface Cache{
  String getId();
  int getSize();
  void putObject(Object key, Object value);
  Object getObject(Object key);
  boolean hasKey(Object key);
  Object removeObject(Object key);
  void clear();
}
```

为了对缓存进行配置,只需要简单地在缓存实现中添加公有的 JavaBean 属性,然后通过 cache 元素传递属性值。例如,下面的例子将在缓存实现上调用一个名为 setCacheFile(String file)的方法:

```
<cache type="com.domain.something.MyCustomCache">
  <property name="cacheFile" value="/tmp/my-custom-cache.tmp"/>
</cache>
```

可以使用所有的简单类型作为 JavaBean 属性的类型,MyBatis 会进行转换。也可以使用占位符(如 ${cache.file}),以便替换成在配置文件属性中定义的值。

从版本 3.4.2 开始,MyBatis 已经支持在所有属性设置完毕之后调用一个初始化方法。如果想使用这个特性,则在自定义缓存类里实现 org.apache.ibatis.builder.InitializingObject 接口。

```
public interface InitializingObject {
    void initialize() throws Exception;
}
```

提示:对缓存的配置(如清除策略、可读或可读写等),不能应用于自定义缓存。

注意：缓存的配置和缓存实例会被绑定到 SQL 映射文件的命名空间中。因此，同一命名空间中的所有语句和缓存将通过命名空间绑定在一起。每条语句可以自定义与缓存交互的方式，或将它们完全排除于缓存之外，这可以通过在每条语句上使用两个简单属性达成。默认情况下，语句会这样配置：

```
<select... flushCache="false" useCache="true"/>
<insert... flushCache="true"/>
<update... flushCache="true"/>
<delete... flushCache="true"/>
```

鉴于这是默认行为，显然不应该以这样的方式显式配置一条语句。但是，如果想改变默认的行为，只需要设置 flushCache 和 useCache 属性。例如，某些情况下可能希望特定 select 语句的结果排除于缓存之外，或希望一条 select 语句清空缓存。类似地，你可能希望某些 update 语句执行时不要刷新缓存。

6. cache-ref

对某一命名空间的语句，MyBatis 只会使用该命名空间的缓存进行缓存或刷新。但你可能会想在多个命名空间中共享相同的缓存配置和实例。要实现这种需求，可以使用 cache-ref 元素引用另一个缓存。

```
<cache-ref namespace="com.someone.application.data.SomeMapper"/>
```

根据上面的介绍，看一个 Student 表的增、删、改、查的案例（Mybatis_CRUD_demo）。

1）在 MySQL 数据库中建立 student 表

```
USE `test`;

/* Table structure for table `student` */

DROP TABLE IF EXISTS `student`;

CREATE TABLE `student`(
  `id` int NOT NULL AUTO_INCREMENT,
  `name`varchar(255) DEFAULT NULL,
  `age` int DEFAULT NULL,
  `description` varchar(255) CHARACTER SET latin1 COLLATE latin1_swedish_ci
DEFAULT NULL,
  PRIMARY KEY (`id`)
) ENGINE=InnoDB AUTO_INCREMENT=4 DEFAULT CHARSET=latin1;

/* Data for the table `student` */

insert  into `student`(`id`,`name`,`age`,`description`) values (1,'Gary',25,
'Software Engineer'),(2,'Linda',35,'Marketing'),(3,'Kevin',22,'Computer Science');
```

2）新建 Java 项目，导入 MyBatis 所依赖 jar 包，建立项目目录

```
src/config                          //存放 mybatis-config.xml 和 db.conf 文件，数据
                                      库连接信息配置
src/com/ascent/mybatis/mapper       //存放 Java API 文件 StudentMapper.class
src/com/ascent/mybatis/mapping      //存放 SQL XML 映射文件 StudentMapper.xml
src/com/ascent/util                 //存放工具文件 SqlSessionFactoryUtil
src/com/ascent/po                   //存放 Java POJO 类，映射查询 Student 实体
src/com/ascent/test                 //存放测试运行主方法类 StudentTest
```

3）在 config 目录下新建 db.conf 文件

```
jdbc.driver=com.mysql.jdbc.Driver
jdbc.url=jdbc:mysql://localhost:3306/test
jdbc.username=root
jdbc.password=root
```

注意：这里的变量要加上 jdbc.作为前缀，否则 MyBatis 可能会无法识别。具体的值根据数据库设置。

4）在 config 目录下新建 mybatis-config 文件

```
<?xml version="1.0" encoding="UTF-8"?>
<!DOCTYPE configuration
  PUBLIC "-//mybatis.org//DTD Config 3.0//EN"
  "http://mybatis.org/dtd/mybatis-3-config.dtd">
<configuration>

  <properties resource="config/db.conf"/>

  <typeAliases>
    <typeAlias alias="Student" type="com.ascent.po.Student"/>
  </typeAliases>

  <environments default="development">
      <environment id="development">
          <transactionManager type="JDBC" />
          <dataSource type="POOLED">
              <property name="driver" value="${jdbc.driver}" />
              <property name="url" value="${jdbc.url}" />
              <property name="username" value="${jdbc.username}" />
              <property name="password" value="${jdbc.password}" />
          </dataSource>
      </environment>
    </environments>
  <mappers>
    <mapper resource="com/ascent/mapping/StudentMapper.xml"/>
  </mappers>
```

```
</configuration>
```

5）在 util 中创建 SqlSessionFactoryUtil 类

```
package com.ascent.util;

import java.io.IOException;
import java.io.InputStream;
import org.apache.ibatis.io.Resources;
import org.apache.ibatis.session.SqlSessionFactory;
import org.apache.ibatis.session.SqlSessionFactoryBuilder;
public class SqlSessionFactoryUtil{

  private staticSqlSessionFactory sqlSessionFactory=null;

  static{
    InputStream input=null;
    try{
      input=Resources.getResourceAsStream("config/mybatis-config.xml");
      sqlSessionFactory=new SqlSessionFactoryBuilder().build(input);
    } catch (IOException e) {
      e.printStackTrace();
    }finally{
      if(input !=null){
        try{
          input.close();
        } catch (IOException e) {
          e.printStackTrace();
        }
      }
    }
  }

  public static SqlSessionFactory getSqlSessionFactory(){
    return sqlSessionFactory;
  }
}
```

6）在 po 目录新建 Student 类

```
package com.ascent.po;

public class Student {
    private int id;
    private String name;                    //姓名
    private int age;                        //年龄
    private String description;             //描述
```

```
    public int getId() {
        return id;
    }
    public void setId(int id) {
        this.id=id;
    }
    public String getName() {
        return name;
    }
    public void setName(String name) {
        this.name=name;
    }
    public int getAge() {
        return age;
    }
    public void setAge(int age) {
        this.age=age;
    }
    public String getDescription() {
        return description;
    }
    public void setDescription(String description) {
        this.description=description;
    }
}
```

7）在 mapping 目录下新建 StudentMapper.xml

```
<?xml version="1.0" encoding="UTF-8"?>
<!DOCTYPE mapper PUBLIC "-//mybatis.org//DTD Mapper 3.0//EN"
"http://mybatis.org/dtd/mybatis-3-mapper.dtd">
<mapper namespace="com.ascent.mapper.StudentMapper">
  <!--定义 Java Bean 的属性与数据库的列之间的映射 -->
  <resultMap type="Student" id="studentResultMap">
    <id column="id" property="id" />
    <result column="name" property="name" />
    <result column="age" property="age" />
    <result column="description" property="description" />
  </resultMap>

  <!--SQL 语句中以"#{}"的形式引用参数 -->
  <!--新增记录 -->
  <insert id="insertStudent" parameterType="Student">
    insert into student(name,age,description)
    values(#{name},#{age},#{description})
```

```
  </insert>

  <!--查询单条记录 -->
  <select id="getById" parameterType="int" resultMap="studentResultMap">
    select * from student where id=#{id}
  </select>

  <!--查询所有记录 -->
  <select id="findAll" resultMap="studentResultMap">
    select * from student
  </select>

  <!--修改记录 -->
  <update id="updateStudent" parameterType="Student">
    update student set name=#{name},age=#{age},description=#{description}
    where id=#{id}
  </update>

  <!--删除单条记录 -->
  <delete id="deleteById" parameterType="int">
    delete from student where id=#{id}
  </delete>
</mapper>
```

8）在 mapper 目录下新建 StudentMapper

```
package com.ascent.mapper;
import com.ascent.po.*;
import java.util.List;
import org.apache.ibatis.annotations.*;

public interface StudentMapper {

    public void insert Student(Student student);
    public Student getById(@Param(value="id") Integer id);
    public List<Student>findAll();
    public void updateStudent(Student student);
    public void deleteById(@Param(value="id") Integer id);

}
```

9）在 test 目录下建立 StudentTest 测试类，含 main()方法

```
package com.ascent.test;

import java.util.List;
import org.apache.ibatis.session.SqlSession;
```

```
import org.apache.ibatis.session.SqlSessionFactory;
import com.ascent.po.Student;
import com.ascent.mapper.StudentMapper;
import com.ascent.util.SqlSessionFactoryUtil;

public class StudentTest {
  private SqlSessionFactory factory=SqlSessionFactoryUtil.getSqlSessionFactory();

  /**
   * 测试新增
   * /
  public void insert(){
    SqlSession session=null;
    try{
      Student student=new Student();
      student.setName("Kevin");
      student.setDescription("Computer Science");
      session=factory.openSession();
      StudentMapper mapper=session.getMapper(com.ascent.mapper.
          StudentMapper.class);
      mapper.insertStudent(student);
      session.commit();
    } catch (Exception e) {
      e.printStackTrace();
    }finally{
      session.close();
    }
  }

  /**
    * 查询单条记录
    * /

  public void getById() {
    SqlSession session=null;
    try{
      session=factory.openSession();
      StudentMapper mapper=
      session.getMapper(com.ascent.mapper.StudentMapper.class);
      Student student=mapper.getById(1);
      System.out.println(student.getName()+" "+student.getDescription());
    } catch (Exception e) {
      e.printStackTrace();
    }finally{
      session.close();
```

```
        }
    }

    /**
     *查询所有记录
     */

    public void findAll(){
        SqlSession session=null;
        try{
            session=factory.openSession();

            StudentMapper mapper=session.getMapper(StudentMapper.class);
            List<Student>students=mapper.findAll();
            System.out.println(students.size());
        } catch (Exception e) {
            e.printStackTrace();
        }finally{
            session.close();
        }
    }

    /**
     *更新记录
     */
    public void update(){
        SqlSession session=null;
        try{
            session=factory.openSession();
            StudentMapper mapper=session.getMapper(StudentMapper.class);
            Student student=mapper.getById(1);
            student.setName("Gary");

    mapper.updateStudent(student);
            session.commit();
        } catch (Exception e) {
            e.printStackTrace();
        }finally{
            session.close();
        }
}

    /**
     *删除记录
     */
```

```
  public void delete(){
    SqlSession session=null;
    try{
      session=factory.openSession();
      StudentMapper mapper=session.getMapper(StudentMapper.class);

      mapper.deleteById(1);
      session.commit();
    } catch (Exception e) {
      e.printStackTrace();
    }finally{
      session.close();
    }
  }
  public static void main(String[]args){

    StudentTest testCase=new StudentTest();

    //testCase.insert();
    testCase.getById();
    //testCase.update();
    //testCase.delete();
    testCase.findAll();

  }

}
```

10）运行 StudentTest，得到结果

在增、删、改、查的 4 个操作中，查询的变化最多。

在上面实例的基础上，再看一个支持更多查询功能的实例（Mybatis_MoreQueries_demo）。前面的步骤和上面实例的步骤相同，下面看不一样的地方。

① StudentMapper.xml

```
<?xml version="1.0" encoding="UTF-8"?>
<!DOCTYPE mapper PUBLIC "-//mybatis.org//DTD Mapper 3.0//EN"
"http://mybatis.org/dtd/mybatis-3-mapper.dtd">
<mapper namespace="com.ascent.mapper.StudentMapper">
  <!--定义 Java Bean 的属性与数据库的列之间的映射 -->
  <resultMap type="Student" id="studentResultMap">
    <id column="id" property="id" />
    <result column="name" property="name" />
    <result column="age" property="age" />
    <result column="description" property="description" />
  </resultMap>
```

```
    <!--SQL 语句中以"#{}"的形式引用参数 -->
    <!--新增记录 -->
    <insert id="insertStudent" parameterType="Student">
      insert into student(name,age,description)
      values(#{name},#{age},#{description})
    </insert>

    <!--查询单条记录 -->
    <select id="getById" parameterType="int" resultMap="studentResultMap">
      select * from student where id=#{id}
    </select>

    <!--查询所有记录 -->
    <select id="findAll" resultMap="studentResultMap">
      select * from student
    </select>

    <!--修改记录 -->
    <update id="updateStudent" parameterType="Student">
      update student set name=#{name},age=#{age},description=#{description}
      where id=#{id}
    </update>

    <!--删除单条记录 -->
    <delete id="deleteById" parameterType="int">
      delete from student where id=#{id}
    </delete>

    <!--模糊查询记录 -->
      <select id="listStudentsByName" parameterType="string" resultType=
          "Student">
          select * from student where name likeconcat('%',#{0},'%')
    </select>

      <!--多条件查询记录 -->
    <select id="listStudentsByIdAndName" parameterType="map" resultType=
        "Student">
        select * from student where id>#{id} and name likeconcat('%',#{name},'%')
    </select>

</mapper>
```

这里增加了模糊查询记录和多条件查询记录两个 SQL 语句。

② StudentMapper 类

```
package com.ascent.mapper;
```

```
import com.ascent.po.*;
import java.util.List;
import java.util.Map;

import org.apache.ibatis.annotations.*;

public interface StudentMapper {

    public void insertStudent(Student student);
    public Student getById(@Param(value="id") Integer id);
    public List<Student>findAll();
    public void updateStudent(Student student);
    public void deleteById(@Param(value="id") Integer id);

    //add more queries
    public List<Student>listStudentsByName(String s);
    public List<Student>listStudentsByIdAndName(Map m);

}
```

这里增加了两个查询方法 listStudentsByName()和 listStudentsByIdAndName()。

③ StudentTest 测试类

```
package com.ascent.test;

import java.util.List;
import java.util.Map;
import java.util.HashMap;

import org.apache.ibatis.session.SqlSession;
import org.apache.ibatis.session.SqlSessionFactory;
import com.ascent.po.Student;
import com.ascent.mapper.StudentMapper;
import com.ascent.util.SqlSessionFactoryUtil;

public classStudentTest {
  privateSqlSessionFactory factory=SqlSessionFactoryUtil.getSqlSessionFactory();

  /**
   * 测试新增记录
   */
  public void insert(){
    SqlSession session=null;
    try{
      Student student=new Student();
      student.setName("Kevin");
```

```
            student.setDescription("Computer Science");
            session=factory.openSession();
            StudentMapper mapper=session.getMapper(com.ascent.mapper.
                StudentMapper.class);
            mapper.insertStudent(student);
            session.commit();
        } catch (Exception e) {
            e.printStackTrace();
        }finally{
            session.close();
        }
    }

    /**
     *查询单条记录
     */

    public voidgetById(){
        SqlSession session=null;
        try{
            session=factory.openSession();
            StudentMapper mapper=
            session.getMapper(com.ascent.mapper.StudentMapper.class);
            Student student=mapper.getById(1);
            System.out.println(student.getName()+" "+student.getDescription());
        } catch (Exception e) {
            e.printStackTrace();
        }finally{
            session.close();
        }
    }

    /**
     *查询所有记录
     */

    public void findAll(){
        SqlSession session=null;
        try{
            session=factory.openSession();

            StudentMapper mapper=
            session.getMapper(StudentMapper.class);
            List<Student>students=mapper.findAll();
            System.out.println(students.size());
```

```
    } catch (Exception e) {
      e.printStackTrace();
    }finally{
      session.close();
    }
  }

  /**
   * 更新记录
   * /
  public void update(){
    SqlSession session=null;
    try{
      session=factory.openSession();
      StudentMapper mapper=session.getMapper(StudentMapper.class);
      Student student=mapper.getById(1);
      student.setName("Gary");

  mapper.updateStudent(student);
      session.commit();
    } catch (Exception e) {
      e.printStackTrace();
    }finally{
      session.close();
    }
  }

  /**
   * 删除记录
   * /
  public void delete(){
    SqlSession session=null;
    try{
      session=factory.openSession();
      StudentMapper mapper=session.getMapper(StudentMapper.class);

      mapper.deleteById(1);
      session.commit();
    } catch (Exception e) {
      e.printStackTrace();
    }finally{
      session.close();
    }
  }
```

```
/**
 *
 * 模糊查询
 */
public void listStudentsByName(){
  SqlSession session=null;
  try {
    session=factory.openSession();
  StudentMapper mapper=session.getMapper(StudentMapper.class);

    List<Student>students=mapper.listStudentsByName("in");
    if(students.size()>0){
      for(Student s: students){
        System.out.println(s.getName());
      }
    }
    session.commit();
  }catch (Exception e) {
    e.printStackTrace();
  }finally{
    session.close();
  }
}
/**
 *
 * 多条件查询
 */
public void listStudentsByIdAndName(){
    SqlSession session=null;
    try {
      session=factory.openSession();
      StudentMapper mapper=session.getMapper(StudentMapper.class);
      Map<String,Object>params=new HashMap<>();
      params.put("id", 2);
      params.put("name", "in");
      List<Student>students=mapper.listStudentsByIdAndName(params);
      if(students.size()>0){
          for(Student s: students){
            System.out.println(s.getName()+" "+s.getDescription());
          }
      }
      session.commit();
    } catch (Exception e) {
      e.printStackTrace();
    }finally{
      session.close();
```

```
        }
    }

    public static void main(String[]args){

        StudentTest testCase=new StudentTest();

        //testCase.insert();
        testCase.getById();
        //testCase.update();
        //testCase.delete();
        testCase.listStudentsByName();
        testCase.listStudentsByIdAndName();
        testCase.findAll();

    }

}
```

这里增加了两个测试方法 listStudentsByName()和 listStudentsByIdAndName()。

7.4 项目案例

7.4.1 学习目标

通过本章的学习，了解 MyBatis 的工作原理，掌握 MyBatis 的核心配置文件 mybatis-config 以及 MyBatis 中关于 SQL 映射与封装的 mapper 接口和 mpping XML 配置文件，以及如何使用 MyBatis 完成基于关系型数据库 SQL 的增、删、改、查操作。

7.4.2 案例描述

本案例以 eGov 的已经开发的 Spring MVC 高级部分中未实现的发布新闻功能为例，完成基于 MyBatis 的新闻发布。

7.4.3 案例要点

本案例需要导入所依赖的 MyBatis 包，编写 MyBatis 的数据库连接配置文件 mybatis-config.xml，并编写保存 SQL 语句的 news-mapping.xml，定义保存新闻信息的 Java 接口 NewsMapper.java，还要编写获取 MyBatis SessionFactory 的 MyBatisSessionFactoryUtils.java，并被 IndexController.java 中用于保存新闻的 addNews()方法调用。

7.4.4 案例实施

(1) 添加 MyBatis 库包，从 MyBatis 官方网站下载，导入 mybatis-3.4.2.jar 以及 lib 目录下所有的包，要求先在 MyEclipse 中建立 User Library (用户自定义的库)，如图 7-11 所示。

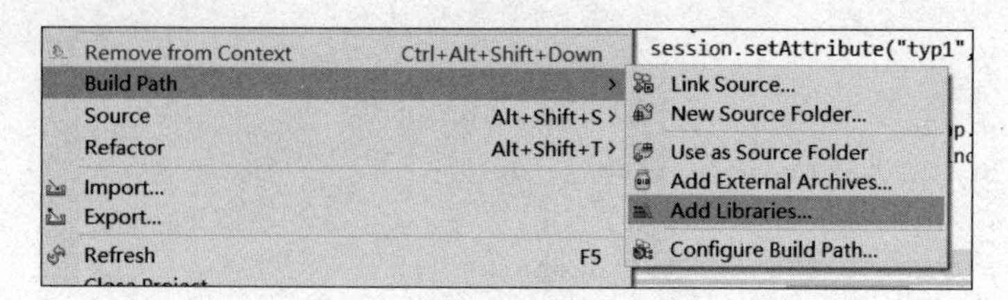

图 7-11 建立用户自定义的库界面 1

单击 Add Libraries,出现图 7-12。

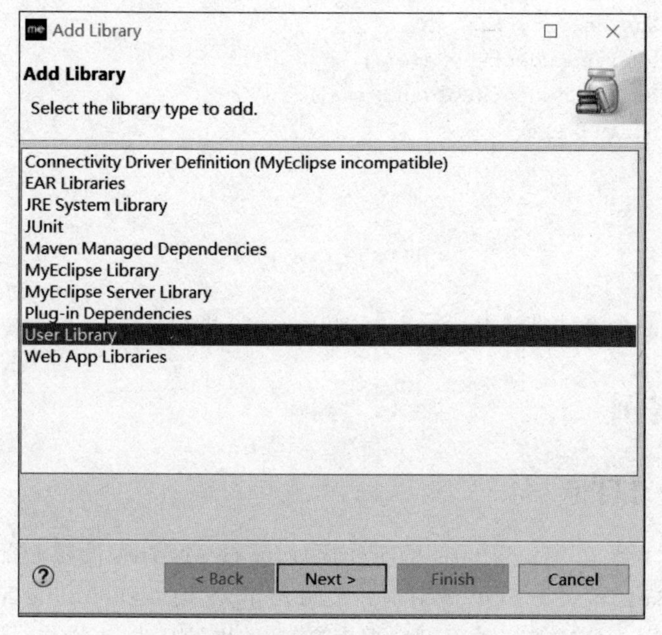

图 7-12 建立用户自定义的库界面 2

选择 User Library,单击 Next 按钮,出现图 7-13。

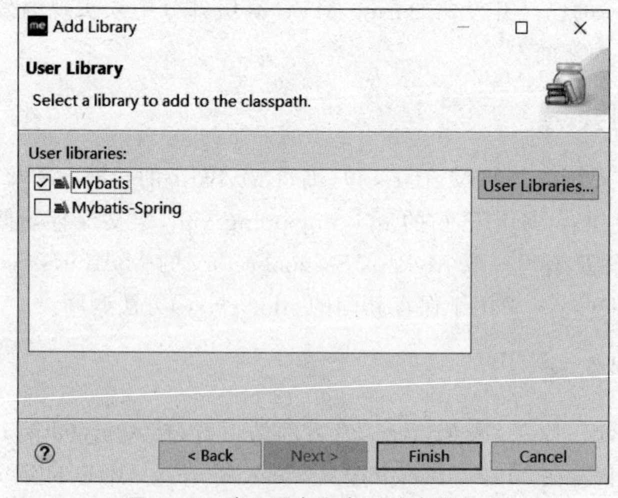

图 7-13 建立用户自定义的库界面 3

勾选 MyBatis 库(如果用户没有该库,则需要新建),如图 7-14 所示,然后单击 Finish 按钮。

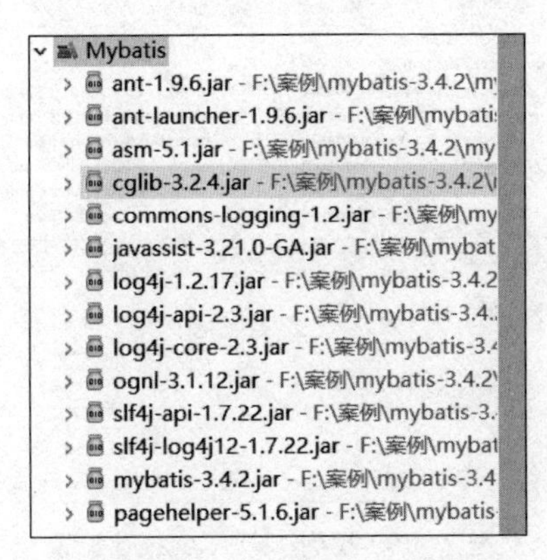

图 7-14 建立用户自定义的库界面 4

(2) 在 MyEclipse 中注册 MyBatis 配置文件需要的 DTD。

解压缩 mybatis-3.4.2.jar 包到指定目录,找到路径\mybatis-3.4.2\org\apache\ibatis\builder\xml,如图 7-15 所示。

新加卷 (F:) › 案例 › mybatis-3.4.2 › mybatis-3.4.2 › mybatis-3.4.2 › org › apache › ibatis › builder › xml 搜索"xm

名称	修改日期	类型	大小
mybatis-3-config.dtd	2017/1/19 11:37	XML Document ...	3 KB
mybatis-3-mapper.dtd	2017/1/19 11:37	XML Document ...	8 KB
XMLConfigBuilder.class	2017/1/19 11:37	CLASS 文件	19 KB
XMLIncludeTransformer.class	2017/1/19 11:37	CLASS 文件	6 KB
XMLMapperBuilder.class	2017/1/19 11:37	CLASS 文件	21 KB
XMLMapperEntityResolver.class	2017/1/19 11:37	CLASS 文件	3 KB
XMLStatementBuilder.class	2017/1/19 11:37	CLASS 文件	11 KB

图 7-15 解压缩 mybatis-3.4.2.jar 包到指定目录

复制两个 DTD 文件到指定目录,如图 7-16 所示。

图 7-16 复制两个 DTD 文件到指定目录

返回 MyEclipse,单击 Window,选择 Preferences。

找到 XML Catalog 配置,如图 7-17 所示。

单击 Add 按钮,添加 mybatis-3-config.dtd,如图 7-18 所示。

单击 OK 按钮。再次单击 Add 按钮,添加 mybatis-3-mapper.dtd,如图 7-19 所示。

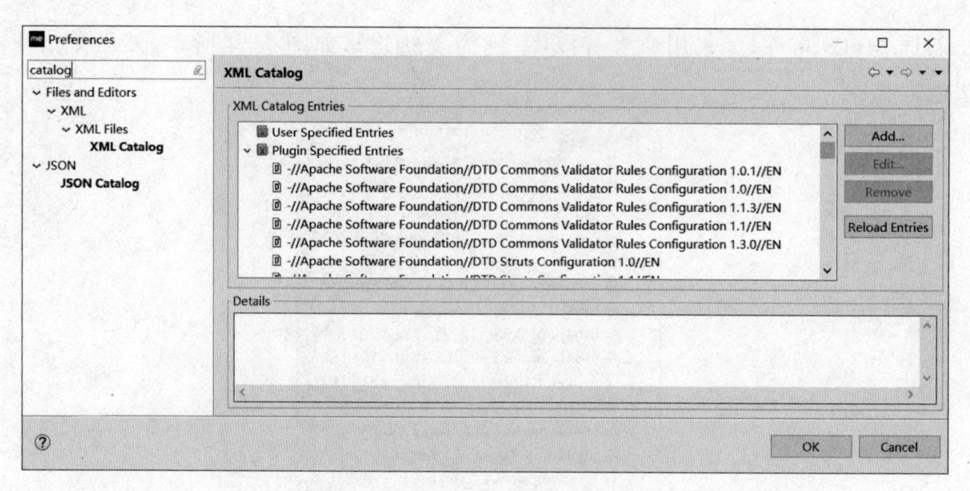

图 7-17　找到 XML Catalog 配置

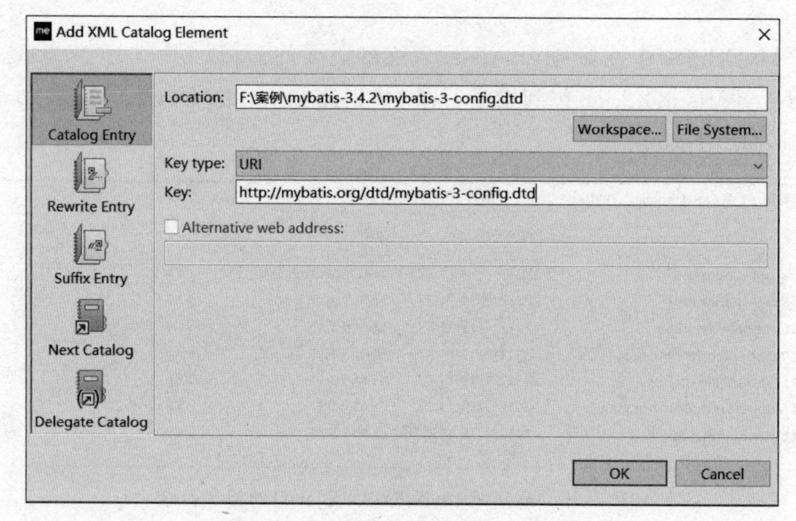

图 7-18　添加 mybatis-3-config.dtd

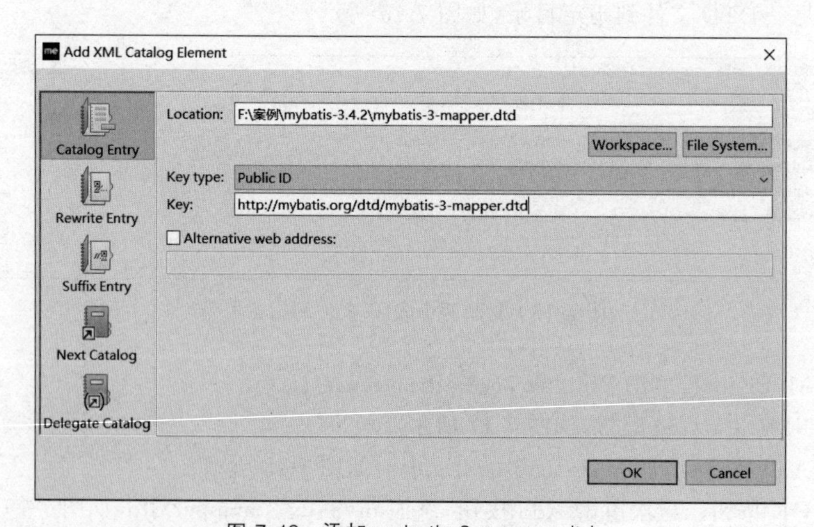

图 7-19　添加 mybatis-3-mapper.dtd

单击 OK 按钮，出现图 7-20。

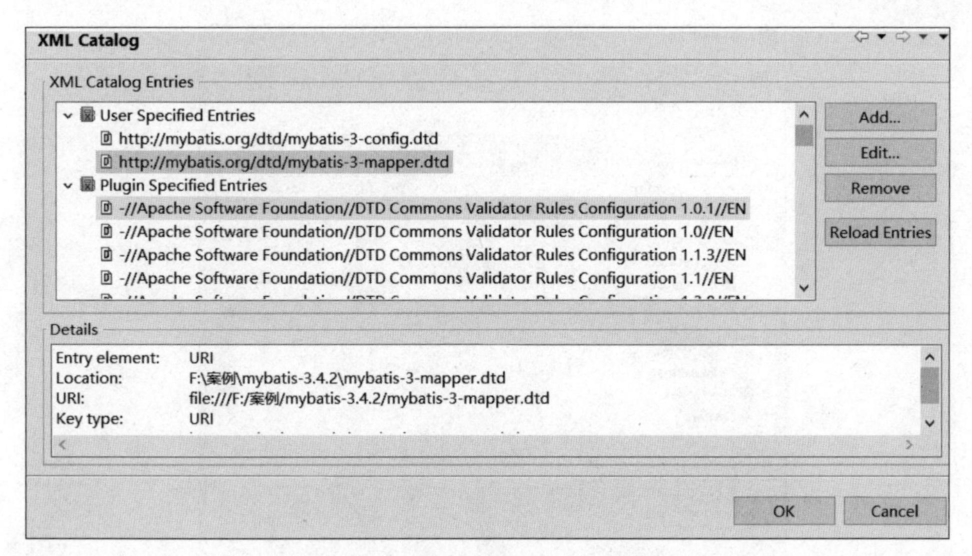

图 7-20　完成添加 DTD 文件

最后单击 OK 按钮。

(3) 编写 MyBatis SqlSessionFactory 配置文件 mybatis-config。

右击 src/config 目录，选择 News→Others 命令，出现图 7-21。

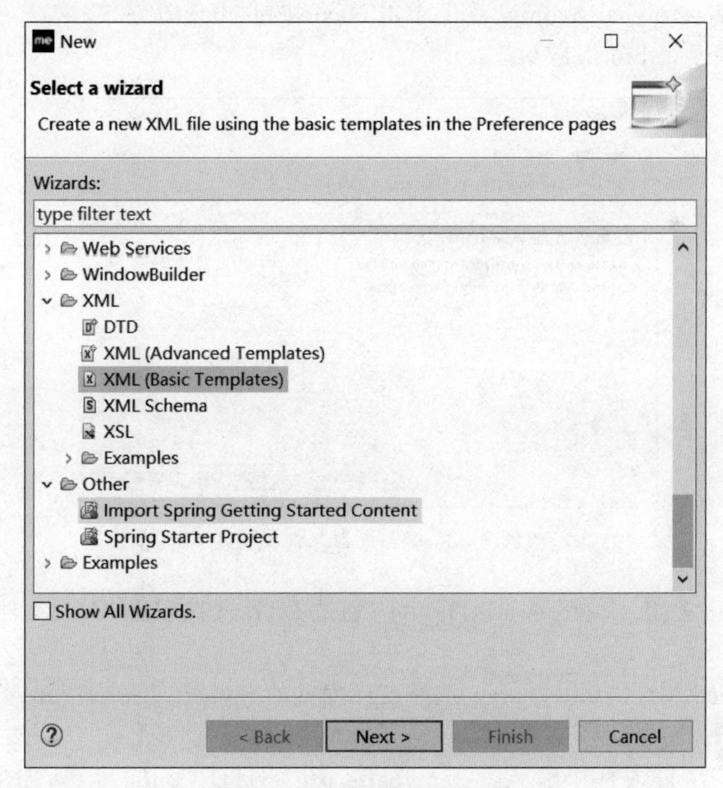

图 7-21　编写配置文件界面 1

选择 XML→XML(Basic Templates)命令,出现图 7-22。

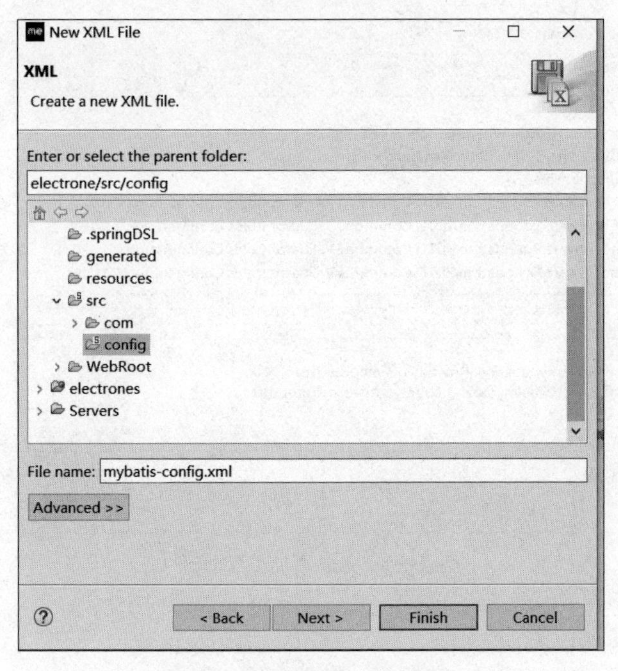

图 7-22 编写配置文件界面 2

输入文件名称 mybatis-config.xml,单击 Next 按钮,出现图 7-23。

注意:不要单击 Finish 按钮。

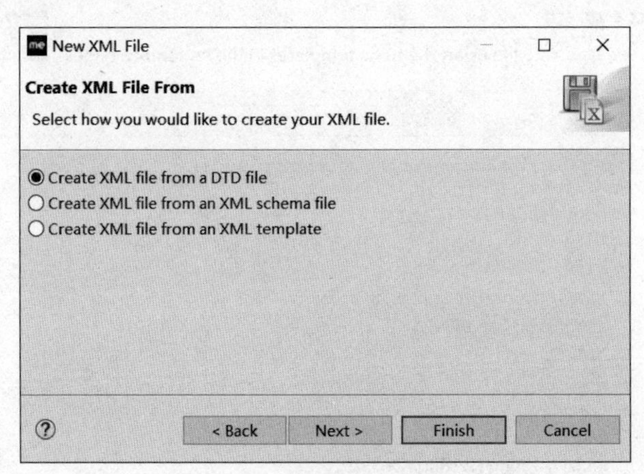

图 7-23 编写配置文件界面 3

勾选 Create XML file from a DTD file (通过 DTD 文件创建 XML 文件),单击 Next 按钮,出现图 7-24。

选中 Select XML Catalog entry,选择已经配置的 mybatis-3-config.dtd,单击 Next 按钮,出现图 7-25。

在 Public ID 输入框中输入:-//mybatis.org//DTD Config 3.0//EN,单击 Finish 按钮。

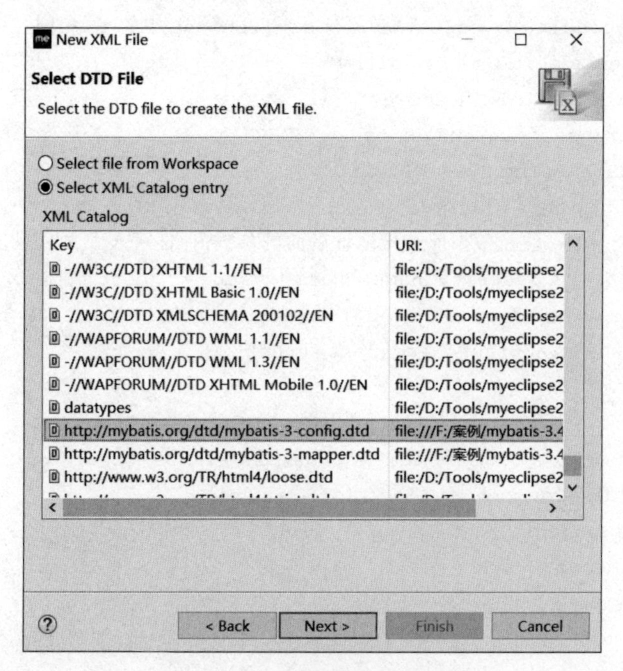

图 7-24 编写配置文件界面 4

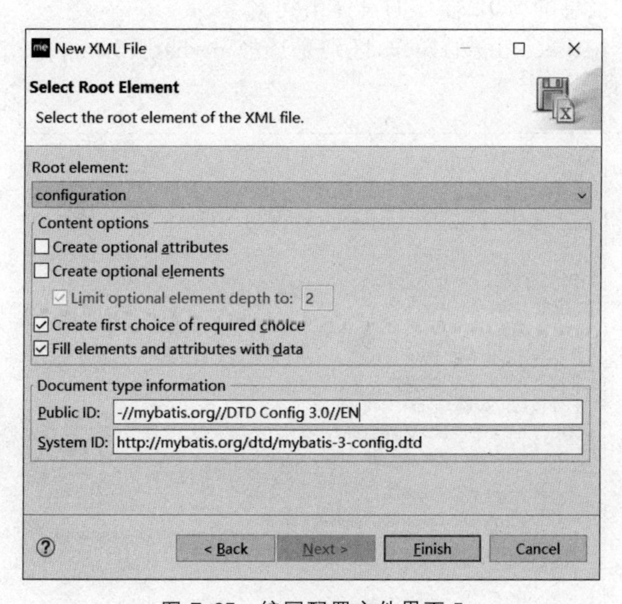

图 7-25 编写配置文件界面 5

（4）编写 mybatis-config.xml。

```
<?xml version="1.0" encoding="UTF-8"?>
<!DOCTYPE configuration PUBLIC "-//mybatis.org//DTD Config 3.0//EN" "http://
mybatis.org/dtd/mybatis-3-config.dtd" >
<configuration>
    <!--配置 SqlFactory 环境以及数据库连接信息,事务处理信息 -->
```

```
<!--注意 default="mysqlDev"必须与 environment 中的 id 相同 -->
<environments default="mysqlDev">
    <environment id="mysqlDev">
        <transactionManager type="JDBC" />
        <dataSource type="POOLED">
    <property name="driver" value="com.mysql.jdbc.Driver" />
            <!--注意数据库链接地址后面的参数,保证写入中文时不会出现乱码 -->
                < property name = "url" value = "jdbc: mysql://localhost: 3306/
                electrones?useUnicode=true&amp;characterEncoding=utf-8
                &amp;autoReconnect=true" />

                <property name="username" value="root" />
                <property name="password" value="root" />
        </dataSource>
    </environment>
</environments>
<!--添加 mapper.xml -->
</configuration>
```

(5) 新建 com/ascent/mybatis/mapping 包,在该目录下新建 NewsMapping.xml 文件,用来保存对新闻信息的 SQL 操作的具体语句。

操作同创建 mybatis-config 的操作,区别是选择 mybatis-3-mapper.dtd,以及 public id 不同,如图 7-26 所示。

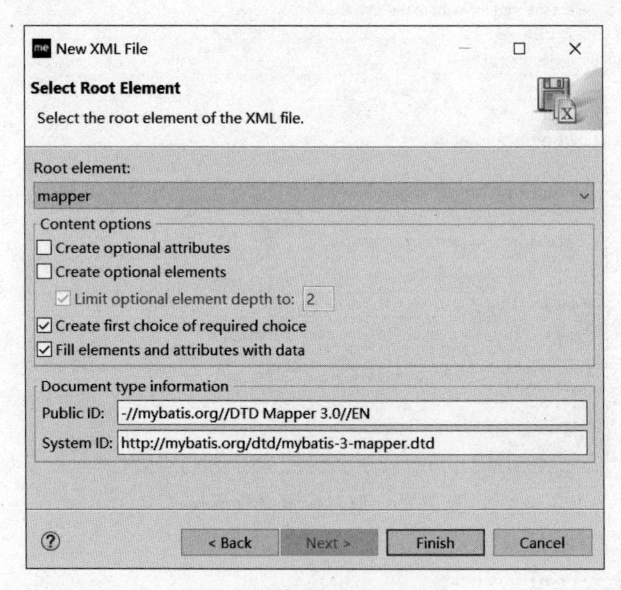

图 7-26 编写 mapper 文件

最后单击 Finish 按钮。

编写 NewsMapping.xml:

```
<?xml version="1.0" encoding="UTF-8"?>
```

```
<!DOCTYPE mapper PUBLIC "-//mybatis.org//DTD Mapper 3.0//EN" "http://mybatis.
org/dtd/mybatis-3-mapper.dtd" >
<mapper namespace="com.ascent.mybatis.mapper.NewsMapper">
  <!--定义查询结果封装实体类 -->
  <resultMap id="BaseResultMap" type="com.ascent.po.News">
    <id column="id"jdbcType="INTEGER" property="id" />
    <result column="title"jdbcType="VARCHAR" property="title" />
    <result column="author"jdbcType="VARCHAR" property="author" />
    <result column="deptid" jdbcType="INTEGER" property="deptid" />
    <result column="content"jdbcType="VARCHAR" property="content" />
    <result column="type"jdbcType="INTEGER" property="type" />
    <result column="checkopinion" jdbcType="VARCHAR" property="checkopinion" />
    <result column="checkopinion2" jdbcType="VARCHAR" property="checkopinion2" />
    <result column="checkstatus" jdbcType="INTEGER" property="checkstatus" />
    <result column="crosscolumn" jdbcType="INTEGER" property="crosscolumn" />
    <result column="crossstatus" jdbcType="INTEGER" property="crossstatus" />
    <result column="picturepath" jdbcType="VARCHAR" property="picturepath" />
    <result column="publishtime" jdbcType="DATE" property="publishtime" />
    <result column="crosspubtime" jdbcType="DATE" property="crosspubtime" />
    <result column="preface"jdbcType="INTEGER" property="preface" />
    <result column="status"jdbcType="INTEGER" property="status" />
    <result column="userid"jdbcType="INTEGER" property="userid" />
  </resultMap>
  <!--将多条 SQL 语句使用的字段定义为一个可以公用的字段集合 -->
  <sql id="Base_Column_List">
    id, title, author, deptid, content, type, checkopinion, checkopinion2,
    checkstatus, crosscolumn, crossstatus, picturepath, publishtime, crosspubtime,
    preface, status, userid
  </sql>
  <!--定义保存新闻信息的 SQL 操作 -->
  <!--parameterType 是指输入参数类型为 com.ascnet.po.News -->
  <!--keyColumn 为数据库 News 表的数据库主键字段 -->
  <!--keyProperty 为对应于 News 表的 News.java 类中用来标识对象唯一性的属性字段 id -->
    <insert id="insert"parameterType="com.ascent.po.News" useGeneratedKeys=
    "true" keyColumn="id" keyProperty="id">
    insert into news(title, author,
      deptid, content, type,
      checkopinion, checkopinion2, checkstatus,
      crosscolumn, crossstatus, picturepath,
      publishtime, crosspubtime, preface,
      status,userid)
    values(#{title,jdbcType=VARCHAR}, #{author,jdbcType=VARCHAR},
       #{deptid, jdbcType = INTEGER}, #{content, jdbcType = VARCHAR}, #{type,
       jdbcType=INTEGER},
      #{checkopinion,jdbcType=VARCHAR}, #{checkopinion2,jdbcType=VARCHAR}, #
```

```
    {checkstatus,jdbcType=INTEGER},
     #{crosscolumn,jdbcType=INTEGER}, #{crossstatus,jdbcType=INTEGER}, #
    {picturepath,jdbcType=VARCHAR},
    #{publishtime,jdbcType=DATE}, #{crosspubtime,jdbcType=DATE}, #{preface,
    jdbcType=INTEGER},
    #{status,jdbcType=INTEGER}, #{userid,jdbcType=INTEGER})
  </insert>
</mapper>
```

本案例主要关注<insert>SQL 操作。

(6) 修改 mybatis-config.xml,将 NewsMapping.xml 加入 context 中。

```
<!--增加 NewsMapping.xml -->
    <mappers>
        <mapper resource="com/ascent/mybatis/mapping/NewsMapping.xml" />
    </mappers>
```

(7) 编写 MyBatis Java 接口 com.ascent.mybatis.mapper.NewsMapper.java,定义 insert 操作。

```
package com.ascent.mybatis.mapper;

import com.ascent.po.News;

public interface NewsMapper {

    public int insert(News news);
}
```

方法 insert 对应 NewsMapper.xml 中<insert>标签的 id="insert"。

(8) 编写 MybatisSqlSessionFactoryUtil.java。

```
package com.ascent.mybatis.utils;

import java.io.IOException;
import java.io.InputStream;

import org.apache.ibatis.io.Resources;
import org.apache.ibatis.session.SqlSession;
import org.apache.ibatis.session.SqlSessionFactory;
import org.apache.ibatis.session.SqlSessionFactoryBuilder;

public class MybatisSqlSessionFactoryUtil{
    //定义配置 mybatis-config.xml 文件的路径
    private static final StringinputResource="config/mybatis-config.xml";
    //定义 SqlSessionFactory 工厂
    private static SqlSessionFactory sqlSessionFactory=null;
```

```
    static {

        try {
            InputStream inputStream=Resources.getResourceAsStream
                (inputResource);
            sqlSessionFactory=new SqlSessionFactoryBuilder().build
                (inputStream);
        } catch (IOException e) {
            //TODO Auto-generated catch block
            e.printStackTrace();
        }

    }

    public static SqlSession getSqlSession(){
        if(null!=sqlSessionFactory)
            return sqlSessionFactory.openSession();
        else
            return null;
    }

}
```

(9) 编写 INewsService.java 接口,定义保存新闻的抽象方法。

```
package com.ascent.service;

import com.ascent.po.News;

public interface INewsService {

    public boolean addNews(News news);

}
```

(10) 编写 INewsServiceImpl.java,实现 INewsService.java 中抽象的 addNews()方法,该方法使用 MyBatisSqlSessionFactoryUtil 类完成 MyBatis 的操作。

```
package com.ascent.service.impl;

import java.util.Date;

import org.apache.ibatis.session.SqlSession;

import com.ascent.mybatis.mapper.NewsMapper;
import com.ascent.mybatis.utils.MybatisSqlSessionFactoryUtil;
import com.ascent.po.News;
```

```
import com.ascent.service.INewsService;

public class INewsServiceImpl implements INewsService {

    /**
     * 定义保存新闻的业务逻辑
     * 如果类型为头版头条新闻,则设置不跨栏目
     * 假设用户 id 为 9,部门 id 为 1
     */
    @Override
    public boolean addNews(News news) {

        SqlSession sqlSession=MybatisSqlSessionFactoryUtil.getSqlSession();

        NewsMapper newsMapper=sqlSession.getMapper(NewsMapper.class);

        Integer deptid=1;
        Integer usrid=9;
        boolean isSuccess=false;

        try {
            //创建新闻对象
            Date dt=new Date();
            news.setAuthor("测试 3");
            news.setPublishtime(dt);
            news.setCheckstatus(0);
            if ("1".equals(news.getType())) {
                news.setCrossstatus(1);
            }
            news.setStatus(0);
            //目前静态设置,实际项目需要从登录用户信息里获取 deptid 和 usrid
            news.setUserid(usrid);
            news.setDeptid(deptid);

            newsMapper.insert(news);

            sqlSession.commit();          //提交事务

            isSuccess=true;

        } catch (NumberFormatException e) {
            //TODO Auto-generated catch block
            e.printStackTrace();
            isSuccess=false;
```

```
            sqlSession.rollback();          //回滚事务
        }

        return isSuccess;

    }

}
```

(11) 修改 IndexController.java 中的 addNews()方法,增加调用 NewsMapper 的代码,完成保存新闻数据的功能。

保存新闻时需要保存发布新闻者的 id、部门 id、本案例选择部门 id 为 1,用户 id 为 9。

```
IndexContrller.java ->addNews
@RequestMapping(value="addNewsnewsAction")
    public  ModelAndView  addNews  (@ Validated  News  news, BindingResult
bindingResult){

        ModelAndView mv=new ModelAndView();
        if(bindingResult.hasErrors()){

            //mv=new ModelAndView("jsp/issue","typ",news.getType());
            mv.addObject("typ",news.getType());
            mv.setViewName("jsp/issue");
            return mv;
        }
        //调用相关 Service,将数据保存到数据库中
        //由于 MyBatis 没有和 Spring 整合,所以直接创建新闻业务对象
        INewsService newsService=new INewsServiceImpl();
        boolean isSuccess=newsService.addNews(news);
        if(isSuccess){
        //保存成功,回到管理界面 selectIssue.jsp
        mv.setViewName("jsp/selectIssue");
    }
    else{
        //保存失败,退回到新闻录入界面
        mv.addObject("typ",news.getType());
        mv.setViewName("jsp/issue");
    }
    return mv;
}
```

(12) 右击项目,从弹出的快捷菜单中选择 Preferences→Web Deployment Assembly 命令,将 MyBatis 库加入部署环境中,如图 7-27 所示。

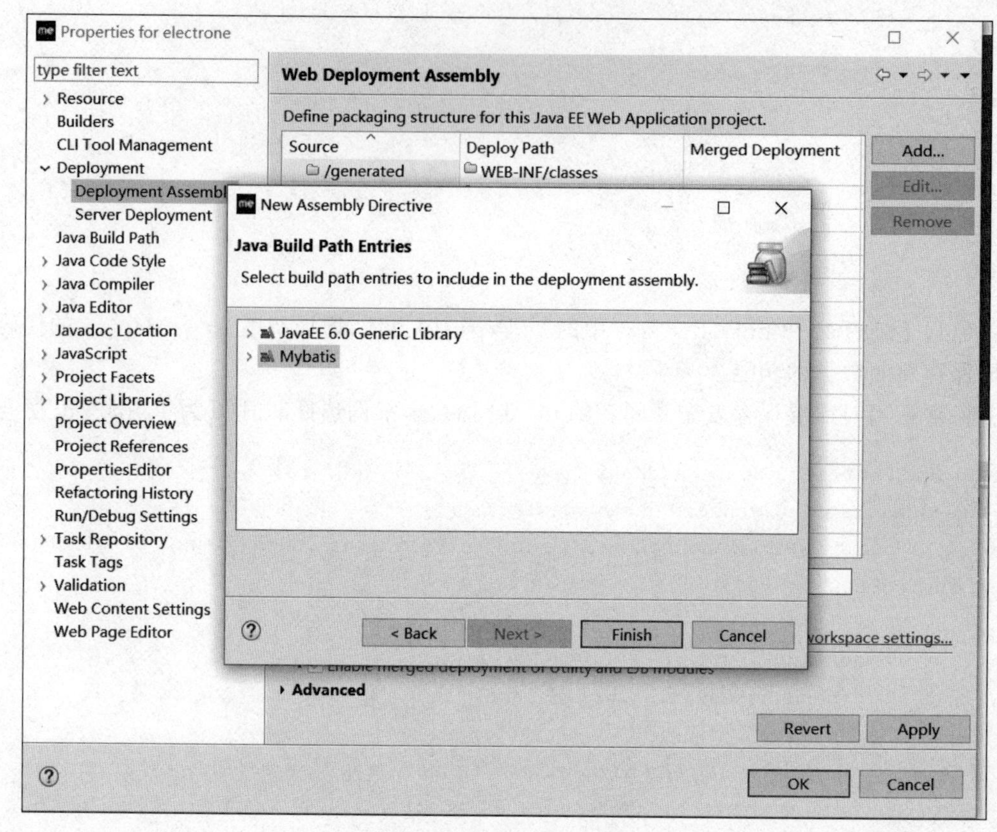

图 7-27　将 MyBatis 库加入部署环境中

(13) 部署项目，进入 selectIssue.jsp，单击其中一个链接，进入录入新闻界面，输入相关内容，单击“提交”按钮，如图 7-28 所示。

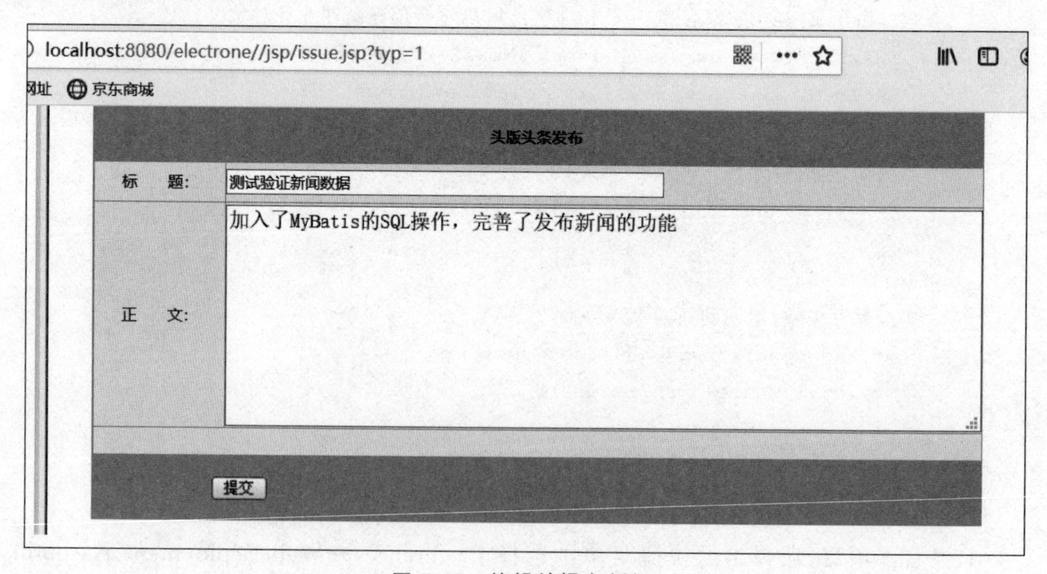

图 7-28　编辑并提交新闻

观察数据库 News 表中的记录，如图 7-29 所示。

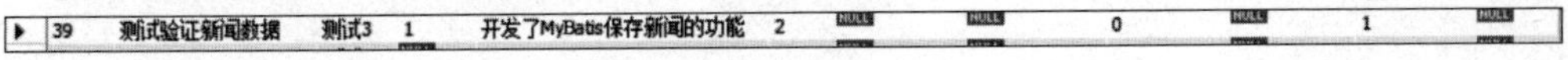

39 测试验证新闻数据 测试3 1 开发了MyBatis保存新闻的功能 2 NULL NULL 0 NULL 1 NULL

图 7-29 数据库结果

至此，使用 MyBatis 完善发布新闻功能开发完毕。

7.4.5 特别提示

MyEclipse 没有提供 MyBatis 集成功能，为了能正常编写 mybatis-config 和 mybatis-mapper 文件，需要在 MyEclipse 中手动添加相应的 DTD 文件，并且选择 URI 类型，不是 public 类型。

MyEclipse 可以使用 MyBatis Generator 插件，该插件不能覆盖上次配置，只是追加，建议初学者在熟悉 MyBatis 之后使用该插件进行创建。

保存发布新闻需要一些逻辑判断，使用 INewsService.java 的实现类封装业务逻辑，同时需要登录用户的部门编号、姓名以及用户编号。本案例采用假设方法设置了 3 个值：1、“测试 3”、9。

本案例没有结合 Spring 框架，因此在 INewsService 中可以直接调用 SqlSession，之后调用 NewsMapper，完成了保存数据的功能。

7.4.6 拓展与提高

根据本案例，请完成一般用户浏览的头版头条新闻和综合新闻查询功能，使用 MyBatis 框架替代 Spring JdbcTemplate 功能。

习题

1. MyBatis 框架总体上可分为哪 3 层？分别描述一下。
2. MyBatis 的主要工作流程是什么？
3. 描述 MyBatis 对象的作用域与生命周期。
4. MyBatis XML 配置文件包含哪些常用设置和属性？
5. MyBatis SQL 映射 XML 文件有哪些基本元素？
6. MyBatis 框架如何实现数据表的增、删、改、查操作？

第8章 MyBatis实体关系映射

学习目的与学习要求

学习目的：了解 MyBatis 对表关系的映射，掌握多对一/一对多、一对一、多对多关系的映射。

学习要求：重点掌握一对一、多对一/一对多和多对多的映射关系，包括表的设计及 MyBatis 对关系表的映射过程和配置。

本章主要内容

本章重点介绍 MyBatis 框架的关系映射，其中包括多对一/一对多、一对一和多对多 3 种关系映射，每种关系都配有实例，掌握每个关系表对应的持久化类和映射文件的设计和开发。

前面介绍的是单表映射的实例，现实中我们可能遇到更多的情况是联表操作。表和表之间通过主键/外键建立联系。接下来介绍如何处理联表关系。

8.1 一对一关系

下面以 customer(客户)和 address(地址)为例，介绍一对一关系的处理。

1. 数据库脚本语句

```
USE `test`;

/* Table structure for table `address` */

DROP TABLE IF EXISTS `address`;

CREATE TABLE `address`(
```

```
  `id` int NOT NULL AUTO_INCREMENT,
  `description`varchar(255) DEFAULT NULL,
  PRIMARY KEY (`id`)
) ENGINE=InnoDB AUTO_INCREMENT=14 DEFAULT CHARSET=latin1;

/* Data for the table `address` */

insert into `address`(`id`,`description`) values (1,'Lixin address'),(2,'Linda
      address'),(3,'Linda address'),(4,'Linda address'),(5,'Linda address'),
      (6,'Kevin address'),(7,'Linda address'),(13,NULL);

/* Table structure for table `customer` */

DROP TABLE IF EXISTS `customer`;

CREATE TABLE `customer`(
  `id` int NOT NULL AUTO_INCREMENT,
  `name`varchar(255) DEFAULT NULL,
  `aid` int DEFAULT NULL,
  PRIMARY KEY (`id`),
  KEY `fk_4` (`aid`),
  CONSTRAINT `fk_4` FOREIGN KEY (`aid`) REFERENCES `address` (`id`)
) ENGINE=InnoDB AUTO_INCREMENT=13 DEFAULT CHARSET=latin1;

/* Data for the table `customer` */

insert into `customer`(`id`,`name`,`aid`) values (5,'Lixin',1),(6,'Peter',
      NULL),(7,'Peter',NULL),(8,'Peter',NULL),(9,'Peter',NULL),(10,'Peter',
      NULL),(11,'Peter',NULL),(12,'Peter',NULL);
```

2. CustomerMapper.xml 文件

```
<?xml version="1.0" encoding="UTF-8"?>
<!DOCTYPE mapper PUBLIC "-//mybatis.org//DTD Mapper 3.0//EN"
"http://mybatis.org/dtd/mybatis-3-mapper.dtd">
<mapper namespace="com.ascent.mapper.CustomerMapper">
  <resultMap type="Customer" id="customerResultMap">
    <id column="id" property="id" />
    <result column="name" property="name" />
    <!--一对一关联映射:association -->
    <association property="address" column="aid" javaType="Address"
       resultMap="addressResultMap" />
  </resultMap>
  <!--定义 Java Bean 的属性与数据库的列之间的映射 -->
  <resultMap type="Address" id="addressResultMap" >
    <id property="id" column="id"/>
```

```
        <result property="description" column="description" />
    </resultMap>

    <!--查询记录 -->
    <select id="selectCustomerById" parameterType="int" resultMap=
        "customerResultMap">
        select * from customer c, address a where c.aid=a.id and c.id=#{id}
    </select>

</mapper>
```

持久化对象之间一对一的关联关系是通过联合(association)元素定义的,详细介绍可以参考第7章。

```
<resultMap type="Customer" id="customerResultMap">
    <id column="id" property="id" />
    <result column="name" property="name" />
    <!--一对一关联映射: association -->
    <association property="address" column="aid" javaType="Address"
        resultMap="addressResultMap" />
</resultMap>
<!--定义 Java Bean 的属性与数据库的列之间的映射 -->
<resultMap type="Address" id="addressResultMap" >
    <id property="id" column="id"/>
    <result property="description" column="description" />
</resultMap>
```

在这里需要一对一关联查询映射的是地址(Address)信息,使用 association 将地址信息映射到客户(Customer)对象的地址属性中。

association:表示进行关联查询单条记录。

property:表示关联查询的结果存储在 address 属性中。

column:数据表的列名。

javaType:表示关联查询的结果类型。

这里使用一个 resultMap 元素映射联合。这使 addressResultMap 可重复使用。

对应的查询语句如下:

```
<!--查询记录 -->
<select id="selectCustomerById" parameterType="int" resultMap="customerResultMap">
    select * from customer c, address a where c.aid=a.id and c.id=#{id}
</select>
```

3. PO类

```
Customer.java

package com.ascent.po;
```

```
public class Customer {
    private int id;
    private String name;
    private Address address;

    public int getId() {
      return id;
    }
    public void setId(int id) {
      this.id=id;
    }
    public String getName() {
      return name;
    }
    public void setName(String name) {
      this.name=name;
    }
    public Address getAddress() {
        return address;
    }
    public void setAddress(Address address) {
      this.address=address;
    }
}
```

注意：Customer 中添加了 private Address address 语句。

```
Address.java

package com.ascent.po;

public class Address {
    private int id;
    private String description;            //描述

    public int getId() {
      return id;
    }
    public void setId(int id) {
      this.id=id;
    }
    public String getDescription() {
      return description;
    }
    public void setDescription(String description) {
      this.description=description;
```

```
    }
}
```

4. Mapper 类

```
package com.ascent.mapper;
import com.ascent.po.*;

public interface CustomerMapper {

    public Customer selectCustomerById(Integer id);

}
```

5. Test 类

```
package com.ascent.test;

import java.util.List;
import org.apache.ibatis.session.SqlSession;
import org.apache.ibatis.session.SqlSessionFactory;
import com.ascent.po.*;
import com.ascent.mapper.CustomerMapper;
import com.ascent.util.SqlSessionFactoryUtil;

public class CustomerTest {
  private SqlSessionFactory factory=SqlSessionFactoryUtil.getSqlSessionFactory();

  /**
   *查询单条记录
   */

  public void selectCustomerById(){
    SqlSession session=null;
    try{
      session=factory.openSession();
      CustomerMapper mapper=session.getMapper(com.ascent.mapper.
          CustomerMapper.class);
      Customer c=mapper.selectCustomerById(5);
      System.out.println(c.getName()+" "+c.getAddress().getDescription());
    } catch (Exception e) {
       e.printStackTrace();
    }finally{
      session.close();
  }
}
```

```
  public static void main(String[]args){
    CustomerTest testCase=new CustomerTest();
    testCase.selectCustomerById();
  }
}
```

8.2 一对多关系

下面以 department(部门)和 employee(员工)为例,介绍一对多关系的处理。

1. 数据库脚本语句

```
USE `test`;

/* Table structure for table `department` */

DROP TABLE IF EXISTS `department`;

CREATE TABLE `department`(
  `id` int NOT NULL AUTO_INCREMENT,
  `name`varchar(255) DEFAULT NULL,
  `description`varchar(255) DEFAULT NULL,
  PRIMARY KEY (`id`)
) ENGINE=InnoDB AUTO_INCREMENT=3 DEFAULT CHARSET=latin1;

/* Data for the table `department` */

insert into `department`(`id`,`name`,`description`) values (1,'IT','computer'),
                      (2,'Marketing','sales');

/* Table structure for table `employee` */

DROP TABLE IF EXISTS `employee`;

CREATE TABLE `employee`(
  `id` int NOT NULL AUTO_INCREMENT,
  `name`varchar(255) DEFAULT NULL,
  `did` int DEFAULT NULL,
  PRIMARY KEY (`id`),
  KEY `FK2` (`did`),
  CONSTRAINT `FK2` FOREIGN KEY (`did`) REFERENCES `department` (`id`)
```

```
) ENGINE=InnoDB AUTO_INCREMENT=6 DEFAULT CHARSET=latin1;

/* Data for the table `employee` */

insert into `employee`(`id`,`name`,`did`) values (1,'Lixin',1),(2,'Admin',1),
        (3,'Linda',2),(4,'Gary',1),(5,'Henry',2);
```

2. DepartmentMapper.xml 文件

```
<?xml version="1.0" encoding="UTF-8"?>
<!DOCTYPE mapper PUBLIC "-//mybatis.org//DTD Mapper 3.0//EN"
"http://mybatis.org/dtd/mybatis-3-mapper.dtd">
<mapper namespace="com.ascent.mapper.DepartmentMapper">
  <!--定义 Java Bean 的属性与数据库的列之间的映射 -->
  <resultMap type="Department" id="departmentResultMap">
    <id column="did" property="id" />
    <result column="dname" property="name" />
    <result column="description" property="description" />
    <!---一对多的关系 -->
         <!--property: 指的是集合属性的值, ofType: 指的是集合中元素的类型 -->
         <collection property="employees"ofType="Employee">
             <id column="eid" property="id" />
             <result column="ename" property="name" />
         </collection>
    </resultMap>

  <!--查询记录 -->
  <select id="listDepartment" resultMap="departmentResultMap">
    select d.*, e.*, d.id 'did', e.id 'eid', d.name 'dname', e.name 'ename' from
           department d left join employee e on d.id=e.did
  </select>

</mapper>
```

处理“一对多”关系要使用聚集(collection),详细介绍可以参考第 7 章。

```
<!--property: 指的是集合属性的值, ofType: 指的是集合中元素的类型 -->
          <collection property="employees"ofType="Employee">
              <id column="eid" property="id" />
              <result column="ename" property="name" />
          </collection>
>
```

property: 指的是集合属性的值。

ofType: 指的是集合中元素的类型,用来区别 JavaBean(或者字段)的属性类型和集合所包括的类型。

对应的查询语句如下:

```
<select id="listDepartment" resultMap="departmentResultMap">
  select d.*, e.*, d.id 'did', e.id 'eid', d.name 'dname', e.name 'ename' from
        department d left join employee e on d.id=e.did
</select>
```

这里已经连接了部门和员工表,并且让列名标签也做了明确的处理,以方便映射。

3. PO 类

Department.java 类:

```
package com.ascent.po;

import java.util.List;

public class Department {
    private int id;
    private String name;
    private String description;
    private List<Employee>employees;

    public int getId() {
      return id;
    }
    public void setId(int id) {
      this.id=id;
    }
    public String getName() {
        return name;
    }
    public void setName(String name) {
        this.name=name;
    }
    public String getDescription() {
      return description;
    }
    public void setDescription(String description) {
      this.description=description;
    }
    public List<Employee>getEmployees() {
        return employees;
    }
    public void setProducts(List<Employee>employees) {
        this.employees=employees;
    }
}
```

注意：Department 中添加了员工集合 List<Employee>employees。

Employee.java 如下：

```
package com.ascent.po;

public class Employee {
    private int id;
    private String name;

    public int getId() {
      return id;
    }
    public void setId(int id) {
      this.id=id;
    }
    public String getName() {
      return name;
    }
    public void setName(String name) {
      this.name=name;
    }

}
```

4. Mapper 类

```
package com.ascent.mapper;
import java.util.List;

import com.ascent.po.*;

public interface DepartmentMapper {

    public List<Department>listDepartment();

}
```

5. Test 类

```
package com.ascent.test;

import java.util.List;
import org.apache.ibatis.session.SqlSession;
import org.apache.ibatis.session.SqlSessionFactory;
import com.ascent.po.*;
import com.ascent.mapper.DepartmentMapper;
import com.ascent.util.SqlSessionFactoryUtil;
```

```
public class DepartmentTest {
  private SqlSessionFactory factory=SqlSessionFactoryUtil.getSqlSessionFactory();

  /**
   * 查询记录
   * /

  public void listDepartment(){
    SqlSession session=null;
    try{
      session=factory.openSession();
      DepartmentMapper mapper=session.getMapper(com.ascent.mapper.
          DepartmentMapper.class);
      List<Department>dl=mapper.listDepartment();
      for(Department d : dl){
        System.out.println("\t"+d.getName());
        List<Employee>es=d.getEmployees();
        for (Employee e : es){
          System.out.println("\t"+e.getName());
        }
      }
    } catch (Exception e) {
      e.printStackTrace();
    }finally{
      session.close();
    }
  }

  public static void main(String[]args){

    DepartmentTest testCase=new DepartmentTest();

    testCase.listDepartment();

  }

}
```

8.3 多对多关系

最后以 student(学生)和 course(课程)为例,介绍多对多关系。

一般来说,在数据库设计阶段,对于有多对多关系的两个表,我们会增加一个代表关系的中间表。这个例子在 student 和 course 之间建立了中间表 student_course。

1. 数据库脚本语句

```
USE `test`;

/* Table structure for table `course` */

DROP TABLE IF EXISTS `course`;

CREATE TABLE `course`(
  `id` int NOT NULL AUTO_INCREMENT,
  `name`varchar(255) DEFAULT NULL,
  `grade`varchar(255) DEFAULT NULL,
  PRIMARY KEY (`id`)
) ENGINE=InnoDB AUTO_INCREMENT=4 DEFAULT CHARSET=latin1;

/* Data for the table `course` */

insert into `course`(`id`,`name`,`grade`) values (1,'Java','90'),(2,'C','95'),
       (3,'Python','100');

/* Table structure for table `student` */

DROP TABLE IF EXISTS `student`;

CREATE TABLE `student`(
  `id` int NOT NULL AUTO_INCREMENT,
  `name`varchar(255) DEFAULT NULL,
  `age` int DEFAULT NULL,
  `description`varchar(255) CHARACTER SET latin1
       COLLATE latin1_swedish_ci DEFAULT NULL,
  PRIMARY KEY (`id`)
) ENGINE=InnoDB AUTO_INCREMENT=4 DEFAULT CHARSET=latin1;

/* Data for the table `student` */

insert into `student`(`id`,`name`,`age`,`description`) values (1,'Gary',25,
       'Software Engineer'),(2,'Linda',35,'Marketing'),(3,'Kevin',22,'Computer
       Science');

/* Table structure for table `student_course` */

DROP TABLE IF EXISTS `student_course`;

CREATE TABLE `student_course`(
  `id` int NOT NULL AUTO_INCREMENT,
```

```
  `sid` int DEFAULT NULL,
  `cid` int DEFAULT NULL,
  PRIMARY KEY (`id`),
  KEY `FK_1` (`sid`),
  KEY `Fk_3` (`cid`),
  CONSTRAINT `FK_1` FOREIGN KEY (`sid`) REFERENCES `student` (`id`),
  CONSTRAINT `Fk_3` FOREIGN KEY (`cid`) REFERENCES `course` (`id`)
) ENGINE=InnoDB AUTO_INCREMENT=6 DEFAULT CHARSET=latin1;

/* Data for the table `student_course` */

insert into `student_course`(`id`,`sid`,`cid`) values (1,1,1),(2,1,2),(3,2,1),
      (5,1,3);
```

2. CustomerMapper.xml 文件

```
<?xml version="1.0" encoding="UTF-8"?>
<!DOCTYPE mapper PUBLIC "-//mybatis.org//DTD Mapper 3.0//EN"
"http://mybatis.org/dtd/mybatis-3-mapper.dtd">
<mapper namespace="com.ascent.mapper.StudentAndCourseMapper">
     <resultMap id="studentCourseMap" type="Student">
       <id property="id" column="s_id"/>
       <result property="name" column="s_name"/>
       <result property="age" column="age"/>
       <result property="description" column="description"/>
       <!--多对多关联映射: collection -->
       <collection property="courses"ofType="Course">
            <id property="id" column="c_id"/>
            <result property="name" column="c_name"/>
            <result property="grade" column="grade"/>
       </collection>
     </resultMap>

     <resultMap id="courseStudentMap" type="Course">
       <id property="id" column="c_id"/>
       <result property="name" column="c_name"/>
       <result property="grade" column="grade"/>
       <!--多对多关联映射: collection -->
       <collection property="students"ofType="Student">
            <id property="id" column="s_id"/>
            <result property="name" column="s_name"/>
            <result property="age" column="age"/>
            <result property="description" column="description"/>
       </collection>
     </resultMap>
```

```
        <select id="getStudentCourse" parameterType="int" resultMap=
           "studentCourseMap" >
             select s.name 's_name',c.name 'c_name', c.grade
             from student s,course c,student_course sc
             where
             s.id=#{id}
             and
             s.id=sc.sid
             and
             sc.cid=c.id
        </select>

        <select id="getCourseStudent" parameterType="int" resultMap=
           "courseStudentMap" >
             select c.name 'c_name',s.name 's_name', s.description
             from student s,course c,student_course sc
             where
             c.id=#{id}
             and
             s.id=sc.sid
             and
             sc.cid=c.id
        </select>

    </mapper>
```

在 MyBaits 中,一对多查询是多对多查询的特例,1 : n 中使用一个 collectin 元素完成关联映射,而 n : n 中可以使用两个 collection。

```
<!--多对多关联映射: collection -->
      <collection property="courses"ofType="Course">
          <id property="id" column="c_id"/>
          <result property="name" column="c_name"/>
          <result property="grade" column="grade"/>
      </collection>


<!--多对多关联映射: collection -->
      <collection property="students"ofType="Student">
          <id property="id" column="s_id"/>
          <result property="name" column="s_name"/>
          <result property="age" column="age"/>
          <result property="description" column="description"/>
      </collection>
```

对应的查询语句如下:

```
<select id="getStudentCourse" parameterType="int" resultMap=
        "studentCourseMap" >
    select s.name 's_name',c.name 'c_name', c.grade
    from student s,course c,student_course sc
    where
    s.id=#{id}
    and
    s.id=sc.sid
    and
    sc.cid=c.id
  </select>

  <select id="getCourseStudent" parameterType="int" resultMap=
          "courseStudentMap" >
    select c.name 'c_name',s.name 's_name', s.description
    from student s,course c,student_course sc
    where
    c.id=#{id}
    and
    s.id=sc.sid
    and
    sc.cid=c.id
  </select>
```

这里已经通过 student_course 连接了 student 和 course 表,并且对列名标签也做了明确的处理,以方便映射。

注意:还有一个方法是使用 collection 嵌套,表示多条记录映射到多条查询。读者可以参考相关文档,这里不再讲解。

3. PO 类

```
Student.java
package com.ascent.po;

import java.util.List;

public class Student {
    private int id;
    private String name;                    //姓名
    private int age;                        //年龄
    private String description;             //描述
    private List<Course> courses;

    public int getId() {
      return id;
    }
```

```
    public void setId(int id) {
      this.id=id;
    }
    public String getName() {
      return name;
    }
    public void setName(String name) {
      this.name=name;
    }
    public int getAge() {
        return age;
    }
    public void setAge(int age) {
        this.age=age;
    }
    public String getDescription() {
      return description;
    }
    public void setDescription(String description) {
      this.description=description;
    }
    public List<Course>getCourses() {
        return courses;
    }
    public void setCourses(List<Course>courses) {
        this.courses=courses;
    }
}
```

注意：Student 中添加了课程集合 List<Course>courses。

```
Course.java
package com.ascent.po;

import java.util.List;

public class Course {
    private int id;
    private String name;
    private int grade;
    private List<Student>students;

    public int getId() {
      return id;
    }
    public void setId(int id) {
      this.id=id;
```

```
    }
    public String getName() {
      return name;
    }
    public void setName(String name) {
      this.name=name;
    }
    public int getGrade() {
        return grade;
    }
    public void setGrade(int grade) {
        this.grade=grade;
    }

    public List<Student>getStudents() {
        return students;
    }
    public void setStudents(List<Student>students) {
        this.students=students;
    }
}
```

注意：Course 中添加了学生集合 List＜Student＞students。

4. Mapper 类

```
package com.ascent.mapper;
import java.util.List;

import com.ascent.po.*;

public interface StudentAndCourseMapper{

    public List<Student>getStudentCourse(int id);
    public List<Course>getCourseStudent(int id);

}
```

5. Test 类

```
package com.ascent.test;

import java.util.List;
import org.apache.ibatis.session.SqlSession;
import org.apache.ibatis.session.SqlSessionFactory;
import com.ascent.po.*;
import com.ascent.mapper.StudentAndCourseMapper;
import com.ascent.util.SqlSessionFactoryUtil;
```

```
public class StudentAndCourseTest{
  private SqlSessionFactory factory=SqlSessionFactoryUtil.getSqlSessionFactory();

  /**
   * 查询记录
   * /

  public void listCourse(){
    SqlSession session=null;
    try{
      session=factory.openSession();
      StudentAndCourseMapper mapper=session.getMapper(com.ascent.mapper.
          StudentAndCourseMapper.class);
    List<Student>ss=mapper.getStudentCourse(1);
      for (Student s: ss){
         System.out.println("\t"+s.getName());
        List<Course>cs=s.getCourses();
        for (Course c : cs){
           System.out.println("\t"+c.getName()+" "+c.getGrade());
        }
    }
    } catch (Exception e) {
      e.printStackTrace();
    }finally{
      session.close();
    }
  }

  /**
   * 查询记录
   * /

  public void listStudent(){
    SqlSession session=null;
    try{
      session=factory.openSession();
      StudentAndCourseMapper mapper=session.getMapper(com.ascent.mapper.
          StudentAndCourseMapper.class);
      List<Course>cc=mapper.getCourseStudent(1);
      for (Course c: cc){
       System.out.println("\t"+c.getName());
      List<Student>ss=c.getStudents();
      for (Student s : ss){
           System.out.println("\t"+s.getName()+" "+s.getDescription());
        }
    }
  } catch (Exception e) {
```

```
        e.printStackTrace();
      }finally{
        session.close();
      }
    }

    public static void main(String[]args){

      StudentAndCourseTest testCase=new StudentAndCourseTest();

      testCase.listCourse();
      testCase.listStudent();
    }

  }
```

8.4 项目案例

8.4.1 学习目标

通过本章的学习了解 MyBatis 的实体关系映射，一对一、一对多以及多对多关系，掌握如何在 mapper.xml 文件中使用特定标签描述实体关系，以及完成实体的增、删、改、查操作。

8.4.2 案例描述

由于 eGov 项目没有使用到实体关系映射，所以新建一个 department 项目，该项目包含 department 部门表和 usr 用户表，一个用户只能属于一个部门，而同样的部门可以包含多条 usr 用户数据，为一对多的项目案例。

8.4.3 案例要点

该案例仅使用 MyBatis 技术，不适用 Web 项目和 Spring MVC、Spring 模块，所以需要建立 Java Project。该项目需要建立 mybatis-config. xml、depart-mapping. xml、usr-mappring.mxl、Department.java、Usr.java、DepartmentMapper.java 接口、UsrMapper.java 接口，以及一个测试类 DepartmentTestCase。

8.4.4 案例实施

(1) 在 MyEclipse 中单击 New→Java Project，如图 8-1 所示。

新建 department 项目，单击 Next 按钮，在出现的对话框中单击 Libraries 选项卡，选择 Mybatis 用户自定义库，如图 8-2 所示。

最后单击 Finish 按钮。

右击 department 项目，选择 Java Build Path，单击 Add External JARs，添加 MySQL 驱动包 mysql-connector，如图 8-3 所示。

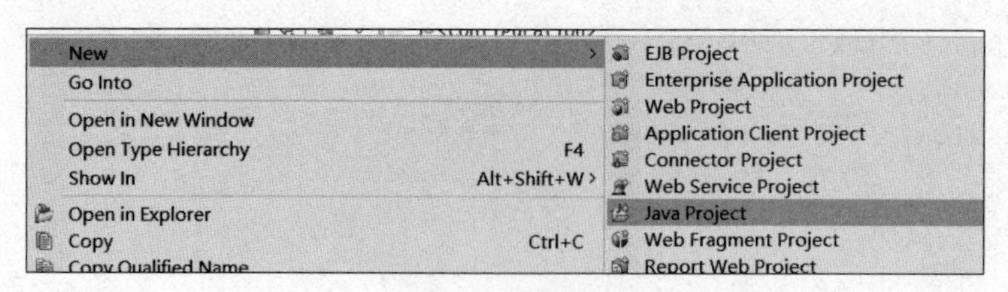

图 8-1 新建项目

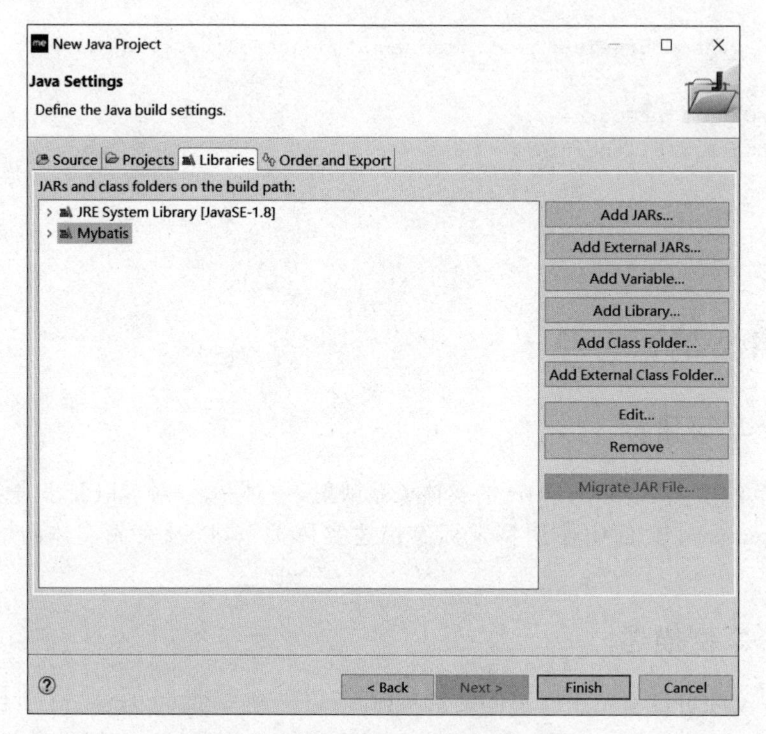

图 8-2 选择 MyBatis 用户自定义库

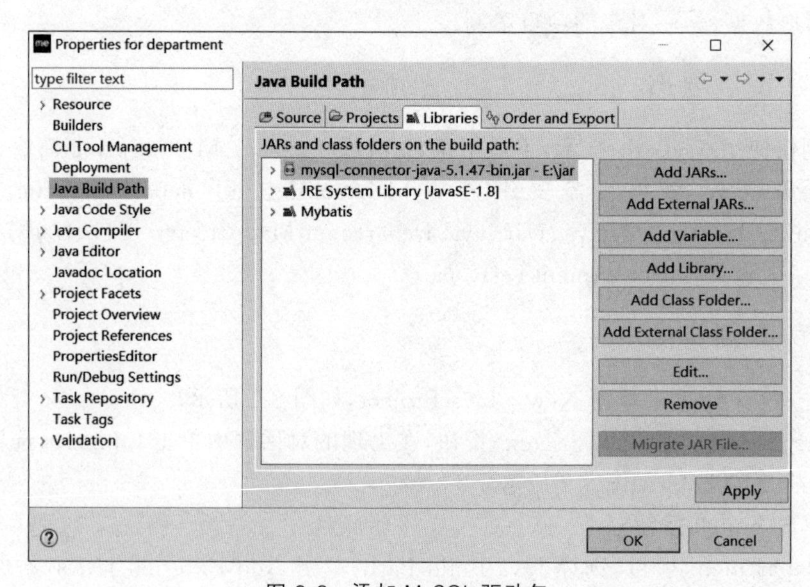

图 8-3 添加 MySQL 驱动包

之后单击 OK 按钮，至此 MyBatis 开发环境准备完毕。

(2) 创建对应于 Usr 表和 Department 表的实体类。

创建 Usr.java：

```
package com.ascent.po;
public class Usr {
    private Integer id;

    private String name;

    private String password;

    private String address;

    private Integer phone;

    private String title;

    private String power;

    private String auth;

    private Integer deptid;

    private Integer homephone;

    private String superauth;

    private Integer groupid;

    private String birthdate;

    private String gender;

    private String email;

    private String nickname;

    public Integer getId() {
        return id;
    }

    public void setId(Integer id) {
        this.id=id;
    }
```

```
public String getName() {
    return name;
}

public void setName(String name) {
    this.name=name==null ? null : name.trim();
}

public String getPassword() {
    return password;
}

public void setPassword(String password) {
    this.password=password==null ? null : password.trim();
}

public String getAddress() {
    return address;
}

public void setAddress(String address) {
    this.address=address==null ? null : address.trim();
}

public Integer getPhone() {
    return phone;
}

public void setPhone(Integer phone) {
    this.phone=phone;
}

public String getTitle() {
    return title;
}

public void setTitle(String title) {
    this.title=title==null ? null : title.trim();
}

public String getPower() {
    return power;
}

public void setPower(String power) {
```

```
        this.power=power==null ? null : power.trim();
    }

    public String getAuth() {
        return auth;
    }

    public void setAuth(String auth) {
        this.auth=auth==null ? null : auth.trim();
    }

    public Integer getDeptid() {
        return deptid;
    }

    public void setDeptid(Integer deptid) {
        this.deptid=deptid;
    }

    public Integer getHomephone() {
        return homephone;
    }

    public void setHomephone(Integer homephone) {
        this.homephone=homephone;
    }

    public String getSuperauth() {
        return superauth;
    }

    public void setSuperauth(String superauth) {
        this.superauth=superauth==null ? null : superauth.trim();
    }

    public Integer getGroupid() {
        return groupid;
    }

    public void setGroupid(Integer groupid) {
        this.groupid=groupid;
    }

    public String getBirthdate() {
        return birthdate;
```

```
    }

    public void setBirthdate(String birthdate) {
        this.birthdate=birthdate==null ? null : birthdate.trim();
    }

    public String getGender() {
        return gender;
    }

    public void setGender(String gender) {
        this.gender=gender==null ? null : gender.trim();
    }

    public String getEmail() {
        return email;
    }

    public void setEmail(String email) {
        this.email=email==null ? null : email.trim();
    }

    public String getNickname() {
        return nickname;
    }

    public void setNickname(String nickname) {
        this.nickname=nickname==null ? null : nickname.trim();
    }
}
```

创建 Department.java：

```
package com.ascent.po;

import java.util.HashSet;
import java.util.Set;

public class Department{
    private Integer id;

    private String name;

    private String description;

    private String status;
```

```
private String goal;
<!--使用 Set 集合包含多条 usr 对象,表示一对多关系-->
private Set<Usr>usrs=new HashSet<Usr>();

public Set<Usr>getUsrs() {
    return usrs;
}

public void setUsrs(Set<Usr>usrs) {
    this.usrs=usrs;
}

public Integer getId() {
    return id;
}

public void setId(Integer id) {
    this.id=id;
}

public String getName() {
    return name;
}

public void setName(String name) {
    this.name=name==null ? null : name.trim();
}

public String getDescription() {
    return description;
}

public void setDescription(String description) {
    this.description=description==null ? null : description.trim();
}

public String getStatus() {
    return status;
}

public void setStatus(String status) {
    this.status=status==null ? null : status.trim();
}

public String getGoal() {
```

```
        return goal;
    }

    public void setGoal(String goal) {
        this.goal=goal==null ? null : goal.trim();
    }
}
```

(3) 创建并编辑 mybatis-config.xml 文件。

```
mybatis-config.xml:
<?xml version="1.0" encoding="UTF-8"?>
<!DOCTYPE configuration PUBLIC "-//mybatis.org//DTD Config 3.0//EN" "http://
mybatis.org/dtd/mybatis-3-config.dtd" >
<configuration>
    <!--配置 SqlFactory 环境以及数据库连接、事务处理信息 -->
    <environments default="mybatisDev">
        <environment id="mybatisDev">
            <transactionManager type="JDBC" />
            <dataSource type="POOLED">
                <property name="driver" value="com.mysql.jdbc.Driver" />
                <property name="url" value="jdbc:mysql://localhost:3306/
                  electrones?useUnicode=true&amp;characterEncoding=utf-8
                  &amp;autoReconnect=true" />
                <property name="username" value="root" />
                <property name="password" value="root" />
            </dataSource>
        </environment>
    </environments>

    <!--增加 mapping.xml -->
    <mappers>
        <!--添加 usr 实体映射文件 -->
        <mapper resource="com/ascent/mybatis/mapping/UsrMapper.xml" />
        <!--添加 Department 实体映射文件 -->
        <mapper resource="com/ascent/mybatis/mapping/DepartmentMapper.xml" />
          </mappers>
</configuration>
```

(4) 编写 Usr.xml。

```
Usr.xml:
<?xml version="1.0" encoding="UTF-8"?>
<!DOCTYPE mapper PUBLIC "-//mybatis.org//DTD Mapper 3.0//EN" "http://mybatis.
org/dtd/mybatis-3-mapper.dtd">
<mapper namespace="com.ascent.mybatis.mapper.UsrMapper">
  <resultMap id=" usrResult" type="com.ascent.po.Usr">
```

```
    <id column="id"jdbcType="INTEGER" property="id" />
    <result column="name"jdbcType="VARCHAR" property="name" />
    <result column="password"jdbcType="VARCHAR" property="password" />
    <result column="address"jdbcType="VARCHAR" property="address" />
    <result column="phone"jdbcType="INTEGER" property="phone" />
    <result column="title"jdbcType="VARCHAR" property="title" />
    <result column="power"jdbcType="VARCHAR" property="power" />
    <result column="auth"jdbcType="VARCHAR" property="auth" />
    <result column="deptid" jdbcType="INTEGER" property="deptid" />
    <result column="homephone"jdbcType="INTEGER" property="homephone" />
    <result column="superauth" jdbcType="VARCHAR" property="superauth" />
    <result column="groupid" jdbcType="INTEGER" property="groupid" />
    <result column="birthdate" jdbcType="VARCHAR" property="birthdate" />
    <result column="gender"jdbcType="VARCHAR" property="gender" />
    <result column="email"jdbcType="VARCHAR" property="email" />
    <result column="nickname"jdbcType="VARCHAR" property="nickname" />
  </resultMap>
</mapper>
```

注意：namespace＝"com.ascent.mybatis.mapper.UsrMapper"很重要，需要在 Department 中引用。

（5）编写包含多条 Usr 实体对象的与 Usr 具有一对多关系的 DepartmentMapper.xml。

```
<?xml version="1.0" encoding="UTF-8"?>
<!DOCTYPE mapper PUBLIC "-//mybatis.org//DTD Mapper 3.0//EN" "http://mybatis.
org/dtd/mybatis-3-mapper.dtd">
<mapper namespace="com.ascent.mybatis.mapper.DepartmentMapper">
  <resultMap id="BaseResultMap" type="com.ascent.po.Department">
    <id column="id"jdbcType="INTEGER" property="id" />
    <result column="name"jdbcType="VARCHAR" property="name" />
    <result column="description"jdbcType="VARCHAR" property="description" />
    <result column="status"jdbcType="VARCHAR" property="status" />
    <result column="goal"jdbcType="VARCHAR" property="goal" />
    <!--定义 Department 类中 usrs 属性,它是一个集合,对应 UsrMapper 中的 id 为
        usrResult 的 resultMap -->
    <collection property="usrs" resultMap="com.ascent.mybatis.mapper.
        UsrMapper.usrResult"></collection>
  </resultMap>

  <!--定义查询所有部门以及该部门下所有用户信息的 select 语句 -->
  <!--使用左外链接查询 -->
  <select id="findAllDepartAndUsrs"resultMap="deptResult">
     select d.id, d.name, d.description, d.status, d.goal,
     u.id, u.name, u.password, u.address, u.phone, u.title,
      u.power, u.auth, u.homephone, u.superauth,
      u.groupid, u.birthdate, u.gender, u.email, u.nickname
```

```
        from department d
        left join usr u
        on d.id=u.deptid
        order by d.id ;
    </select>

    <!--定义根据部门 ID,查询该部门信息以及包含的用户信息 -->
    <!--#{deptid}为获取方法传入的整型参数 deptid -->
    <select id="findDeptAndUsrsById" resultMap="deptResult" parameterType=
        "java.lang.Integer">
        select d.id, d.name, d.description, d.status, d.goal,
      u.id, u.name, u.password, u.address, u.phone, u.title,
        u.power, u.auth, u.homephone, u.superauth,
        u.groupid, u.birthdate, u.gender, u.email, u.nickname
        from department d
        left join usr u
        on d.id=#{deptid}
        and d.id=u.deptid
        order by d.id ;
    </select>
</mapper>
```

（6）编写 DepartmentMapper.java 接口，定义 findAllDepartAndUsrs 以及 findDeptAndUsrsById 的 Java 抽象方法。

```
package com.ascent.mybatis.mapper;

import java.util.List;

import com.ascent.po.Department;

public interface DepartmentMapper {
    /**
     * 查询所有部门以及其下属用户
     * @return
     */
    public List<Department> findAllDepartAndUsrs();
    /**
     * 根据 ID 查询一个部门及其下属用户
     * @param deptid
     * @return
     */
    public Department findDeptAndUsrsById(Integer deptid);

}
```

（7）编写 MybatisSqlSessionFactoryUtil.java。

```
package com.ascent.mybatis.utils;

import java.io.IOException;
import java.io.InputStream;

import org.apache.ibatis.io.Resources;
import org.apache.ibatis.session.SqlSession;
import org.apache.ibatis.session.SqlSessionFactory;
import org.apache.ibatis.session.SqlSessionFactoryBuilder;

public class MybatisSqlSessionFactoryUtil{

    private static final StringinputResource="config/mybatis-config.xml";
    private static SqlSessionFactory sqlSessionFactory=null;

    static {

        try {
            InputStream inputStream=Resources.getResourceAsStream
                (inputResource);
            sqlSessionFactory=new SqlSessionFactoryBuilder().build
                (inputStream);
        } catch (IOException e) {
            //TODO Auto-generated catch block
            e.printStackTrace();
        }

    }

    public static SqlSession getSqlSession(){
        if(null!=sqlSessionFactory)
            return sqlSessionFactory.openSession();
        else
            return null;
    }
}
```

（8）编写 DepartmentTestCase.java。

```
package com.ascent.mybatis.test;

import java.util.List;
import java.util.Set;

import org.apache.ibatis.session.SqlSession;
```

```
import com.ascent.mybatis.mapper.DepartmentMapper;
import com.ascent.mybatis.utils.MybatisSqlSessionFactoryUtil;
import com.ascent.po.Department;
import com.ascent.po.Usr;

public class DepartmentTestCase {

    public static void main(String[] args) {

        SqlSession sqlSession=MybatisSqlSessionFactoryUtil.getSqlSession();

         DepartmentMapper deptMapper = sqlSession. getMapper (DepartmentMapper.
         class);
        //测试 findAllDepartAndUsrs()方法
        List<Department>departments=deptMapper.findAllDepartAndUsrs();

    for (Department department : departments) {
System.out.println("部门信息: "+department.getId()+"|"+department.getName()+"
|"+department.getDescription());

            Set<Usr>usrs=department.getUsrs();

            for (Usr usr : usrs) {
System.out.println("用户信息: "+usr.getId()+">>"+usr.getName()+">>"+usr.
getNickname());
            }
        }

    }
    //调用 findDeptAndUsrsById()方法
}
```

运行该类，看到图 8-4 所示的输出，表示配置实体关系成功。

```
部门信息: 1|研发部|null
用户信息: 1>>研发部>>测试6
```

图 8-4　运行结果

8.4.5　特别提示

MyBatis 实体关系映射常用于封装查询结果，因此本案例采用连表查询封装查询结果作为案例。注意数据库驱动包的导入以及 Mybatis 库的引入。

8.4.6 拓展与提高

开发新闻与用户的多对多的查询功能。

习题

1. 写出客户和地址的持久化类和映射文件。
2. 映射文件中如何使用联合表示一对一映射关系?
3. 写出部门和员工的持久化类和映射文件。
4. 映射文件中如何使用聚集表示一对多映射关系?
5. 如何设计学生和课程的多对多关系表?写出持久化类和映射文件。

第9章 MyBatis高级特性

学习目的与学习要求

学习目的：了解 MyBatis 动态 SQL、MyBatis 注解和日志的使用。

学习要求：重点掌握 MyBatis 动态 SQL、MyBatis 注解和日志等高级特性。

本章主要内容

本章重点介绍 MyBatis 动态 SQL、MyBatis 注解和日志的使用，以及高级特性的设计和开发。

9.1 MyBatis 动态 SQL

MyBatis 最强大的特性之一是它的动态语句功能。如果以前使用 JDBC 或者类似的框架，就会明白把 SQL 语句条件连接在一起是多么困难，空格和逗号等都不能疏忽。动态语句能解决这些烦恼。

尽管使用动态 SQL 也不是那么简单，但是 MyBatis 确实能通过使用强大的动态 SQL 解决许多问题。

动态 SQL 元素对于任何使用过 JSTL 或者类似于 XML 之类的文本处理的人来说，都是非常熟悉的。在上一版本中(iBatis 2)，动态 SQL 有许多元素需要学习和了解，但在 MyBatis 3 中，这些都有了许多改进，现在只剩下不到原先一半的元素。MyBatis 使用基于强大的 OGNL (Object-Graph Navigation Language 的缩写，它是一种功能强大的表达式语言)表达式避免了大部分其他的元素。动态 SQL 元素主要包括：

- if；
- choose (when，otherwise)；

- trim (where,set);
- foreach;
- bind。

1. if 语句

动态 SQL 中最常做的事情就是有条件地包含一个 where 子句。例如：

```
<select id="findActiveBlogWithTitleLike" parameterType="Blog" resultType="
Blog">
  SELECT * FROM BLOG
  WHERE state='ACTIVE'
  <if test="title !=null">
    AND title like #{title}
  </if>
</select>
```

这条语句提供一个带功能性的可选的文字。如果没有传入标题,那么所有的博客都会被返回,如果传入了一个标题,就会寻找相似的标题。

如果想可选地使用标题或者作者查询怎么办？需要将语句的名称稍改一下,使得语句看起来更直观,然后简单地加上另外一个条件。

```
<select id="findActiveBlogLike" parameterType="Blog" resultType="Blog">
  SELECT * FROM BLOG WHERE state='ACTIVE'
  <if test="title !=null">
    AND title like #{title}
  </if>
  <if test="author !=null and author.name !=null">
    AND title like #{author.name}
  </if>
</select>
```

2. choose(when,otherwise)

有时候并不想应用所有的条件,只是想从多个选项中选择一个。类似于 Java 中的 switch 语句,MyBatis 提供了 choose 元素。继续以上面的例子举例,只是现在搜索有查询标题的,或者只搜索有查询作者的。如果都没有提交,则只选有特性的(比如说是管理员加精的,或者是置顶的)。

```
<select id="findActiveBlogLike" parameterType="Blog" resultType="Blog">
  SELECT * FROM BLOG WHERE state='ACTIVE'
  <choose>
    <when test="title !=null">
      AND title like #{title}
    </when>
    <when test="author !=null and author.name !=null">
      AND title like #{author.name}
```

```
    </when>
    <otherwise>
      AND featured=1
    </otherwise>
  </choose>
</select>
```

3. trim(where,set)

考虑上面提到的 if 的例子,如果把 ACTIVE=1 也作为条件,会发生什么情况?

```
<select id="findActiveBlogLike" parameterType="Blog" resultType="Blog">
  SELECT * FROM BLOG
  WHERE
  <if test="state !=null">
    state=#{state}
  </if>
  <if test="title !=null">
    AND title like #{title}
  </if>
  <if test="author !=null and author.name !=null">
    AND title like #{author.name}
  </if>
</select>
```

如果一个条件都不给出,会怎样?语句可能会变成 SELECT * FROM BLOG WHERE,这会有问题。或者如果仅获得第二个条件,语句又会变成:

```
SELECT * FROM BLOG
WHERE
AND title like'someTitle'
```

这同样很糟糕。这个问题仅用条件很难简单地解决,MyBatis 有一个简单的方案可以解决这里面 90%的问题。如果 where 没有出现,可以自定义一个。稍稍修改一下,就能完全解决:

```
<select id="findActiveBlogLike" parameterType="Blog" resultType="Blog">
  SELECT * FROM BLOG
  <where>
    <if test="state !=null">
      state=#{state}
    </if>
    <if test="title !=null">
      AND title like #{title}
    </if>
    <if test="author !=null and author.name !=null">
      AND title like #{author.name}
    </if>
```

```
  </where>
</select>
```

这个 where 元素会知道如果它包含的标签中有返回值,它就插入一个 where。此外,如果标签返回的内容是以 AND 或 OR 开头的,则它会剔除掉。如果 where 元素并没有完全达到预期的目标效果,那也可以使用 trim 元素自定义它。下面的 trim 也相当于 where:

```
<trim prefix="WHERE"prefixOverrides="AND |OR ">
...
</trim>
```

prefixOverrides 元素使用一个管理分隔符组成的文字进行覆写,空白符也是不能忽略的。这样的结果是移出了所有指定在 overrides 里的字符,使用 with 属性里的字符覆写。在动态 update 语句里相似的情形叫作 set。这个 set 元素被用作动态囊括列名来更新,而忽略其他的元素。例如:

```
<update id="updateAuthorIfNecessary"parameterType="domain.blog.Author">
  update Author
  <set>
    <if test="username !=null">username=#{username},</if>
    <if test="password !=null">password=#{password},</if>
    <if test="email !=null">email=#{email},</if>
    <if test="bio !=null">bio=#{bio}</if>
  </set>
  where id=#{id}
</update>
```

set 元素将动态地配置 SET 关键字,也用来剔除追加到条件末尾的任何不相关的逗号,例如:

```
<trim prefix="SET"suffixOverrides=",">
...
</trim>
```

注意:这里只写了一个前缀,同样也可以追加一个后缀。

4. foreach

动态 SQL 的另外一个常用的必要操作是需要对一个集合进行遍历,通常是在构建 IN 条件语句的时候。例如:

```
<select id="selectPostIn" resultType="domain.blog.Post">
  SELECT *
  FROM POST P
  WHERE ID in
  <foreach item="item" index="index" collection="list"
      open="(" separator="," close=")">
      #{item}
  </foreach>
```

```
</select>
```

foreach 元素非常强大，允许指定一个集合，申明一个项和一个索引变量，用在这个元素的方法体内，也允许指定开始和结束的字符，还可以在两个迭代器之间加入一个分隔符。它的智能之处在于它不会追加额外的分隔符。

注意：可以传入一个 List 实例或者一个数组给 MyBatis 作为一个参数对象。如果这么做，MyBatis 会自动将它包装成一个 Map，使用它的名称作为 key。List 实例将使用 list 作为键，数组实例以 array 作为键。

5. bind

bind 标签可以使用 OGNL 表达式创建一个变量，并将其绑定到上下文中。例如：

```
<select id="getEmpsTest" resultType=" Employee">
    <!--bind:可以将 OGNL 表达式的值绑定到一个变量中，方便后来引用这个变量的值 -->
    <bind name="bindName" value="'%'+eName+'%'"/> eName 是 employee 中的一个属
性值
      SELECT * FROM emp
    <if test="_parameter!=null">
      where ename like #{bindName}
  </if>
</select>
```

bind 标签的两个属性都是必选项，name 为绑定到上下文的变量名，value 为 OGNL 表达式。创建一个 bind 标签的变量后，就可以在下面直接使用。使用 bind 拼接字符串不仅可以避免因更换数据库而修改 SQL，也能预防 SQL 注入。

接下来看一个 Dynamic SQL 的实例(Mybatis_dynamicSQL_demo)。

1）在 MySQL 数据库中建立 student 表

```
USE `test`;

/* Table structure for table `student` */

DROP TABLE IF EXISTS `student`;

CREATE TABLE `student`(
  `id` int NOT NULL AUTO_INCREMENT,
  `name`varchar(255) DEFAULT NULL,
  `age` int DEFAULT NULL,
  `description` varchar(255) CHARACTER SET latin1 COLLATE latin1_swedish_ci
DEFAULT NULL,
  PRIMARY KEY (`id`)
) ENGINE=InnoDB AUTO_INCREMENT=4 DEFAULT CHARSET=latin1;

/* Data for the table `student` */

insert into `student`(`id`,`name`,`age`,`description`) values (1,'Gary',25,
```

```
'Software Engineer'), (2, 'Linda', 35, 'Marketing'), (3, 'Kevin', 22, 'Computer
Science');
```

2）新建 Java 项目，导入 MyBatis 所依赖的 jar 包，建立项目目录

```
  src/config                          //存放 mybatis-config.xml 和 db.conf 文件，数据
                                        库连接信息配置
src/com/ascent/mybatis/mapper       //存放 Java API 接口文件 StudentMapper.class
src/com/ascent/mybatis/mapping      //存放 SQL XML 映射文件 StudentMapper.xml
src/com/ascent/util                 //存放工具文件 SqlSessionFactoryUtil
src/com/ascent/po                   //存放 Java POJO 类，映射查询 Student 实体
src/com/ascent/test                 //存放测试运行主方法类 StudentTest
```

3）在 config 目录下新建 db.conf 文件

```
jdbc.driver=com.mysql.jdbc.Driver
jdbc.url=jdbc:mysql://localhost:3306/test
jdbc.username=root
jdbc.password=root
```

注意：这里的变量要加上 jdbc.作为前缀，否则 MyBatis 可能无法识别。具体的值根据自己的数据库设置。

4）在 config 目录下新建 mybatis-config 文件

```
<?xml version="1.0" encoding="UTF-8"?>
<!DOCTYPE configuration
  PUBLIC "-//mybatis.org//DTD Config 3.0//EN"
  "http://mybatis.org/dtd/mybatis-3-config.dtd">
<configuration>

  <properties resource="config/db.conf"/>

  <typeAliases>
    <typeAlias alias="Student" type="com.ascent.po.Student"/>
  </typeAliases>

  <environments default="development">
        <environment id="development">
            <transactionManager type="JDBC" />
            <dataSource type="POOLED">
                <property name="driver" value="${jdbc.driver}" />
                <property name="url" value="${jdbc.url}" />
                <property name="username" value="${jdbc.username}" />
                <property name="password" value="${jdbc.password}" />
            </dataSource>
        </environment>
    </environments>
  <mappers>
```

```
    <mapper resource="com/ascent/mapping/StudentMapper.xml"/>
  </mappers>
</configuration>
```

5）在 util 中创建 SqlSessionFactoryUtil 类

```
package com.ascent.util;

import java.io.IOException;
import java.io.InputStream;
import org.apache.ibatis.io.Resources;
import org.apache.ibatis.session.SqlSessionFactory;
import org.apache.ibatis.session.SqlSessionFactoryBuilder;
public class SqlSessionFactoryUtil{

  private static SqlSessionFactory sqlSessionFactory=null;

  static{
    InputStream input=null;
    try{
      input=Resources.getResourceAsStream("config/mybatis-config.xml");
      sqlSessionFactory=new SqlSessionFactoryBuilder().build(input);
    } catch (IOException e) {
      e.printStackTrace();
    }finally{
      if(input !=null){
          try{
            input.close();
          } catch (IOException e) {
            e.printStackTrace();
        }
      }
    }
  }

  public static SqlSessionFactory getSqlSessionFactory(){
    return sqlSessionFactory;
  }
}
```

6）在 po 目录下新建 Student 类

```
package com.ascent.po;

public class Student {
    private int id;
    private String name;                //姓名
```

```
    private int age;                    //年龄
    private String description;         //描述

    public int getId() {
      return id;
    }
    public void setId(int id) {
      this.id=id;
    }
    public String getName() {
      return name;
    }
    public void setName(String name) {
      this.name=name;
    }
    public int getAge() {
        return age;
    }
    public void setAge(int age) {
        this.age=age;
    }
    public String getDescription() {
      return description;
    }
    public void setDescription(String description) {
      this.description=description;
    }
}
```

7）在 mapping 目录下新建 StudentMapper.xml

```
<?xml version="1.0" encoding="UTF-8"?>
<!DOCTYPE mapper PUBLIC "-//mybatis.org//DTD Mapper 3.0//EN"
"http://mybatis.org/dtd/mybatis-3-mapper.dtd">
<mapper namespace="com.ascent.mapper.StudentMapper">
  <!--定义 Java Bean 的属性与数据库的列之间的映射 -->
  <resultMap type="Student" id="studentResultMap">
    <id column="id" property="id" />
    <result column="name" property="name" />
    <result column="age" property="age" />
    <result column="description" property="description" />
  </resultMap>

  <!--SQL 语句中以"#{}"的形式引用参数 -->

  <!--Dynamic If 查询记录 -->
```

```
    < select  id =" searchByStudentSelectiveAndIf "  parameterType =" Student "
resultMap="studentResultMap">
            select * from student
            <if test="name!=null">
               where name like concat('%',#{name},'%')
            </if>
   </select>

    <!--Dynamic where 查询记录 -->
   <select id="searchByStudentSelectiveAndWhere" parameterType="Student"
   resultMap="studentResultMap">
            select * from student
            <where>
              <if test="name!=null">
                and name likeconcat('%',#{name},'%')
              </if>
              <if test="age!=null and age!=0">
                and age>#{age}
              </if>
            </where>
   </select>

    <!--Dynamic choose 查询记录 -->
   <select id="searchByStudentSelectiveAndChoose" parameterType="Student"
   resultMap="studentResultMap">
            select * from student
            <where>
              <choose>
                <when test="name !=null">
              and name likeconcat('%',#{name},'%')
                </when>
                <when test="age !=null and age !=0">
                  and age>#{age}
                </when>
                <otherwise>
                  and id >1
                </otherwise>
              </choose>
            </where>
   </select>

    <!--Dynamic foreach 查询记录 -->
    <select id="searchByStudentSelectiveAndForeach" resultMap="studentResultMap">
              select * from student
              where id in
```

```
            <foreach item="item" index="index" collection="list"
                open="(" separator="," close=")">
                #{item}
            </foreach>
  </select>

    <!--Dynamic bind 查询记录 -->
  <select id="searchByStudentSelectiveAndBind" parameterType="Student"
  resultMap="studentResultMap">
          <bind name="likename" value="'%'+name+'%'" />
          select * from student where name like #{likename}
    </select>

</mapper>
```

8）在 mapper 目录下新建 StudentMapper

```
package com.ascent.mapper;
import com.ascent.po.*;
import java.util.List;
import org.apache.ibatis.annotations.*;

public interface StudentMapper {

    //dynamic SQL if
    public List<Student>searchByStudentSelectiveAndIf(Student student);
    //dynamic SQL where
    public List<Student>searchByStudentSelectiveAndWhere(Student student);
    //dynamic SQL choose
    public List<Student>searchByStudentSelectiveAndChoose(Student student);
    //dynamic SQL foreach
    public List<Student>searchByStudentSelectiveAndForeach(List<Integer>ids);
    //dynamic SQL bind
    public List<Student>searchByStudentSelectiveAndBind(Student student);

}
```

9）在 test 目录下建立 StudentTest 测试类，含 main()方法

```
package com.ascent.test;

import java.util.HashMap;
import java.util.LinkedList;
import java.util.List;
import java.util.Map;

import org.apache.ibatis.session.SqlSession;
```

```
import org.apache.ibatis.session.SqlSessionFactory;
import com.ascent.po.Student;
import com.ascent.mapper.StudentMapper;
import com.ascent.util.SqlSessionFactoryUtil;

public class StudentTest {
  private SqlSessionFactory factory=SqlSessionFactoryUtil.getSqlSessionFactory();

  /**
   * Dynamic If 查询记录
   */

  public void searchByStudentSelectiveAndIf(){
    System.out.println("This is searchByStudentSelectiveAndIf...");
    SqlSession session=null;
    try{
      session=factory.openSession();
      Student st=new Student();
      System.out.println("no name...");
      StudentMapper mapper=session.getMapper(StudentMapper.class);
      List<Student>students1=mapper.searchByStudentSelectiveAndIf(st);
      if(students1.size()>0){
            for(Student s: students1){
              System.out.println(s.getName());
            }
      }
      st.setName("in");
      System.out.println("with name...");
      List<Student>students2=mapper.searchByStudentSelectiveAndIf(st);
      if(students2.size()>0){
            for(Student s: students2){
              System.out.println(s.getName());
            }
      }
    } catch (Exception e) {
      e.printStackTrace();
    }finally{
      session.close();
    }
  }

  /**
   * Dynamic Where 查询记录
   */
```

```
public void searchByStudentSelectiveAndWhere(){
  System.out.println("This is searchByStudentSelectiveAndWhere...");
  SqlSession session=null;
  try{
    session=factory.openSession();
    Student st=new Student();
    System.out.println("no name, no age...");
    StudentMapper mapper=session.getMapper(StudentMapper.class);
    List<Student>students1=mapper.searchByStudentSelectiveAndWhere(st);
    if(students1.size()>0){
          for(Student s: students1){
            System.out.println(s.getName());
          }
    }
    st.setName("in");
    System.out.println("with name, no age...");
    List<Student>students2=mapper.searchByStudentSelectiveAndWhere(st);
    if(students2.size()>0){
          for(Student s: students2){
            System.out.println(s.getName());
          }
    }
    st.setName(null);
    st.setAge(26);
    System.out.println("no name, with age...");
    List<Student>students3=mapper.searchByStudentSelectiveAndWhere(st);
    if(students3.size()>0){
          for(Student s: students3){
            System.out.println(s.getName());
          }
    }
  } catch (Exception e) {
    e.printStackTrace();
  }finally{
    session.close();
  }
}

/**
 * Dynamic Choose 查询记录
 * /

public void searchByStudentSelectiveAndChoose(){
  System.out.println("This is searchByStudentSelectiveAndChoose...");
  SqlSession session=null;
```

```
    try{
      session=factory.openSession();
      Student st=new Student();
      System.out.println("no name, no age...");
      StudentMapper mapper=session.getMapper(StudentMapper.class);
      List<Student>students1=mapper.searchByStudentSelectiveAndChoose(st);
      if(students1.size()>0){
            for(Student s: students1){
              System.out.println(s.getName());
            }
      }
      st.setName("in");
      System.out.println("with name, no age...");
      List<Student>students2=mapper.searchByStudentSelectiveAndChoose(st);
      if(students2.size()>0){
            for(Student s: students2){
              System.out.println(s.getName());
            }
      }
      st.setName(null);
      st.setAge(26);
      System.out.println("no name, with age...");
      List<Student>students3=mapper.searchByStudentSelectiveAndChoose(st);
      if(students3.size()>0){
            for(Student s: students3){
              System.out.println(s.getName());
            }
      }
    } catch (Exception e) {
      e.printStackTrace();
    }finally{
      session.close();
    }
  }

  /**
   * Dynamic Foreach 查询记录
   * /

  public void searchByStudentSelectiveAndForeach(){
    System.out.println("This is searchByStudentSelectiveAndForeach...");
    SqlSession session=null;
    try{
      session=factory.openSession();
      List<Integer>ids=newLinkedList<>();
```

```
            ids.add(1);
            ids.add(3);
            StudentMapper mapper=session.getMapper(StudentMapper.class);
            List<Student>students1=mapper.searchByStudentSelectiveAndForeach(ids);
            if(students1.size()>0){
                for(Student s: students1){
                    System.out.println(s.getName());
                }
            }
        } catch (Exception e) {
            e.printStackTrace();
        }finally{
            session.close();
        }
    }
    /**
     * Dynamic Bind 查询记录
     */

    public void searchByStudentSelectiveAndBind(){
        System.out.println("This is searchByStudentSelectiveAndBind...");
        SqlSession session=null;
        try{
            session=factory.openSession();
            Student st=new Student();
            st.setName("in");
            System.out.println("with name...");
            StudentMapper mapper=session.getMapper(StudentMapper.class);
            List<Student>students2=mapper.searchByStudentSelectiveAndBind(st);
            if(students2.size()>0){
                for(Student s: students2){
                    System.out.println(s.getName());
                }
            }
        } catch (Exception e) {
            e.printStackTrace();
        }finally{
            session.close();
        }
    }

    public static void main(String[]args){

        StudentTest testCase=new StudentTest();
```

```
        testCase.searchByStudentSelectiveAndIf();
        //testCase.searchByStudentSelectiveAndWhere();
        //testCase.searchByStudentSelectiveAndChoose();
        //testCase.searchByStudentSelectiveAndForeach();
        //testCase.searchByStudentSelectiveAndBind();
    }

}
```

10）运行 StudentTest，得到结果

9.2　MyBatis 注解

最开始的时候，MyBatis 是一个 XML 驱动的框架，配置文件是 XML，语句映射定义在 XML 文件中。之后，MyBatis 也支持基于注解方式的配置，可以简便地映射 SQL 语句。

注意：*Java 注解在表达式和弹性上有一些限制，一些强大的 MyBatis 映射是无法用注解生成的。*

1. MyBatis 基本注解

MyBatis 提供的一些基本注解见表 9-1。

表 9-1　MyBatis 提供的一些基本注解

注　解	目标	相应的 XML	描　述
@CacheNamespace	类	<cache>	为给定的命名空间(如类)配置缓存。 属性：implemetation、eviction、flushInterval、size 和 readWrite
@CacheNamespaceRef	类	<cacheRef>	参照另外一个命名空间的缓存使用。 属性：value，也就是类的完全限定名
@ConstructorArgs	方法	<constructor>	收集一组结果传递给对象构造方法。 属性：value，是形式参数的数组
@Arg	方法	<arg> <idArg>	单独的构造方法参数，是 ConstructorArgs 集合的一部分。 属性：id、column、javaType、typeHandler id 属性是一个布尔值，用来标识比较的属性，和<idArg>XML 元素相似
@TypeDiscriminator	方法	<discriminator>	一组实例值被用来决定结果映射的表现。 属性：Column、javaType、jdbcType typeHandler、cases。cases 属性就是实例的数组
@Case	方法	<case>	单独实例的值和它对应的映射。 属性：value、type、results。 Results 属性是结果数组，因此这个注解和实际的 ResultMap 很相似，由下面的 Results 注解指定

续表

注　　解	目标	相应的 XML	描　　述
@Results	方法	<resultMap>	结果映射的列表，包含一个特别结果列如何被映射到属性或字段的详情。 属性：value，是 Result 注解的数组
@Result	方法	<result> <id>	在列和属性或字段之间的单独结果映射。 属性：id、column、property、javaType、jdbcType、type Handler、one、many。id 属性是一个布尔值，表示被比较的属性。one 属性是单独的联系，和<association>相似。many 属性是对集合而言的，和<collection>相似
@One	方法	<association>	复杂类型的单独属性值映射。 属性：select 已映射语句(也就是映射器方法)的完全限定名，它可以加载合适类型的实例。注意，联合映射在注解 API 中是不支持的
@Many	方法	<collection>	复杂类型的集合属性映射。 属性：select 是映射器方法的完全限定名，它可加载合适类型的一组实例。注意，联合映射在 Java 注解中是不支持的
@Options	方法	映射语句的属性	这个注解提供访问交换和配置选项的宽广范围，它们通常在映射语句上作为属性出现，而不是将每条语句注解变复杂。Options 注解提供连贯清晰的方式访问它们。 属性：useCache=true， flushCache=false， resultSetType=FORWARD_ONLY， statementType=PREPARED， fetchSize=-1，timeout=-1， useGeneratedKeys=false， keyProperty="id"。 理解 Java 注解很重要，因为没有办法指定 null 作为值。因此，一旦使用了 Options 注解，语句就受所有默认值的支配。要注意使用什么样的默认值，以避免不期望的行为
@Insert @Update @Delete	方法	<insert> <update> <delete>	代表了执行的真实 SQL。每一个注解都使用字符串数组(或单独的字符串)。如果传递的是字符串数组，它们由每个分隔它们的单独空间串联起来。 属性：value，这是字符串数组用来组成单独的 SQL 语句
@InsertProvider @UpdateProvider @DeleteProvider @SelectProvider	方法	<insert> <update> <delete> <select> 允许创建动态 SQL	这些可选的 SQL 注解允许指定一个类名和一个方法在执行时返回运行的 SQL。基于执行的映射语句，MyBatis 会实例化这个类，然后执行由 provider 指定的方法。这个方法可以选择性地接受参数对象作为它的唯一参数，但是必须指定该参数或者没有参数。 属性：type、method。type 属性是类的完全限定名。method 属性是该类中的方法名

续表

注　　解	目标	相应的 XML	描　　述
@Param	参数	N/A	当映射器方法需要多个参数时,这个注解可以被应用于映射器方法参数给每个参数一个名字。否则,多参数将会以它们的顺序位置被命名。则如 #{1},#{2} 等,这是默认的。 使用@Param("person"),SQL 中的参数应该被命名为 #{person}

这些注解都是运用到传统意义上映射器接口中的方法、类或者方法参数中的。

这里主要介绍两种使用注解的方式:一种是直接在映射器接口中写 SQL 语句;一种是利用 SQLProvider 创建 SQL,再由映射器接口调用。

2. 在映射器接口中写 SQL 语句

先采用第一种方式,看一个 CRUD 注解的实例(Mybatis_CRUD_Annotation_demo)。

1) 在 MySQL 数据库中建立 student 表

```
USE `test`;

/* Table structure for table `student` */

DROP TABLE IF EXISTS `student`;

CREATE TABLE `student`(
  `id` int NOT NULL AUTO_INCREMENT,
  `name`varchar(255) DEFAULT NULL,
  `age` int DEFAULT NULL,
  `description` varchar (255) CHARACTER SET latin1 COLLATE latin1_swedish_ci
DEFAULT NULL,
  PRIMARY KEY (`id`)
) ENGINE=InnoDB AUTO_INCREMENT=4 DEFAULT CHARSET=latin1;

/* Data for the table `student` */

insert  into `student`(`id`,`name`,`age`,`description`) values (1,'Gary',25,
'Software Engineer'),(2,'Linda',35,'Marketing'),(3,'Kevin',22,'Computer
Science');
```

2) 新建 Java 项目,导入 MyBatis 所依赖的 jar 包,建立项目目录

```
src/config                          //存放 mybatis-config.xml 和 db.conf 文件,数据
                                      库连接信息配置
src/com/ascent/mybatis/mapper       //存放 Java API 接口文件 StudentMapper.class
src/com/ascent/mybatis/mapping      //存放 SQL XML 映射文件 StudentMapper.xml
src/com/ascent/util                 //存放工具文件 SqlSessionFactoryUtil
src/com/ascent/po                   //存放 Java POJO 类,映射查询 Student 实体
```

```
src/com/ascent/test                //存放测试运行主方法类 StudentTest
```

3）在 config 目录下新建 db.conf 文件

```
jdbc.driver=com.mysql.jdbc.Driver
jdbc.url=jdbc:mysql://localhost:3306/test
jdbc.username=root
jdbc.password=root
```

注意：这里的变量要加上 jdbc.作为前缀，否则 MyBatis 可能无法识别。具体的值根据自己的数据库设置。

4）在 config 目录下新建 mybatis-config 文件

```
<?xml version="1.0" encoding="UTF-8"?>
<!DOCTYPE configuration
  PUBLIC "-//mybatis.org//DTD Config 3.0//EN"
  "http://mybatis.org/dtd/mybatis-3-config.dtd">
<configuration>

  <properties resource="config/db.conf"/>

  <typeAliases>
    <typeAlias alias="Student" type="com.ascent.po.Student"/>
  </typeAliases>

  <environments default="development">
        <environment id="development">
            <transactionManager type="JDBC" />
            <dataSource type="POOLED">
                <property name="driver" value="${jdbc.driver}" />
                <property name="url" value="${jdbc.url}" />
                <property name="username" value="${jdbc.username}" />
                <property name="password" value="${jdbc.password}" />
            </dataSource>
        </environment>
    </environments>
<mappers>
  <!--mapper resource="com/ascent/mapping/StudentMapper.xml"/-->
  <mapper class="com.ascent.mapper.StudentMapper"/>
</mappers>
    </configuration>
```

注意：mapper 由 resource 改为 class。

5）在 util 中创建 SqlSessionFactoryUtil 类

```
package com.ascent.util;

import java.io.IOException;
import java.io.InputStream;
```

```
import org.apache.ibatis.io.Resources;
import org.apache.ibatis.session.SqlSessionFactory;
import org.apache.ibatis.session.SqlSessionFactoryBuilder;
public class SqlSessionFactoryUtil{

  private static SqlSessionFactory sqlSessionFactory=null;

  static{
    InputStream input=null;
    try{
      input=Resources.getResourceAsStream("config/mybatis-config.xml");
      sqlSessionFactory=new SqlSessionFactoryBuilder().build(input);
    } catch (IOException e) {
      e.printStackTrace();
    }finally{
      if(input !=null){
        try{
          input.close();
        } catch (IOException e) {
          e.printStackTrace();
        }
      }
    }
  }

  public staticSql SessionFactory getSqlSessionFactory(){
    return sqlSessionFactory;
  }
}
```

6）在 po 目录下新建 Student 类

```
package com.ascent.po;

public class Student {
    private int id;
    private String name;                    //姓名
    private int age;                        //年龄
    private String description;             //描述

    public int getId() {
      return id;
    }
    public void setId(int id) {
      this.id=id;
    }
```

```
    public String getName() {
      return name;
    }
    public void setName(String name) {
      this.name=name;
    }
    public int getAge() {
        return age;
    }
    public void setAge(int age) {
        this.age=age;
    }
    public String getDescription() {
      return description;
    }
    public void setDescription(String description) {
      this.description=description;
    }
}
```

7）在 mapping 目录下新建 StudentMapper.xml

```
<?xml version="1.0" encoding="UTF-8"?>
<!DOCTYPE mapper PUBLIC "-//mybatis.org//DTD Mapper 3.0//EN"
"http://mybatis.org/dtd/mybatis-3-mapper.dtd">

<mapper namespace="com.ascent.mapper.StudentMapper">
  <!--定义 Java Bean 的属性与数据库的列之间的映射 -->
  <resultMap type="Student" id="studentResultMap">
    <id column="id" property="id" />
    <result column="name" property="name" />
    <result column="age" property="age" />
    <result column="description" property="description" />
  </resultMap>
    <!--删除了 SQL 语句 -->
</mapper>
```

注意：这里删除了 SQL 语句，将其放到 StudentMapper 的注解中了。

8）在 mapper 目录下新建 StudentMapper

```
package com.ascent.mapper;
import com.ascent.po.*;
import java.util.List;
import org.apache.ibatis.annotations.*;

public interfaceStudentMapper {
```

```
    @Insert("insert into student(name,age,description) values(#{name},#{age},
#{description})")
    public void insertStudent(Student student);
    @Select("select * from student where id=#{id}")
    public Student getById(@Param(value="id") Integer id);
    @Select("select * from student")
    public List<Student>findAll();
    @Update("update student set name=#{name},age=#{age},description=#
{description} where id=#{id}")
    public void updateStudent(Student student);
    @Delete("delete from student where id=#{id}")
    public void deleteById(@Param(value="id") Integer id);

}
```

注意：注解部分映射了 SQL 操作。

9）在 test 目录下建立 StudentTest 测试类，含 main()方法

```
package com.ascent.test;

import java.util.List;
import org.apache.ibatis.session.SqlSession;
import org.apache.ibatis.session.SqlSessionFactory;
import com.ascent.po.Student;
import com.ascent.mapper.StudentMapper;
import com.ascent.util.SqlSessionFactoryUtil;

public class StudentTest {
  private SqlSessionFactory factory=SqlSessionFactoryUtil.getSqlSessionFactory();

  /**
   * 测试新增
   */
  public void insert(){
  SqlSession session=null;
    try{
      Student student=new Student();
      student.setName("Kevin");
      student.setDescription("Computer Science");
      session=factory.openSession();
      StudentMapper mapper=session.getMapper(com.ascent.mapper.
          StudentMapper.class);
      mapper.insertStudent(student);
      session.commit();
    } catch (Exception e) {
      e.printStackTrace();
```

```
    }finally{
        session.close();
    }
}

/**
 * 查询单条记录
 */

public void getById(){
    SqlSession session=null;
    try{
        session=factory.openSession();
        StudentMapper mapper=session.getMapper(com.ascent.mapper.
            StudentMapper.class);
        Student student=mapper.getById(3);
        System.out.println(student.getName()+" "+student.getDescription());
    } catch (Exception e) {
        e.printStackTrace();
    }finally{
        session.close();
    }
}

/**
 * 查询所有记录
 */

public void findAll(){
SqlSession session=null;
    try{
        session=factory.openSession();

        StudentMapper mapper=session.getMapper(StudentMapper.class);
        List<Student>students=mapper.findAll();
        System.out.println(students.size());
    } catch (Exception e) {
        e.printStackTrace();
    }finally{
        session.close();
    }
}

/**
 * 更新记录
```

```
  */
public void update(){
  SqlSession session=null;
  try{
    session=factory.openSession();
    StudentMapper mapper=session.getMapper(StudentMapper.class);
    Student student=mapper.getById(1);
    student.setName("Gary");

    mapper.updateStudent(student);
    session.commit();
  } catch (Exception e) {
    e.printStackTrace();
  }finally{
    session.close();
  }
}

/**
  * 删除记录
  */
public void delete(){
  SqlSession session=null;
  try{
    session=factory.openSession();
    StudentMapper mapper=session.getMapper(StudentMapper.class);

    mapper.deleteById(1);
    session.commit();
  } catch (Exception e) {
    e.printStackTrace();
  }finally{
    session.close();
  }
}
public static void main(String[]args){

  StudentTest testCase=new StudentTest();

  //testCase.insert();
  testCase.getById();
  //testCase.update();
  //testCase.delete();
  //testCase.findAll();
```

```
    }

}
```

10）运行 StudentTest，得到结果

3. 映射器接口调用 SQLProvider 生成的 SQL

先介绍一下相关背景知识，这里主要介绍@SelectProvider、@InsertProvider、@UpdateProvider 和@DeleteProvider 的使用方式。

这几个注解声明在 Mapper 对应的 interface 方法上，注解用于生成增、删、改、查用的 SQL 语句。如果对应的 Mapper 中已使用@Param 注解参数，则在对应的 Prodiver 的方法中无须写参数。

注解中的参数：

- type 参数指定的 Class 类，必须能够通过无参的构造函数初始化；
- method 参数指定的方法，必须是 public 的，返回值必须为 String，可以为 static。

1）@SelectProvider

@ResultMap 注解用于从查询结果集 RecordSet 中取数据，然后拼装实体 bean。

```
public interface UserMapper {
    @SelectProvider(type=SqlProvider.class, method="selectUser")
    @ResultMap("userMap")
    public User getUser(long userId);
}

public class SqlProvider {

    public String selectUser(long userId) {
        SELECT("id, name, email");
        FROM("USER");
        WHERE("ID=#{userId}");

    }
}
```

上例中定义了一个 Mapper 接口，其中定义了一个 getUser()方法，这个方法根据用户 id 获取用户信息，并返回相应的 User，而对应的 SQL 语句则写在 SqlProvider 类中。

2）@InsertProvider

```
public interface UserMapper {
    @InsertProvider(type=SqlProvider.class, method="addUser")
    @Options(useGeneratedKeys=true, keyProperty="id")
    intaddUser(Tutor tutor);
}

public class SqlProvider {
```

```
    public String addUser(User user) {
        return new SQL(){
            {
                INSERT_INTO("USER");
                if (user.getName() !=null) {
                    VALUES("NAME", "#{name}");
                }
                if (user.getEmail() !=null) {
                    VALUES("EMAIL", "#{email}");
                }
            }
        }.toString();
    }
}
```

3) @UpdateProvider

```
public interface UserMapper {
    @UpdateProvider(type=SqlProvider.class, method="updateUser")
    int updateUser(User user);
}

public class SqlProvider {
    public String updateUser(User user) {
        return new SQL(){
            {
                UPDATE("USER");
                if (user.getName() !=null) {
                    SET("NAME=#{name}");
                }
                if (user.getEmail() !=null) {
                    SET("EMAIL=#{email}");
                }
                WHERE("ID=#{id}");
            }
        }.toString();
    }
}
```

4) @DeleteProvider

```
public interface UserMapper {
    @DeleteProvider(type=SqlProvider.class, method="deleteUser")
    intdeleteUser(int id);
}

public class SqlProvider {
```

```
    public String deleteUser(int id) {
        return new SQL(){
            {
                DELETE_FROM("USER");
                WHERE("ID=#{id}");
            }
        }.toString();
    }
}
```

接下来看一个 Dynamic SQL Annotation 的实例(Mybatis_DynamicSQL_Annotation_demo)。

1) 在 MySQL 数据库中建立 student 表

```
USE `test`;

/* Table structure for table `student` */

DROP TABLE IF EXISTS `student`;

CREATE TABLE `student`(
  `id` int NOT NULL AUTO_INCREMENT,
  `name`varchar(255) DEFAULT NULL,
  `age` int DEFAULT NULL,
  `description` varchar(255) CHARACTER SET latin1 COLLATE latin1_swedish_ci
DEFAULT NULL,
  PRIMARY KEY (`id`)
) ENGINE=InnoDB AUTO_INCREMENT=4 DEFAULT CHARSET=latin1;

/* Data for the table `student` */

insert into `student`(`id`,`name`,`age`,`description`) values (1,'Gary',25,
'Software Engineer'),(2,'Linda',35,'Marketing'),(3,'Kevin',22,'Computer
Science');
```

2) 新建 Java 项目,导入 MyBatis 所依赖的 jar 包,建立项目目录

```
src/config                        //存放 mybatis-config.xml 和 db.conf 文件,数据
                                    库连接信息配置
src/com/ascent/mybatis/mapper     //存放 Java API 接口文件 StudentMapper.class
src/com/ascent/mybatis/mapping    //存放 SQL XML 映射文件 StudentMapper.xml
src/com/ascent/util               //存放工具文件 SqlSessionFactoryUtil
src/com/ascent/po                 //存放 Java POJO 类,映射查询 Student 实体
src/com/ascent/test               //存放测试运行主方法类 StudentTest
```

3) 在 config 目录下新建 db.conf 文件

```
jdbc.driver=com.mysql.jdbc.Driver
```

```
jdbc.url=jdbc:mysql://localhost:3306/test
jdbc.username=root
jdbc.password=root
```

注意：这里的变量要加上 jdbc.作为前缀，否则 MyBatis 可能无法识别。具体的值根据自己的数据库设置。

4）在 config 目录下新建 mybatis-config 文件

```
<?xml version="1.0" encoding="UTF-8"?>
<!DOCTYPE configuration
  PUBLIC "-//mybatis.org//DTD Config 3.0//EN"
  "http://mybatis.org/dtd/mybatis-3-config.dtd">
<configuration>

  <properties resource="config/db.conf"/>

<typeAliases>
  <typeAlias alias="Student" type="com.ascent.po.Student"/>
</typeAliases>

<environments default="development">
        <environment id="development">
            <transactionManager type="JDBC" />
            <dataSource type="POOLED">
                <property name="driver" value="${jdbc.driver}" />
                <property name="url" value="${jdbc.url}" />
                <property name="username" value="${jdbc.username}" />
                <property name="password" value="${jdbc.password}" />
            </dataSource>
        </environment>
    </environments>
<mappers>
  <!--mapper resource="com/ascent/mapping/StudentMapper.xml"/ -->
  <mapper class="com.ascent.mapper.StudentMapper"/>
</mappers>
    </configuration>
```

注意：mapper 由 resource 改为 class。

5）在 util 中创建 SqlSessionFactoryUtil 类

```
package com.ascent.util;

import java.io.IOException;
import java.io.InputStream;
import org.apache.ibatis.io.Resources;
import org.apache.ibatis.session.SqlSessionFactory;
import org.apache.ibatis.session.SqlSessionFactoryBuilder;
```

```
public class SqlSessionFactoryUtil{

  private static SqlSessionFactory sqlSessionFactory=null;

  static{
    InputStream input=null;
    try{
      input=Resources.getResourceAsStream("config/mybatis-config.xml");
      sqlSessionFactory=new SqlSessionFactoryBuilder().build(input);
    } catch (IOException e) {
      e.printStackTrace();
      }finally{
      if(input !=null){
      try{
        input.close();
      } catch (IOException e) {
        e.printStackTrace();
      }
    }
  }
}

  public static SqlSessionFactory getSqlSessionFactory(){
    return sqlSessionFactory;
  }
}
```

6）在 util 中创建 StudentDynaSqlProvider 类

```
package com.ascent.util;

import org.apache.ibatis.jdbc.SQL;

public class StudentDynaSqlProvider {
    public String findAll() {
        return new SQL()
                .SELECT("*")
                .FROM("student")
                .toString();

    }
    public String getById() {
        return new SQL()
                .SELECT("*")
                .FROM("student")
                .WHERE("id=#{id}")
```

```
                .toString();
    }

    public String insertStudent(){
        return new SQL()
                .INSERT_INTO("student")
                .VALUES("name", "#{name}")
                .toString();
    }
    public String updateStudent(){
        return new SQL()
                .UPDATE("student")
                .SET("name=#{name}")
                .WHERE("id=#{id}")
                .toString();
    }
    public String deleteById(){
        return new SQL()
                .DELETE_FROM("student")
                .WHERE("id=#{id}")
                .toString();
    }

}
```

7）在 po 目录下新建 Student 类

```
package com.ascent.po;

public class Student {
    private int id;
    private String name;                    //姓名
    private int age;                        //年龄
    private String description;             //描述

    public int getId() {
      return id;
    }
    public void setId(int id) {
      this.id=id;
    }
    public String getName() {
      return name;
    }
    public void setName(String name) {
      this.name=name;
```

```
    }
    public int getAge() {
        return age;
    }
    public void setAge(int age) {
        this.age=age;
    }
    public String getDescription() {
      return description;
    }
    public void setDescription(String description) {
      this.description=description;
    }
}
```

8）在 mapping 目录下新建 StudentMapper.xml

```
<?xml version="1.0" encoding="UTF-8"?>
<!DOCTYPE mapper PUBLIC "-//mybatis.org//DTD Mapper 3.0//EN"
"http://mybatis.org/dtd/mybatis-3-mapper.dtd">

<mapper namespace="com.ascent.mapper.StudentMapper">
  <!--定义 Java Bean 的属性与数据库的列之间的映射 -->
  <resultMap type="Student" id="studentResultMap">
    <id column="id" property="id" />
    <result column="name" property="name" />
    <result column="age" property="age" />
    <result column="description" property="description" />
  </resultMap>
    <!--删除了 SQL 语句 -->
</mapper>
```

注意：这里删除了 SQL 语句，将其放到 StudentMapper 的注解中了。

9）在 mapper 目录下新建 StudentMapper

```
package com.ascent.mapper;
import com.ascent.po.*;
import com.ascent.util.StudentDynaSqlProvider;

import java.util.List;
import org.apache.ibatis.annotations.*;

public interface StudentMapper {

    @InsertProvider(type=StudentDynaSqlProvider.class,method="insertStudent")
    public void insertStudent(Student student);
    @SelectProvider(type=StudentDynaSqlProvider.class,method="getById")
```

```
    public Student getById(@Param(value="id") Integer id);
    @SelectProvider(type=StudentDynaSqlProvider.class,method="findAll")
    public List<Student>findAll();
    @UpdateProvider(type=StudentDynaSqlProvider.class,method="updateStudent")
    public void updateStudent(Student student);
@DeleteProvider(type=StudentDynaSqlProvider.class,method="deleteById")
    public void deleteById(@Param(value="id") Integer id);

}
```

注意：注解部分映射了 SQL 操作。

10）在 test 目录下建立 StudentTest 测试类，含 main()方法

```
package com.ascent.test;

import java.util.List;
import org.apache.ibatis.session.SqlSession;
import org.apache.ibatis.session.SqlSessionFactory;
import com.ascent.po.Student;
import com.ascent.mapper.StudentMapper;
import com.ascent.util.SqlSessionFactoryUtil;

public classStudentTest {
  private SqlSessionFactory factory=SqlSessionFactoryUtil.getSqlSessionFactory();

  /**
   * 测试新增记录
   * /
  public void insert(){
    SqlSession session=null;
    try{
      Student student=new Student();
      student.setName("Kevin");
      student.setDescription("Computer Science");
      session=factory.openSession();
      StudentMapper mapper=session.getMapper(com.ascent.mapper.
          StudentMapper.class);
      mapper.insertStudent(student);
      session.commit();
    } catch (Exception e) {
      e.printStackTrace();
    }finally{
      session.close();
    }
  }
```

```
/**
 * 查询单条记录
 * /

public void getById(){
  SqlSession session=null;
  try{
    session=factory.openSession();
    StudentMapper mapper=session.getMapper(com.ascent.mapper.StudentMapper.
class);
    Student student=mapper.getById(2);
    System.out.println(student.getName()+" "+student.getDescription());
  } catch (Exception e) {
    e.printStackTrace();
  }finally{
    session.close();
  }
}

/**
 * 查询所有记录
 * /

public void findAll(){
  SqlSession session=null;
  try{
    session=factory.openSession();

    StudentMapper mapper=session.getMapper(StudentMapper.class);
    List<Student>students=mapper.findAll();
    System.out.println(students.size());
  } catch (Exception e) {
    e.printStackTrace();
  }finally{
    session.close();
  }
}

/**
 * 更新记录
 * /
public void update(){
  SqlSession session=null;
  try{
    session=factory.openSession();
```

```
            StudentMapper mapper=session.getMapper(StudentMapper.class);
            Student student=mapper.getById(1);
            student.setName("Gary");

            mapper.updateStudent(student);
            session.commit();
        } catch (Exception e) {
            e.printStackTrace();
        }finally{
            session.close();
        }
    }

    /**
     * 删除记录
     * /
    public void delete(){
        SqlSession session=null;
        try{
            session=factory.openSession();
            StudentMapper mapper=session.getMapper(StudentMapper.class);

            mapper.deleteById(1);
            session.commit();
        } catch (Exception e) {
            e.printStackTrace();
        }finally{
            session.close();
        }
    }
    public static void main(String[]args){

        StudentTest testCase=new StudentTest();

        //testCase.insert();
        testCase.getById();
        //testCase.update();
        //testCase.delete();
        testCase.findAll();

    }

}
```

11）运行 StudentTest，得到结果

9.3 日志

1. MyBatis 日志概述

MyBatis 内置日志工厂提供日志功能。内置日志工厂将日志交给以下其中一种工具作代理：

- SLF4J；
- Apache Commons Logging；
- Log4J 2；
- Log4J；
- JDK logging。

MyBatis 内置日志工厂基于运行时自省机制选择合适的日志工具。它会使用第一个查找得到的工具(按上文列举的顺序查找)。如果一个都未找到，日志功能就会被禁用。

不少应用服务器(如 Tomcat 和 WebShpere)的类路径中已经包含 Apache Commons Logging，所以在这一种配置环境下的 MyBatis 会把它作为日志工具，记住这一点非常重要。这将意味着，在诸如 WebSphere 的环境中，它提供了 Apache Commons Logging 的私有实现，你的 Log4J 配置将被忽略。MyBatis 将你的 Log4J 配置忽略掉是相当令人郁闷的(事实上，正是因为在这种配置环境下，MyBatis 才会选择使用 Apache Commons Logging，而不是 Log4J)。如果你的应用部署在一个类路径已经包含 Apache Commons Logging 的环境中，而你又想使用其他日志工具，可以通过在 MyBatis 配置文件 mybatis-config.xml 里添加一项 setting 选择别的日志工具。

```
<configuration>
  <settings>
    ...
    <setting name="logImpl" value="LOG4J"/>
    ...
  </settings>
</configuration>
```

logImpl 可选的值有 SLF4J、LOG4J、LOG4J2、JDK_LOGGING、COMMONS_LOGGING、STDOUT_LOGGING、NO_LOGGING，或者是实现了接口 org.apache.ibatis.logging.Log 的，且构造方法是以字符串为参数的类的完全限定名。也可以调用如下任一方法来使用日志工具：

```
org.apache.ibatis.logging.LogFactory.useSlf4jLogging();
org.apache.ibatis.logging.LogFactory.useLog4JLogging();
org.apache.ibatis.logging.LogFactory.useJdkLogging();
org.apache.ibatis.logging.LogFactory.useCommonsLogging();
org.apache.ibatis.logging.LogFactory.useStdOutLogging();
```

如果决定调用以上某个方法，请在调用其他 MyBatis 方法之前调用它。另外，仅当运

行时类路径中存在该日志工具时，调用与该日志工具对应的方法才会生效，否则 MyBatis 一概忽略。如环境中并不存在 Log4J，却调用了相应的方法，MyBatis 就会忽略这一调用，转而以默认的查找顺序查找日志工具。

2. Log4J 日志配置

可以对包、映射类的全限定名、命名空间或全限定语句名开启日志功能，查看 MyBatis 的日志语句。

再次说明，具体怎么做由使用的日志工具决定，这里以 Log4J 为例。配置日志功能非常简单：添加一个或多个配置文件(如 log4j.properties)，添加 jar 包(如 log4j.jar)。下面的例子使用 Log4J 配置完整的日志服务，共两个步骤：

(1) 添加 Log4J 的 jar 包。

因为使用的是 Log4J，所以要确保它的 jar 包在应用中是可用的。要启用 Log4J，只要将 jar 包添加到应用的类路径中即可。Log4J 的 jar 包可以从上面的链接中下载。

对于 Web 应用或企业级应用，需要将 Log4J 的 jar 包添加到 WEB-INF/lib 目录下；对于独立应用，可以将它添加到 JVM 的-classpath 启动参数中。

(2) 配置 Log4J。

配置 Log4J 比较简单，假如需要记录这个映射器接口的日志：

```
package org.mybatis.example;
public interface BlogMapper {
  @Select("SELECT * FROMblog WHERE id=#{id}")
  Blog selectBlog(int id);
}
```

在应用的类路径中创建一个名称为 log4j.properties 的文件，文件的具体内容如下：

```
#Global logging configuration
log4j.rootLogger=ERROR, stdout
#MyBatis logging configuration...
log4j.logger.org.mybatis.example.BlogMapper=TRACE
#Console output...
log4j.appender.stdout=org.apache.log4j.ConsoleAppender
log4j.appender.stdout.layout=org.apache.log4j.PatternLayout
log4j.appender.stdout.layout.ConversionPattern=%5p [%t]- %m%n
```

添加以上配置后，Log4J 就会记录 org.mybatis.example.BlogMapper 的详细执行操作，且仅记录应用中其他类的错误信息(若有)。

也可以将日志的记录方式从接口级别切换到语句级别，从而实现更细粒度的控制。如下配置只对 selectBlog 语句记录日志：

```
log4j.logger.org.mybatis.example.BlogMapper.selectBlog=TRACE
```

与此相对，可以对一组映射器接口记录日志，只要对映射器接口所在的包开启日志功能即可：

```
log4j.logger.org.mybatis.example=TRACE
```

某些查询可能返回庞大的结果集,此时只想记录其执行的 SQL 语句,而不想记录结果该怎么办?为此,MyBatis 中 SQL 语句的日志级别被设为 DEBUG(JDK 日志被设为 FINE),结果的日志级别为 TRACE(JDK 日志被设为 FINER)。所以,只要将日志级别调整为 DEBUG 即可达到目的:

```
log4j.logger.org.mybatis.example=DEBUG
```

要记录日志的是类似下面的映射器文件,而不是映射器接口又该怎么做呢?

```
<?xml version="1.0" encoding="UTF-8" ?>
<!DOCTYPE mapper
  PUBLIC "-//mybatis.org//DTD Mapper 3.0//EN"
  "http://mybatis.org/dtd/mybatis-3-mapper.dtd">
<mapper namespace="org.mybatis.example.BlogMapper">
  <select id="selectBlog" resultType="Blog">
    select * from Blog where id=#{id}
  </select>
</mapper>
```

如需对 XML 文件记录日志,只对命名空间增加日志记录功能即可:

```
log4j.logger.org.mybatis.example.BlogMapper=TRACE
```

要记录具体语句的日志,可以这样做:

```
log4j.logger.org.mybatis.example.BlogMapper.selectBlog=TRACE
```

可见,为映射器接口和 XML 文件添加日志功能的语句毫无差别。

注意:如果使用的是 SLF4J 或 Log4J 2,MyBatis 将以 MYBATIS 这个值进行调用。

配置文件 log4j.properties 的余下内容是针对日志输出源的,这一内容已经超出本文档范围。关于 Log4J 的更多内容,可以参考 Log4J 的网站。

目前为止,分别介绍了 Spring 和 MyBatis,接下来看一个 Spring 整合 MyBatis 的实例(Spring_Mybatis_demo)。

(1) 在 MySQL 数据库中建立 student 表。

```
USE `test`;

/* Table structure for table `student` */

DROP TABLE IF EXISTS `student`;

CREATE TABLE `student`(
  `id` int NOT NULL AUTO_INCREMENT,
  `name`varchar(255) DEFAULT NULL,
  `age` int DEFAULT NULL,
  `description` varchar(255) CHARACTER SET latin1 COLLATE latin1_swedish_ci
DEFAULT NULL,
  PRIMARY KEY (`id`)
```

```
) ENGINE=InnoDB AUTO_INCREMENT=4 DEFAULT CHARSET=latin1;

/* Data for the table `student` */

insert into `student`(`id`,`name`,`age`,`description`) values (1,'Gary',25,
'Software Engineer'),(2,'Linda',35,'Marketing'),(3,'Kevin',22,'Computer
Science');
```

(2) 新建 Java 项目,导入 Spring 4.1.0 的 jar 包和 MyBatis 的 jar 包,这里还要多导入一个 mybatis-spring-1.3.2 的 jar 包,如图 9-1 所示。

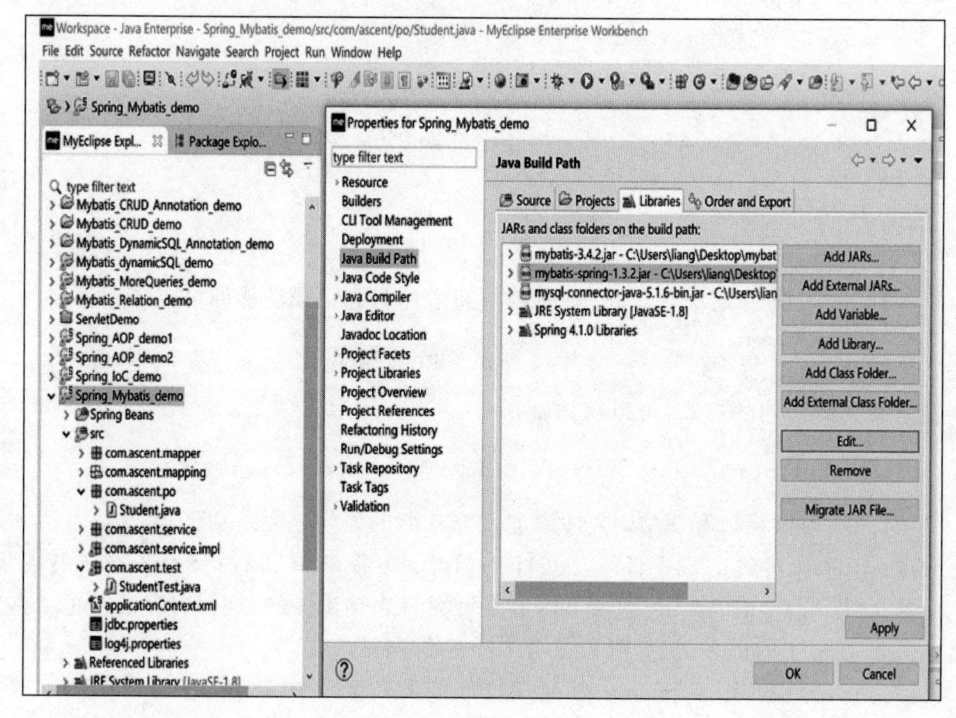

图 9-1 导入 jar 包

(3) 在 src 下建立项目目录。

```
com/ascent/mapper              //存放 Java API 接口文件 StudentMapper.class
com/ascent/mapping             //存放 SQL XML 映射文件 StudentMapper.xml
com/ascent/po                  //存放 Java POJO 类,映射 Student 实体
com/ascent/service             //存放 Java 业务类的接口 IStudentService
com/ascent/service/impl        //存放 Java 业务类的实现 IStudentServiceImpl

com/ascent/test                //存放测试运行主方法类 StudentTest
```

(4) 在 src 下编写 jdbc.properties,存放数据库配置信息。

```
jdbc.url=jdbc:mysql://localhost:3306/test?useUnicode=true&characterEncoding=
utf-8&autoReconnect=true
jdbc.driver=com.mysql.jdbc.Driver
```

```
jdbc.username=root
jdbc.password=root
```

（5）在 src 下编写 log4j.properties，存放日志配置信息。

```
log4j.rootLogger=INFO, stdout
log4j.appender.stdout=org.apache.log4j.ConsoleAppender
log4j.appender.stdout.layout=org.apache.log4j.PatternLayout
log4j.appender.stdout.layout.ConversionPattern=%5p [%t]-%m%n
```

（6）在 src 下编辑 applicationConetxt.xml，它已经把 mybatis-config.xml 中的 dataSource、mappers 等配置整合成 bean 或 peroperty。

```
<?xml version="1.0" encoding="UTF-8"?>
<beans xmlns="http://www.springframework.org/schema/beans"
    xmlns:xsi="http://www.w3.org/2001/XMLSchema-instance"
xmlns:p="http://www.springframework.org/schema/p"
    xmlns:context="http://www.springframework.org/schema/context"
    xmlns:mvc="http://www.springframework.org/schema/mvc"
    xsi:schemaLocation="http://www.springframework.org/schema/beans
    http://www.springframework.org/schema/beans/spring-beans-3.1.xsd
    http://www.springframework.org/schema/context
    http://www.springframework.org/schema/context/spring-context-3.1.xsd
    http://www.springframework.org/schema/mvc
    http://www.springframework.org/schema/mvc/spring-mvc-4.0.xsd">

  <!--自动扫描所有包,加载装配所有的实体类 bean -->
  <context:component-scan base-package="com.ascent"/>

<!--加载配置文件 -->
<context:property-placeholder location="classpath:*.properties" />

<!--配置数据源 -->
<bean id="dataSource"
class="org.springframework.jdbc.datasource.DriverManagerDataSource">
  <property name="driverClassName" value="${jdbc.driver}"/>
  <property name="url" value="${jdbc.url}"/>
  <property name="username" value="${jdbc.username}"/>
  <property name="password" value="${jdbc.password}"/>
</bean>

<!--sqlSessionFactory -->
<bean id="sqlSessionFactory" class="org.mybatis.spring.SqlSessionFactoryBean">
  <!--加载 mapper.xml 的配置文件 -->
  <property name="mapperLocations" value="com/ascent/mapping/*.xml"/>
  <!--添加所需的数据源 -->
  <property name="dataSource" ref="dataSource"/>
```

```
</bean>

<!--扫描 Dao 包,加载接口映射类 -->
<bean class="org.mybatis.spring.mapper.MapperScannerConfigurer">
    <property name="basePackage" value="com.ascent.mapper" />
    <property name="sqlSessionFactoryBeanName" value="sqlSessionFactory" />
</bean>


<!--(事务管理)transaction manager -->
  <bean id="transactionManager"
      class="org.springframework.jdbc.datasource.DataSourceTransactionManager">
      <property name="dataSource" ref="dataSource" />
  </bean>

</beans>
```

(7) 在 po 目录下新建 Student 类。

```
package com.ascent.po;

public class Student {
    private int id;
    private String name;             //姓名
    private int age;                 //年龄
    private String description;      //描述

    public int getId() {
      return id;
    }
    public void setId(int id) {
      this.id=id;
    }
    public String getName() {
      return name;
    }
    public void setName(String name) {
      this.name=name;
    }
    public int getAge() {
        return age;
    }
    public void setAge(int age) {
        this.age=age;
    }
    public String getDescription() {
```

```
        return description;
    }
    public void setDescription(String description) {
        this.description=description;
    }
}
```

(8) 在 mapping 目录下新建 StudentMapper.xml。

```
<?xml version="1.0" encoding="UTF-8"?>
<!DOCTYPE mapper PUBLIC "-//mybatis.org//DTD Mapper 3.0//EN"
"http://mybatis.org/dtd/mybatis-3-mapper.dtd">
<mapper namespace="com.ascent.mapper.StudentMapper">
  <!--定义 Java Bean 的属性与数据库的列之间的映射 -->
  <resultMap type="com.ascent.po.Student" id="studentResultMap">
    <id column="id" property="id" />
    <result column="name" property="name" />
    <result column="age" property="age" />
    <result column="description" property="description" />
  </resultMap>

  <!--查询所有记录 -->
  <select id="findAll" resultMap="studentResultMap">
    select * from student
  </select>

</mapper>
```

(9) 在 mapper 目录下新建 StudentMapper。

```
package com.ascent.mapper;
import com.ascent.po.Student;
import java.util.List;

public interface StudentMapper {

    public List<Student> findAll();

}
```

(10) 在 service 目录下新建 IStudentService。

```
package com.ascent.service;

import java.util.List;

import com.ascent.po.Student;
```

```
public interface IStudentService {

    public List<Student>findAll();

}
```

(11) 在 service/impl 目录下新建 IStudentServiceImpl。

```
package com.ascent.service.impl;

import java.util.List;

import org.springframework.context.ApplicationContext;
import org.springframework.context.support.ClassPathXmlApplicationContext;
import org.springframework.stereotype.Component;
import org.springframework.stereotype.Service;

import com.ascent.mapper.StudentMapper;
import com.ascent.po.Student;
import com.ascent.service.IStudentService;

@Service
public classIStudentServiceImpl implements IStudentService {

    private StudentMapper studentMapper;

    @Override
    public List<Student>findAll() {

        ApplicationContext apc=new ClassPathXmlApplicationContext
            ("applicationContext.xml");
        studentMapper=apc.getBean(StudentMapper.class);
        //使用 StudentMapper 接口完成 SQL 查询操作
        List<Student>students=studentMapper.findAll();

        return students;
    }
}
```

注意:@Service 标注业务层组件。这个注解写在类的上面,标注将这个类交给 Spring 容器管理,相当于 applicationContext.xml 配置中的 bean,Spring 容器要为它创建对象。

(12) 在 test 目录下建立 StudentTest 测试类,含 main()方法。

```
package com.ascent.test;

import java.util.List;
```

```
import org.springframework.context.ApplicationContext;
import org.springframework.context.support.ClassPathXmlApplicationContext;

import com.ascent.po.Student;
import com.ascent.service.IStudentService;
import com.ascent.service.impl.IStudentServiceImpl;

public class StudentTest {

    public static void main(String[] args) {

        ApplicationContext apc=new ClassPathXmlApplicationContext
            ("applicationContext.xml");
        IStudentService stuService=apc.getBean(IStudentServiceImpl.class);
        List<Student> students=stuService.findAll();
        for(Student s : students){
            System.out.println(s.getName());
        }

    }
}
```

(13) 运行 StudentTest 类,得到如下结果:

```
Gary
Linda
Kevin
```

9.4 项目案例

9.4.1 学习目标

了解 MyBatis 的高级特性,掌握外部插件的使用。

9.4.2 案例描述

本案例使用 MyBatis 的 PageHelper 插件,将一般用户浏览的查询头版头条新闻和综合新闻改为使用 MyBatis 实现,在业务代码中使用该插件完成分页查询操作。

9.4.3 案例要点

在已经安装配置了 MyBatis 环境的项目 electrone 中取消 Spring 的 JdbcTemplate,采用独立的 MyBatis 查询方式查询头版头条新闻和综合新闻。使用 PageHelper 安装查询分页插件。修改 IAuthorizationServiceImpl 中获取头版头条新闻和综合新闻的业务方法。

9.4.4 案例实施

(1) 引入 PageHelper 插件包,如图 9-2 所示。

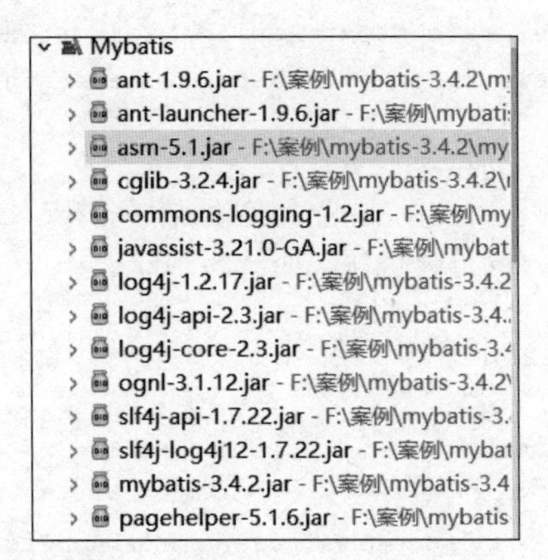

图 9-2 引入 PageHelper 插件包

添加 jsqlparser 的 jar 包,否则 PageHelper 会抛出异常,如图 9-3 所示。

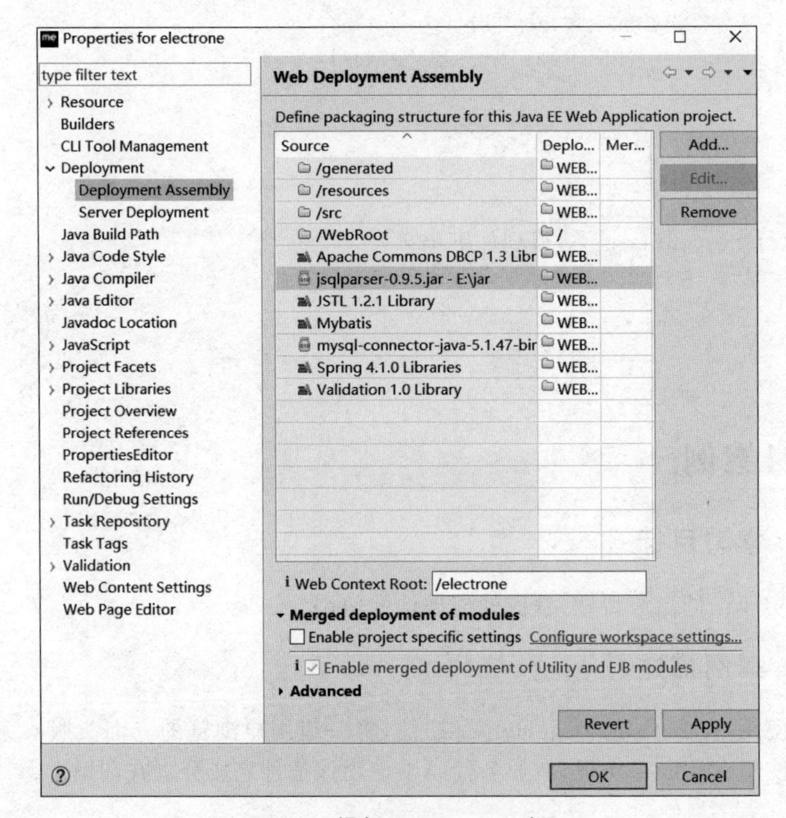

图 9-3 添加 jsqlparser.jar 包

(2) 修改 NewsMapping.xml,增加查询头版头条新闻和综合新闻的查询语句。

```
NewsMappingxml:
增加如下内容
<sql id="Base_Column_List">
    id, title, author,deptid, content, type, checkopinion, checkopinion2,
    checkstatus, crosscolumn, crossstatus, picturepath, publishtime,
```

```
    crosspubtime, preface, status, userid
  </sql>
  <!--查询头版头条新闻-->
  <select id="findHeaderNews" resultType="com.ascent.po.News">
    select
      <include refid="Base_Column_List" />
      from News n where (n.status=1 and n.type=1) or
        (n.type=2 and n.checkstatus=2 and n.status=1) order by n.id desc limit 1
  </select>
  <!--查询综合新闻-->
  <select id="findIndexNews" resultType="com.ascent.po.News">
      select
        <includerefid="Base_Column_List" />
        from News n where n.status=1 and n.type=2 and n.checkstatus!=2 order by n.
        id desc
</select>
```

(3) 修改 NewsMapper.java，增加两个方法。

```
package com.ascent.mybatis.mapper;

import java.util.List;
import com.ascent.po.News;

public interface NewsMapper {

    public int insert(News news);
    public News findHeaderNews();
    public List<News> findIndexNews();
}
```

(4) 修改 IAuthorizationServiceImpl.java。

```
package com.ascent.service.impl;

import java.util.ArrayList;
import java.util.HashMap;
import java.util.List;

import com.ascent.mybatis.mapper.NewsMapper;
import com.ascent.mybatis.utils.MybatisSqlSessionFactoryUtil;
import com.ascent.po.News;
import com.ascent.service.IAuthorizationService;
import com.github.pagehelper.PageHelper;

public class IAuthorizationServiceImpl implements IAuthorizationService{
```

```
    private NewsMapper newsMapper;

    /**
     * 使用 NewsDAO 完成获取头版头条新闻和综合新闻前 6 条的业务过程
     *
     * @return 返回包含了头版头条新闻和综合新闻的 HashMap,
     * 头版头条新闻对应一个 key:headerNews,综合新闻对应一个 Key:indexNews
     */
    @Override
    public HashMap<String, List<News>>findHeaderNews() {

        /*替换原有的 jdbcTemplate 使用*/
        //获取 NewsMapper 对象
        //使用 MyBatis 的 SqlSession
        newsMapper=MybatisSqlSessionFactoryUtil.getSqlSession().getMapper
            (NewsMapper.class);
        //查询头版头条新闻
        News headerNews=newsMapper.findHeaderNews();

        List<News>headerNewsList=new ArrayList<News>(1);
        //构造空的 List<News>集合对象,集合成员必须为 News 对象

        headerNewsList.add(headerNews);    //将查询得到的新闻头版头条保存到 List 中

        //针对查询使用 PageHelper 进行分页,第一个参数 1 表示第一页,第二个参数 6 表示有
          6 条记录
        PageHelper.startPage(1, 6);
        List<News>indexNews=newsMapper.findIndexNews();
        //取消之前取前 6 条记录的业务操作

        //定义并初始化 HashMap<String,List<News>>对象
        HashMap<String, List<News>>headerAndIndexNews=new HashMap<String,
            List<News>>(2);

        //将头版头条新闻保存在 Map 中
        headerAndIndexNews.put("headerNews", headerNewsList);
        //将综合新闻保存在 Map 中
        headerAndIndexNews.put("indexNews", indexNews);

        return headerAndIndexNews;
    }
}
```

(5) 修改 mybatis-config.xml,添加 plugins 插件。

注意:plugins 标签要写在 enviroments 标签之前。

```
<plugins>
    <!--com.github.pagehelper 为 PageHelper 类所在包名 -->
    <plugin interceptor="com.github.pagehelper.PageInterceptor">
        <!--使用 MySQL 的分页 -->
        <property name="helperDialect" value="mysql"/>
        <property name="pageSizeZero" value="true"/>
    </plugin>
</plugins>
```

(6) 由于之前在 Spring 中配置 IAuthorizationServiceImpl 注入了 NewsDAO,现在需要取消注入。

修改 applicationContext.xml 中的以下内容:

```
<!--配置 IAuthorizationService -->
    <bean id="iAuthorizationService" class="com.ascent.service.impl.
        IAuthorizationServiceImpl" >
    </bean>
```

(7) 不需要改动 IndexController 中的方法。

(8) 部署项目,运行 tomcat,输入 http://localhost:8080/electrone/,运行结果如图 9-4 所示,表示 MyBatis 的分页插件配置和运行成功。

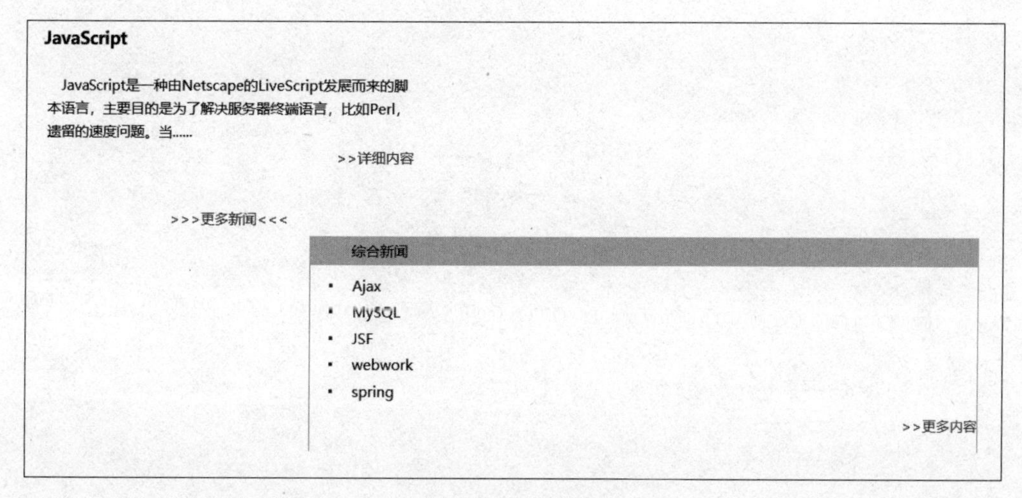

图 9-4 运行结果

9.4.5 特别提示

PageHelper 依赖于 jsqlparser 包,因此必须导入两个 jar 包。该插件主要用于分页查询。配置 mybatis-config.xml 中,plugins 的位置要优先于 enviroments,否则 MyEclipse 校验会出错。

在 Spring MVC 中依然调用 Spring ApplicationContext 对象,获取配置的 IAuthorizationServiceImpl 对象,没有进行整合,需要手动代码获取,IAuthorizationServiceImpl 中使用 MybatisSqlSessionFactoryUtils 获取 sqlSession 对象,再获取 NewsMapper.class 接

口。这一过程没有使用 Spring 的 DI/IoC 特性。在第 10 章的 SSM 整合时,一切都要围绕 Spring 展开配置。

9.4.6 拓展与提高

在 eGov 项目中,有发布新闻权限的用户具有查看自己发布的新闻列表的功能,请自行开发 MyBatis 版,查看发布的新闻列表,可以根据页码翻页查看,注意,要使用 PageHelper 插件。

习题

1. MyBatis 动态 SQL 元素主要有哪些?
2. MyBatis 提供哪些基本注解?
3. MyBatis 有哪两种使用注解的方式?
4. MyBatis 的内置日志工厂可以将日志交给哪些工具作代理?
5. MyBatis 如何进行 Log4J 日志配置?

学习目的与学习要求

学习目的：学会Spring、Spring MVC和MyBatis 3个框架的搭建流程及整合过程。

学习要求：熟练整合Spring、Spring MVC和MyBatis 3个框架，并开发基于Spring＋Spring MVC＋MyBatis的Web应用。

本章主要内容

本章重点学习Spring＋Spring MVC＋MyBatis整合过程。

在介绍完Spring、Spring MVC、MyBatis 3个框架之后，接下来通过用户登录功能讲解Spring＋Spring MVC＋MyBatis的集成原理和开发步骤。

10.1 Spring＋Spring MVC＋MyBatis集成原理和实例

1. 环境版本，以及开发前的准备

IDE：MyEclipse 2017 CI 10

数据库：MySQL 8

Spring MVC 4.1.0/Spring 4.1.0/MyBatis-3.4.2

JDK 1.8/Tomcat 8.5

在使用MyEclipse进行开发之前，需要下载如下组件：mybatis-3.4.2、mybatis-spring-1.3.2、mysql-connector-java-5.1.6-bin.jar、pagehelper-5.1.9，以及mybatis-3.4.2目录下的lib目录中的jar包，如图10-1所示。

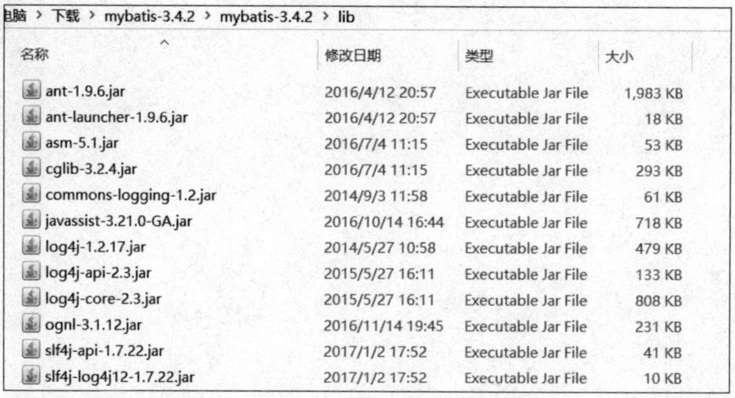

电脑 › 下载 › mybatis-3.4.2 › mybatis-3.4.2 › lib

名称	修改日期	类型	大小
ant-1.9.6.jar	2016/4/12 20:57	Executable Jar File	1,983 KB
ant-launcher-1.9.6.jar	2016/4/12 20:57	Executable Jar File	18 KB
asm-5.1.jar	2016/7/4 11:15	Executable Jar File	53 KB
cglib-3.2.4.jar	2016/7/4 11:15	Executable Jar File	293 KB
commons-logging-1.2.jar	2014/9/3 11:58	Executable Jar File	61 KB
javassist-3.21.0-GA.jar	2016/10/14 16:44	Executable Jar File	718 KB
log4j-1.2.17.jar	2014/5/27 10:58	Executable Jar File	479 KB
log4j-api-2.3.jar	2015/5/27 16:11	Executable Jar File	133 KB
log4j-core-2.3.jar	2015/5/27 16:11	Executable Jar File	808 KB
ognl-3.1.12.jar	2016/11/14 19:45	Executable Jar File	231 KB
slf4j-api-1.7.22.jar	2017/1/2 17:52	Executable Jar File	41 KB
slf4j-log4j12-1.7.22.jar	2017/1/2 17:52	Executable Jar File	10 KB

图 10-1　mybatis-3.4.2 的 lib 目录中的 jar 包

2. 开发步骤

(1) 在 MySQL(注意,这里的 MySQL 数据库的用户名和密码都是 root,读者可以根据需要改为自己的属性值)中创建数据库和 usr 表,并插入两行记录,SQL 语句如下:

```
/*
SQLyog Ultimate v12.09 (64 bit)
MySQL-5.5.60 : Database-test
*********************************************************************
*/

CREATE DATABASE /*!32312 IF NOT EXISTS*/`test` /*!40100 DEFAULT CHARACTER SET
utf8 */;

USE `test`;

/*Table structure for table `usr` */

DROP TABLE IF EXISTS `usr`;

CREATE TABLE `usr` (
  `id` int(11) NOT NULL AUTO_INCREMENT,
  `username`varchar(255) DEFAULT NULL,
  `password`varchar(255) DEFAULT NULL,
  PRIMARY KEY (`id`)
) ENGINE=InnoDB AUTO_INCREMENT=3 DEFAULT CHARSET=utf8;

/*Data for the table `usr` */

insert  into `usr`(`id`,`username`,`password`) values (1,'Lixin','123456'),(2,
'admin','123456');
```

(2) 创建 Web 工程,如图 10-2 所示。

单击 Next 按钮,如图 10-3 所示。

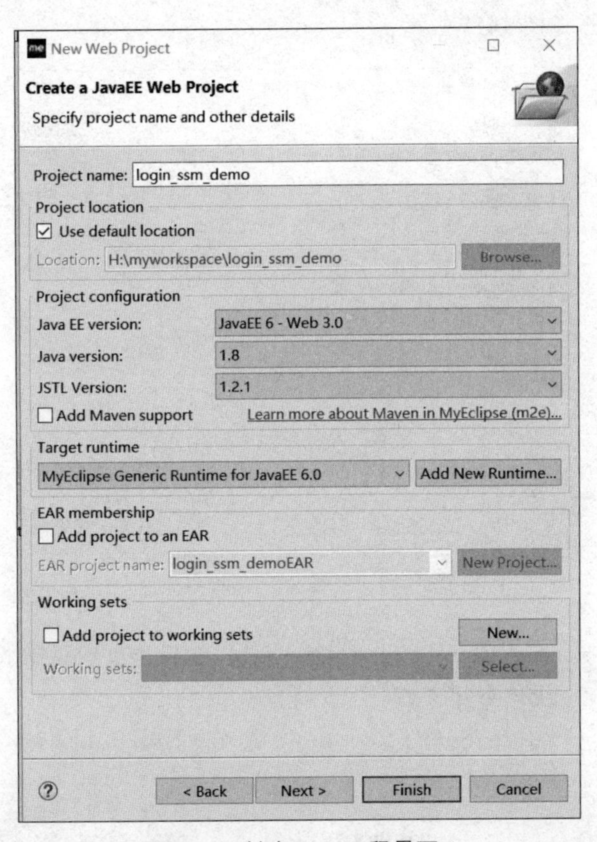

图 10-2 创建 Web 工程界面 1

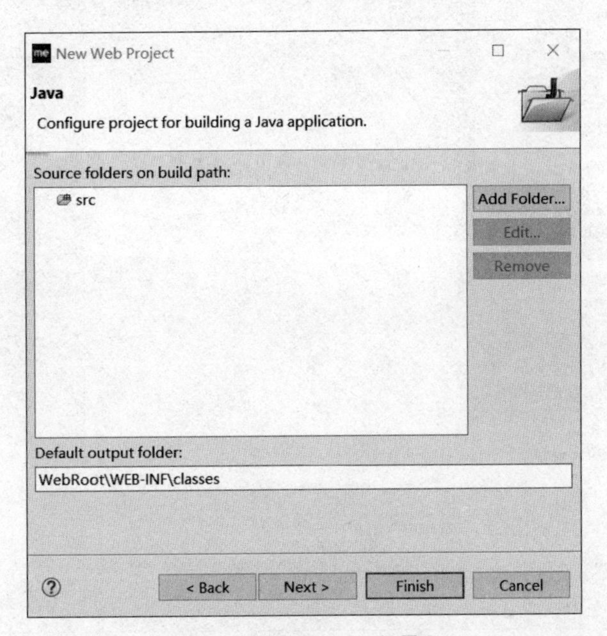

图 10-3 创建 Web 工程界面 2

单击 Next 按钮，如图 10-4 所示。

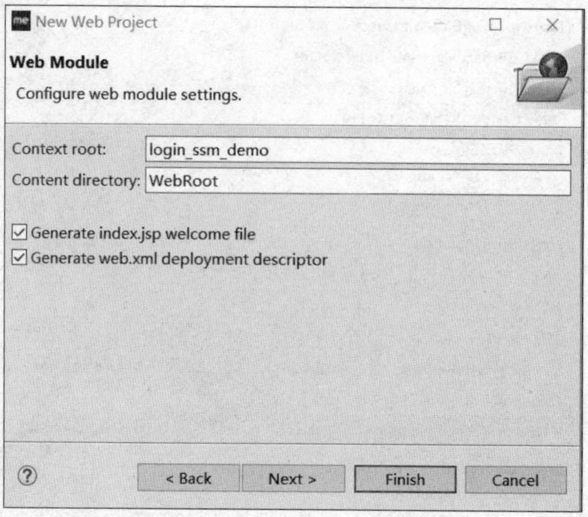

图 10-4 创建 Web 工程界面 3

勾选 Generate web.xml deployment descriptor，之后单击 Finish 按钮。

(3) 创建 MyBatis 用户自定义库。

选择 Windows→Preferences→Java→Build Path→User Libraries，如图 10-5 所示。

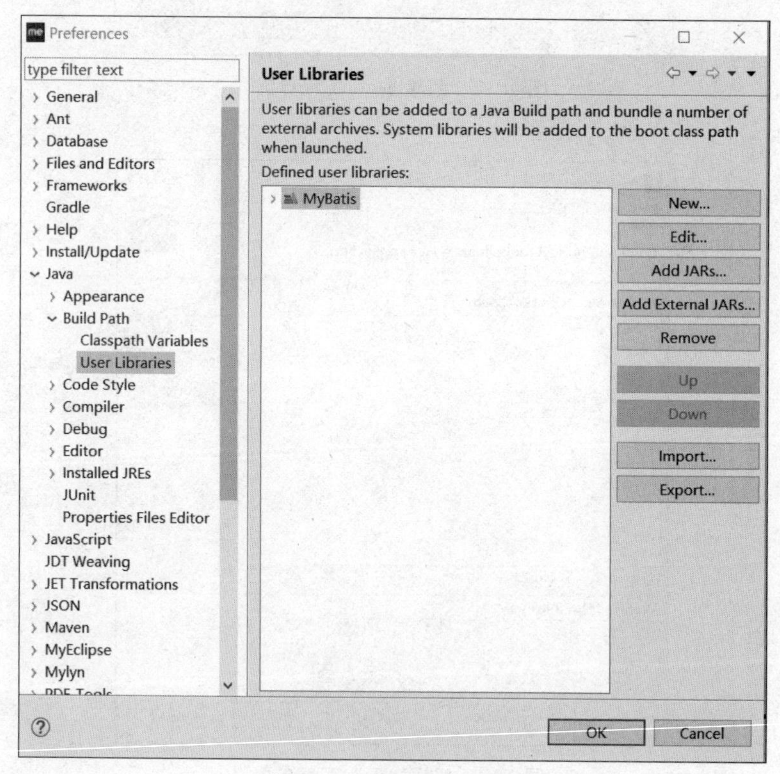

图 10-5 User Libraries

单击 New 按钮，新建 MyBatis 用户库。单击 Add External JARs 按钮，将上文中已经下载并解压缩完成的 jar 文件加入 MyBatis 库，如图 10-6 所示。

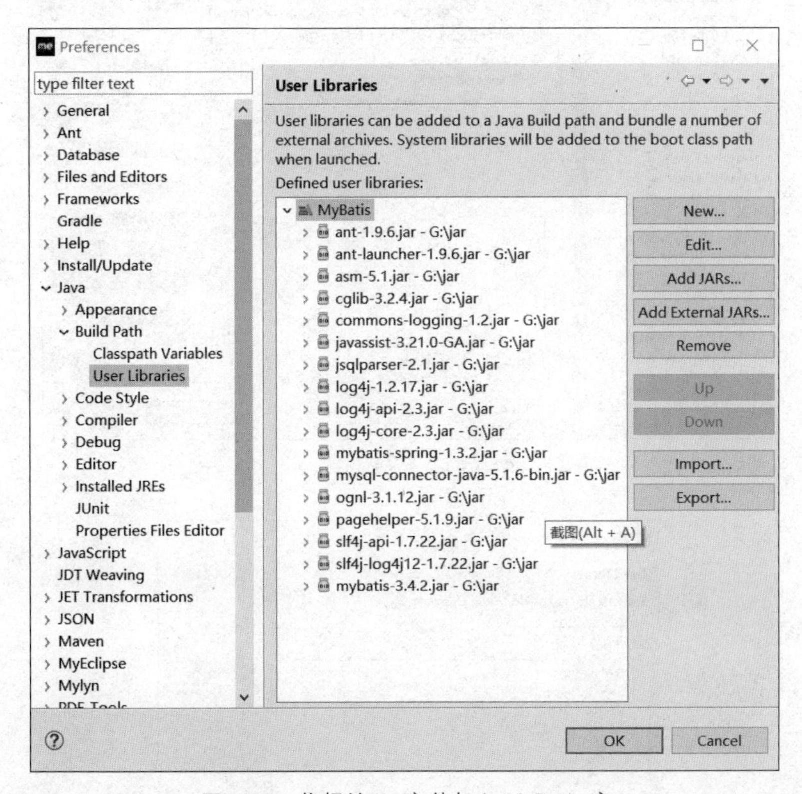

图 10-6　将相关 jar 文件加入 MyBatis 库

单击 OK 按钮，关闭对话框。

(4) 为 login_ssm_demo 项目添加 MyBatis 自定义库

右击 login_ssm_demo 项目，选择 Build Path→Configure Build Path，如图 10-7 所示。

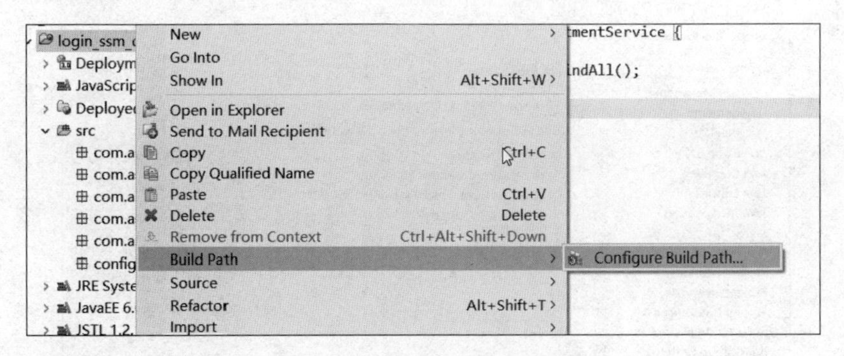

图 10-7　Configure Build Path

选择 Libraries 选项卡，之后单击 Add Library 按钮，如图 10-8 所示。

选择 User Library，勾选 MyBatis，如图 10-9 所示。

单击 Finish 按钮，出现图 10-10。

单击 OK 按钮，完成 MyBatis 库的引用配置。

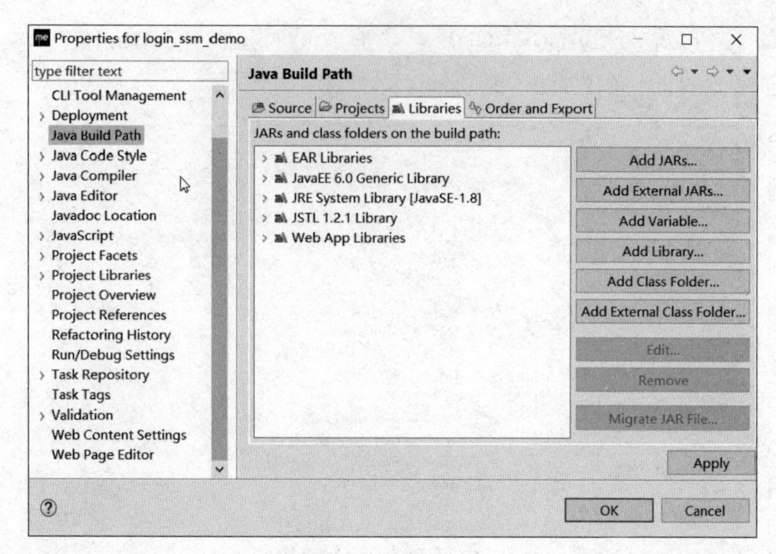

图 10-8　Add Library

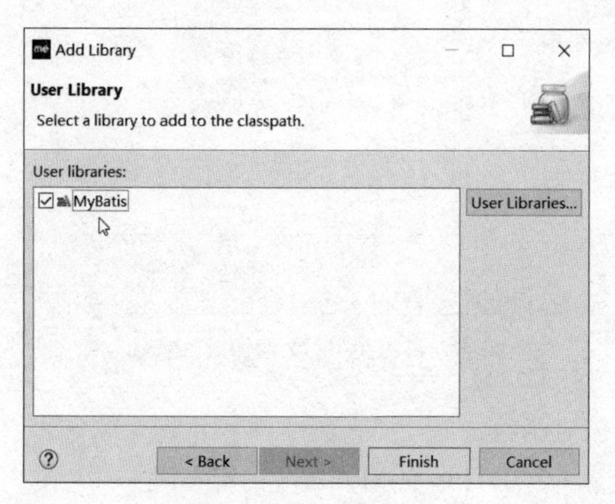

图 10-9　勾选 MyBatis

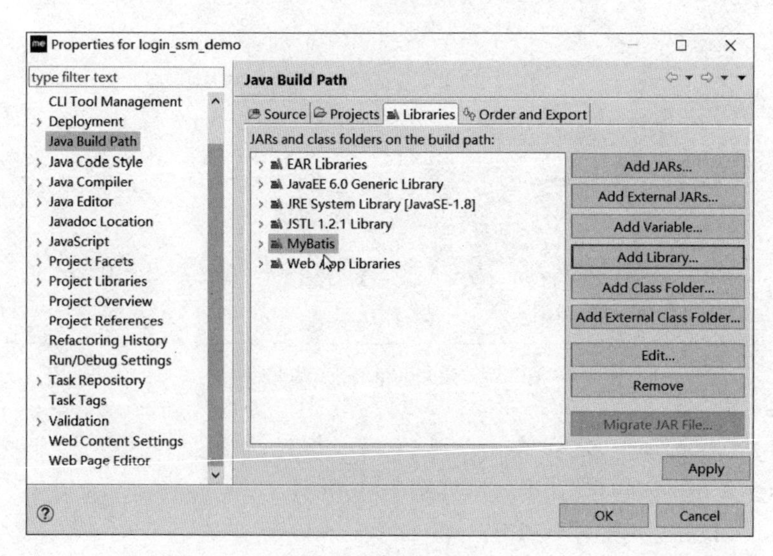

图 10-10　加入 MyBatis

(5) 为 login_ssm_demo 项目添加 Spring Facets 特性。

右击 login_ssm_demo 项目,选择 Configure Facets→Install Spring Facet,如图 10-11 所示。

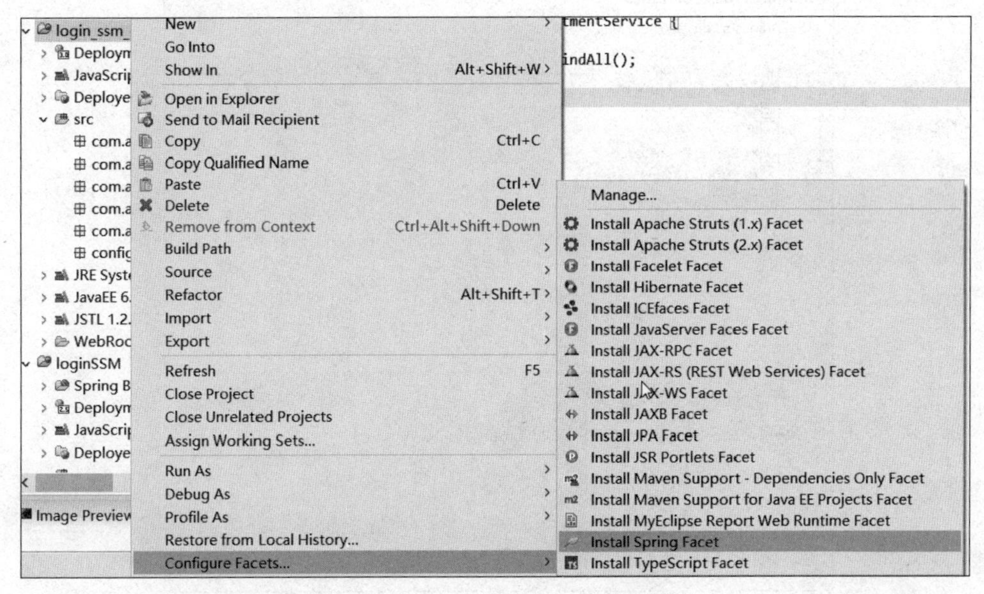

图 10-11 添加 Spring Facet 界面 1

单击 Next 按钮,如图 10-12 所示。

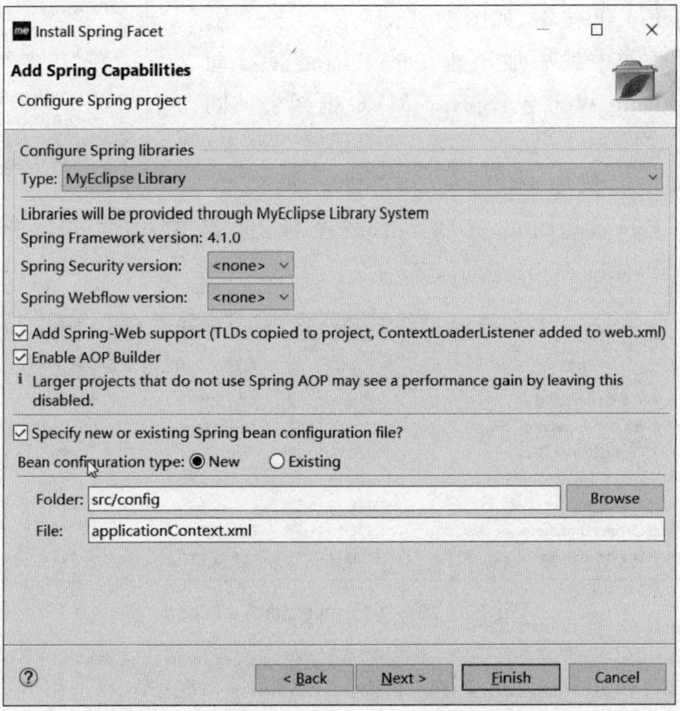

图 10-12 添加 Spring Facet 界面 2

Folder 项选择 login_ssm_demo 下的 config 文件夹，单击 Next 按钮，如图 10-13 所示。

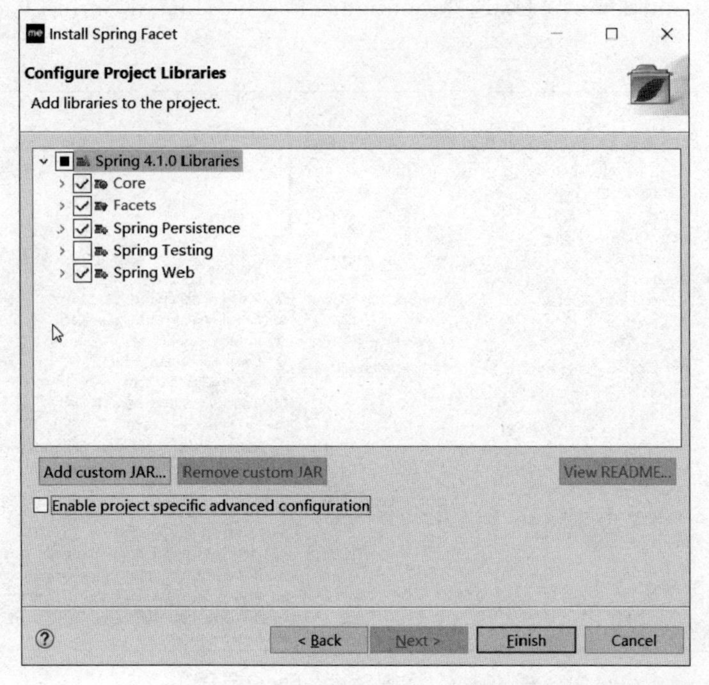

图 10-13　添加 Spring Facet 界面 3

如果使用 Spring 持久化 API，则需要勾选 Spring Persistence，单击 Finish 按钮，如图 10-14 所示。

此时资源管理器中就添加了相关的 Spring 库。同时，由于导入了 Spring Web 包，Spring MVC 框架会一同被引入项目中。

- Spring Beans
- Deployment Descriptor: login_ssı
- JavaScript Resources
- Deployed Resources
- src
- JRE System Library [JavaSE-1.8]
- JavaEE 6.0 Generic Library
- JSTL 1.2.1 Library
- Spring 4.1.0 Libraries
- WebRoot

图 10-14　添加 Spring Facet 界面 4

(6) 为 MyEclipse 添加 MyBatis 的 DTD 文件。

使用解压缩工具解压 mybatis-3.4.2.jar 文件，进入 mybatis 3.4.2 目录，如图 10-15 所示。

此电脑 › 下载 › mybatis-3.4.2 › mybatis-3.4.2 › mybatis-3.4.2 › org › apache › ibatis › builder › xml

名称	修改日期	类型	大小
mybatis-3-config.dtd	2017/1/19 11:37	DTD 文件	3 KB
mybatis-3-mapper.dtd	2017/1/19 11:37	DTD 文件	8 KB
XMLConfigBuilder.class	2017/1/19 11:37	CLASS 文件	19 KB
XMLIncludeTransformer.class	2017/1/19 11:37	CLASS 文件	6 KB
XMLMapperBuilder.class	2017/1/19 11:37	CLASS 文件	21 KB
XMLMapperEntityResolver.class	2017/1/19 11:37	CLASS 文件	3 KB
XMLStatementBuilder.class	2017/1/19 11:37	CLASS 文件	11 KB

图 10-15　解压 mybatis 3.4.2.jar

在 org/apache/ibatis/builder/xml 下找到 mybatis-3-config.dtd 和 mybatis 3-mapper.dtd，并将它们复制到其他位置。

单击 Windows→Preferences→Files and Editors→XML → XML Files→ XML

Catalog，如图 10-16 所示。

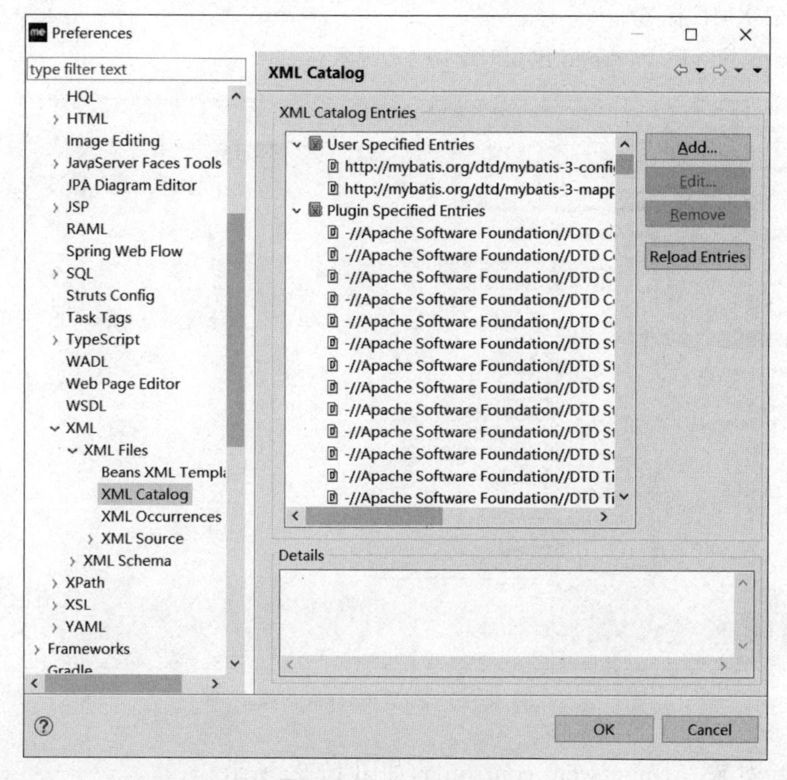

图 10-16 XML Catalog

单击 Add 按钮，添加 config DTD 和 mapping DTD，如图 10-17 和图 10-18 所示。

Location: C:\Users\gary\Downloads\mybatis-3.4.2\mybatis-3-config.d
Workspace... File System...
Key type: URI
Key: http://mybatis.org/dtd/mybatis-3-config.dtd
Alternative web address:
OK Cancel

图 10-17 添加 config DTD

Location: C:\Users\gary\Downloads\mybatis-3.4.2\mybatis-3-mapper.
Workspace... File System...
Key type: URI
Key: http://mybatis.org/dtd/mybatis-3-mapper.dtd
Alternative web address:
OK Cancel

图 10-18 添加 mapping DTD

单击 OK 按钮，关闭对话框。以后在建立 MyBatis 配置文件时，可以使用上述 DTD 辅助编写正确的 XML 标签。

（7）在 src 下新建 Package 包，如图 10-19 所示。

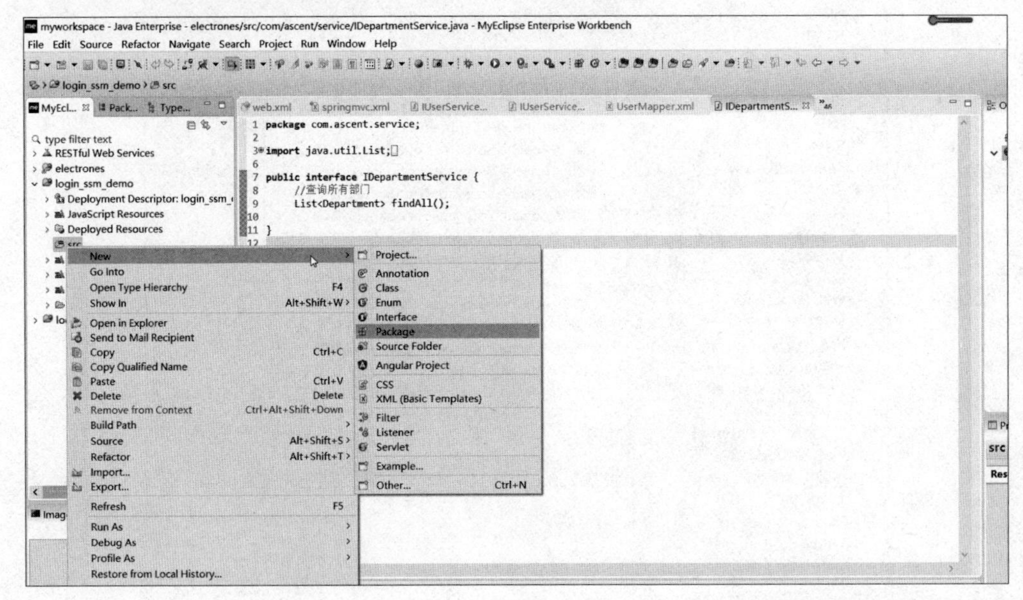

图 10-19 新建 Package 包

在 Name 处输入 com.ascent.controller，如图 10-20 所示。

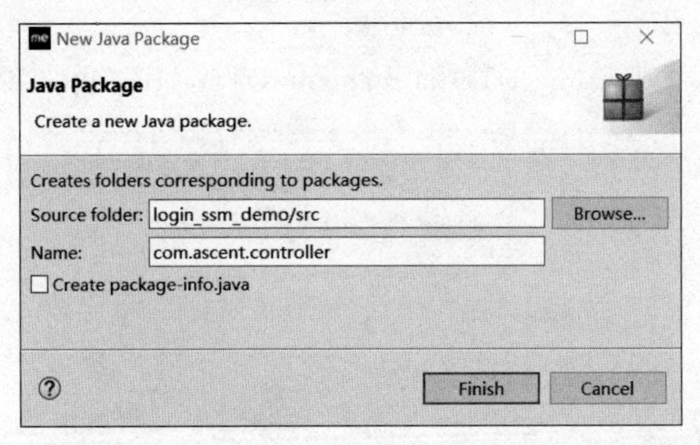

图 10-20 命名 Package

按照上述步骤依次创建

config：存放 Spring 配置文件、Spring MVC 配置文件、数据库 JDBC 链接所需信息文件；

com.ascent.mybatis.po：存放 Java POJO 类，映射 MySQL 关系数据库表；

com.ascent.service：存放业务接口；

com.ascent.service.impl：存放业务接口实现类；

com.ascent.mybatis.mapper：存放 Java 接口，定义 SQL 访问数据方法；

com.ascent.mybatis.mapping：存放 mapper 方法与 SQL 语句映射关系配置文件；

(8) 为 login_ssm_demo 项目编写 log4j.properties 以及 db.conf 数据库信息。

右击 login_ssm_demo 下的 src 目录,新建 log4j.properties 文件,如图 10-21 所示。

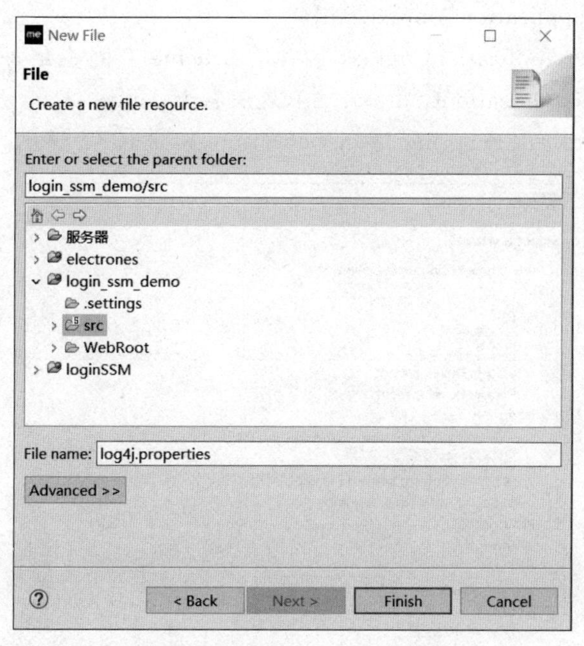

图 10-21 新建 log4j.properties 文件

编辑 log4j.properties,输入以下内容:

```
log4j.rootLogger=INFO, stdout
log4j.appender.stdout=org.apache.log4j.ConsoleAppender
log4j.appender.stdout.layout=org.apache.log4j.PatternLayout
log4j.appender.stdout.layout.ConversionPattern=%5p [%t]-%m%n
```

重复上述步骤,在 config 目录下建立 db.conf 文件,输入以下内容:

```
jdbc.url=jdbc:mysql://localhost:3306/test?useUnicode=true&characterEncoding=
utf-8&autoReconnect=true
jdbc.driver=com.mysql.jdbc.Driver
jdbc.username=root
jdbc.password=root
```

注意:一定要写 jdbc.××××,如果不以 jdbc.开头,则在引入 Spring 之后可能出现无法识别的问题。

(9) 编写 Spring 配置信息。

① 查看 web.xml,确认有如下内容:

```
<listener>
    <listener-class>org.springframework.web.context.ContextLoaderListener</
listener-class>
</listener>
<context-param>
    <param-name>contextConfigLocation</param-name>
```

```
    <param-value>classpath:config/applicationContext.xml</param-value>
</context-param>
```

② 重建 config/applicationContext.xml。

由于 IDE 生成的 applicationContext 包含的 schema 空间不足，所以需要重新建立。删除 config 目录下的 applicationContext.xml，之后右击 config 目录，从弹出的快捷菜单中选择 New→Other→Spring→Spring Bean Definition，如图 10-22 所示。

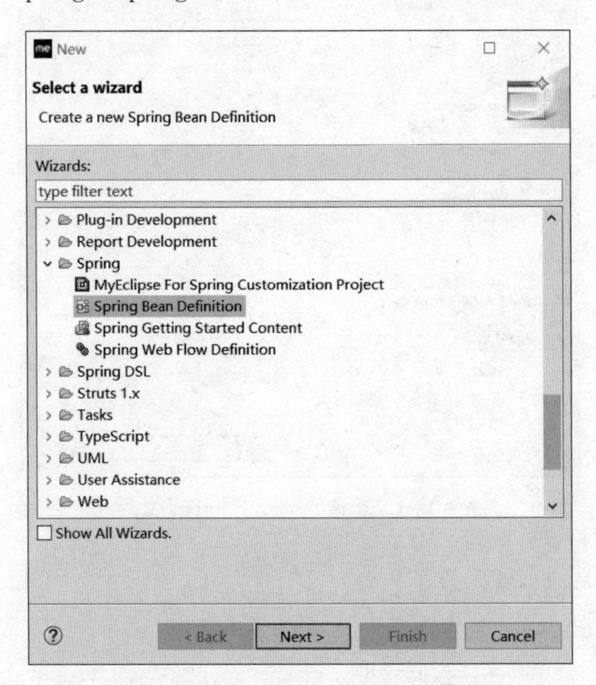

图 10-22　Spring Bean Definition 界面 1

单击 Next 按钮，如图 10-23 所示。

Create a new Spring Bean Definition file
New Spring Bean Definition file
Select the location and give a name for the Spring Bean Definition file
Project: login_ssm_demo
Folder: src/config　Browse
File: applicationContext.xml
< Back　Next >　Finish　Cancel

图 10-23　Spring Bean Definition 界面 2

单击 Next 按钮，选择 aop、beans、context、p、tx、util 标签，并在下方选择对应的版本 4.1，如图 10-24 所示。

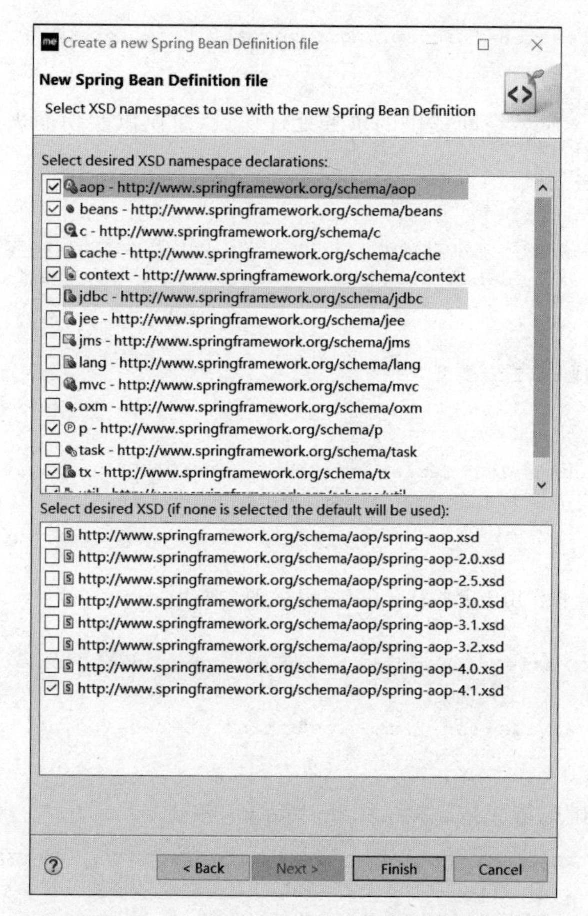

图 10-24 Spring Bean Definition 界面 3

单击 Finish 按钮,之后编辑 config/applicationContext.xml。

a. 定义 DataSource 数据源:

```
<!--读取外部配置文件 -->
<context:property-placeholder location="classpath:config/db.conf"/>
<!--配置数据源 -->
<bean id="dataSource"
  class="org.springframework.jdbc.datasource.DriverManagerDataSource">
    <property name="driverClassName" value="${jdbc.driver}"></property>
    <property name="url" value="${jdbc.url}"/>
    <property name="username" value="${jdbc.username}"></property>
    <property name="password" value="${jdbc.password}"></property>
</bean>
```

b. 配置自动扫描注入声明为 Service 业务 Bean 的包。

```
<!--配置自动扫描 Service -->
<context:component-scan base-package="com.ascent">
  <!--避免与 Spring MVC 配置产生冲突->
    <!--制定扫包规则,扫描除使用@Controller 注解的 Java 类 -->
    <context:exclude-filter type="annotation"
```

```
            expression="org.springframework.stereotype.Controller" />
</context:component-scan>
```

c. 配置 Spring tx 命名空间提供的事务处理切点、建议以及切面表达式。

```
<!--配置业务事务处理 AOP -->
<bean id="txManager"
  class="org.springframework.jdbc.datasource.DataSourceTransactionManager">
    <property name="dataSource" ref="dataSource"></property>
</bean>

<!--配置事务处理 advice 类型  -->
<tx:advice id="txAdvice" transaction-manager="txManager">
  <tx:attributes>
    <tx:method name=" * " isolation="READ_COMMITTED" propagation="REQUIRED"/>
  </tx:attributes>
</tx:advice>

<!--配置 advice 将作用于哪些符合切点表达式的方法上 -->
<aop:config >
  <aop:pointcut expression="execution( * com.ascent.service. * . * (..))" id=
 "pointCut"/>
  <aop:advisor advice-ref="txAdvice" pointcut-ref="pointCut"/>
</aop:config>
```

至此,项目的基础配置完成,如图 10-25 所示。

```
13         http://www.springframework.org/schema/util http://www.springframework.org/schema/util/spring-util-4.1.xsd">
14
15  <!-- 读取外部配置文件 -->
16  <context:property-placeholder location="classpath:config/db.conf"/>
17
18  <!-- 配置数据源 -->
19  <bean id="dataSource" class="org.springframework.jdbc.datasource.DriverManagerDataSource">
20    <property name="driverClassName" value="${jdbc.driver}"></property>
21    <property name="url" value="${jdbc.url}"/>
22    <property name="username" value="${jdbc.username}"></property>
23    <property name="password" value="${jdbc.password}"></property>
24  </bean>
25
26  <!-- 配置自动扫描Service -->
27 <context:component-scan base-package="com.ascent">
28   <!-- 避免与Spring MVC配置产生冲突 -->
29    <!-- 制定扫包规则 ,扫描除去使用@Controller注解的Java类 -->
30        <context:exclude-filter type="annotation"
31            expression="org.springframework.stereotype.Controller" />
32 </context:component-scan>
33
34 <!-- 配置业务事务处理AOP -->
35 <bean id="txManager"  class="org.springframework.jdbc.datasource.DataSourceTransactionManager">
36  <property name="dataSource" ref="dataSource"></property>
37 </bean>
38
39 <!-- 配置事务处理advice类型 -->
40 <tx:advice id="txAdvice" transaction-manager="txManager">
41   <tx:attributes>
42     <tx:method name="*" isolation="READ_COMMITTED" propagation="REQUIRED"/>
43   </tx:attributes>
44 </tx:advice>
45
46 <!-- 配置advice将作用于哪些符合切点表达式的方法上 -->
47 <aop:config >
48   <aop:pointcut expression="execution(* com.ascent.service.*.*(..))" id="pointCut"/>
49   <aop:advisor advice-ref="txAdvice" pointcut-ref="pointCut"/>
50 </aop:config>
51
```

图 10-25 项目的基础配置完成

(10) 编写 Spring MVC 配置信息。

① 修改 web.xml,添加如图 10-26 所示的信息。

```
    <!-- 注册编码过滤器 -->
    <filter>
        <filter-name>charset</filter-name>
        <filter-class>org.springframework.web.filter.CharacterEncodingFilter</filter-class>
        <init-param>
           <param-name>encoding</param-name>
            <param-value>UTF-8</param-value>
        </init-param>
    </filter>
    <filter-mapping>
        <filter-name>charset</filter-name>
        <url-pattern>/*</url-pattern>
    </filter-mapping>

  <!-- 配置Spring MVC Servlet  -->
  <servlet>
    <servlet-name>springMvcServlet</servlet-name>
    <servlet-class>org.springframework.web.servlet.DispatcherServlet</servlet-class>
    <init-param>
        <param-name>contextConfigLocation</param-name>
        <param-value>classpath:config/springmvc.xml</param-value>
    </init-param>
    <load-on-startup>1</load-on-startup>
</servlet>
    <servlet-mapping>
        <servlet-name>springMvcServlet</servlet-name>
        <url-pattern>*.action</url-pattern>
    </servlet-mapping>
```

图 10-26 修改 web.xml

```
<!--配置 Spring MVC 框架中提供的请求编码过滤器 -->
<filter>
    <filter-name>charset</filter-name>
    <filter-class>
      org.springframework.web.filter.CharacterEncodingFilter
    </filter-class>
    <init-param>
      <param-name>encoding</param-name>
      <param-value>UTF-8</param-value>
    </init-param>
</filter>
<filter-mapping>
        <filter-name>charset</filter-name>
        <url-pattern>/*</url-pattern>
</filter-mapping>

<!--配置 Spring MVC 核心 DispatcherServlet -->
<servlet>
    <servlet-name>
      Spring MVCServlet
</servlet-name>
    <servlet-class>
      org.springframework.web.servlet.DispatcherServlet
</servlet-class>
    <init-param>
```

```
    <param-name>contextConfigLocation</param-name>
        <param-value>classpath:config/Spring MVC.xml</param-value>
        </init-param>
    <load-on-startup>1</load-on-startup>
</servlet>
<servlet-mapping>
    <servlet-name>Spring MVCServlet</servlet-name>
    <url-pattern>*.action</url-pattern>
</servlet-mapping>
```

② 配置 Spring MVC.xml。

在 config 目录下建立 Spring Bean Difinition 文件，命名为 spring mvc.xml，如图 10-27 所示。

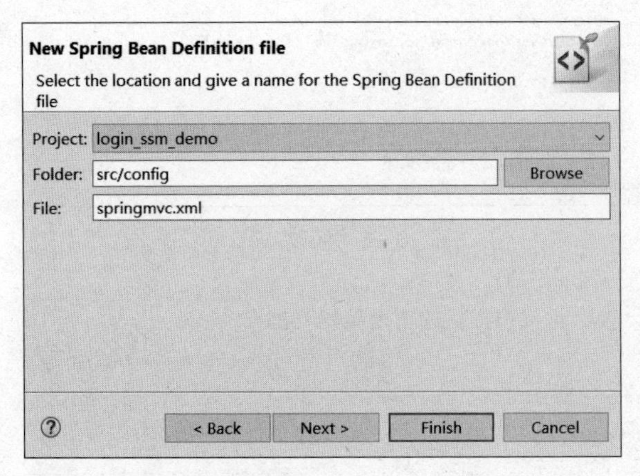

图 10-27　配置 spring mvc.xml 界面 1

单击 Next 按钮，如图 10-28 所示。

Select desired XSD namespace declarations:
☑ beans - http://www.springframework.org/schema/beans
☐ c - http://www.springframework.org/schema/c
☐ cache - http://www.springframework.org/schema/cache
☐ context - http://www.springframework.org/schema/context
☐ jdbc - http://www.springframework.org/schema/jdbc
☐ jee - http://www.springframework.org/schema/jee
☐ jms - http://www.springframework.org/schema/jms
☐ lang - http://www.springframework.org/schema/lang
☑ mvc - http://www.springframework.org/schema/mvc
☐ oxm - http://www.springframework.org/schema/oxm
☑ p - http://www.springframework.org/schema/p
☐ task - http://www.springframework.org/schema/task
☑ tx - http://www.springframework.org/schema/tx
☑ util - http://www.springframework.org/schema/util

Select desired XSD (if none is selected the default will be used):
☐ http://www.springframework.org/schema/mvc/spring-mvc.xsd
☐ http://www.springframework.org/schema/mvc/spring-mvc-3.0.xsd
☐ http://www.springframework.org/schema/mvc/spring-mvc-3.1.xsd
☐ http://www.springframework.org/schema/mvc/spring-mvc-3.2.xsd
☐ http://www.springframework.org/schema/mvc/spring-mvc-4.0.xsd
☑ http://www.springframework.org/schema/mvc/spring-mvc-4.1.xsd

图 10-28　配置 spring mvc.xml 界面 2

选中 aop、beans、context、mvc、p、tx、util，选择相应的 xsd 版本为 4.1，beans、mvc、context 必选，单击 Finish 按钮，完成 spring mvc.xml 文件的创建。编辑该文件，如图 10-29 所示。

```
<!-- spring mvc 框架采用注解 注入 Bean -->
  <context:component-scan
         base-package="com.ascent"/>
   <mvc:annotation-driven />
!-- 配置静态资源映射 -->
mvc:resources location="/static/css" mapping="/css/**"></mvc:resources>

!-- 配置动态视图解析器 -->
   <bean class="org.springframework.web.servlet.view.InternalResourceViewResolver">
         <property name="prefix" value="/WEB-INF/view/"/>
         <property name="suffix" value=".jsp"/>
   </bean>
```

图 10-29　编辑 spring mvc.xml

```
<!--spring mvc 框架采用注解注入 Bean -->
  <context:component-scan
         base-package="com.ascent"/>
   <mvc:annotation-driven />
<!--配置静态资源映射 -->
<mvc:resources location="/static/css" mapping="/css/**"></mvc:resources>

<!--配置动态视图解析器 -->
<bean class="org.springframework.web.servlet.view.InternalResourceViewResolver">
       <property name="prefix" value="/WEB-INF/view/"/>
       <property name="suffix" value=".jsp"/>
</bean>
```

在 WebRoot 下新建 static/css 目录，存放样式表文件，Spring MVC 可以将此目录映射为 /css URL，从而隐藏 static 目录，使得资源更安全。

在 WEB-INF 下新建 view 目录，所有的.jsp 文件将被 Spring MVC 的视图解析器识别，其他的.hml 则不会被识别。WEB-INF 目录结构如图 10-30 所示。

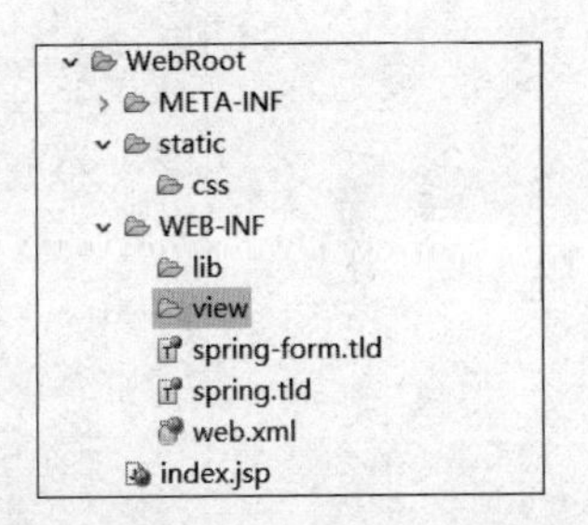

图 10-30　WEB-INF 目录结构

至此，Spring MVC 的核心配置以及视图、扫描策略等配置完毕。

(11) 编写 MyBatis 配置信息。

修改 applicationContext.xml，配置如图 10-31 所示。

```
<!-- 配置 MyBatis SqlSessionFactory -->
<bean id="sqlSessionFactory" class="org.mybatis.spring.SqlSessionFactoryBean" >
   <property name="dataSource" ref="dataSource"></property>
   <property name="mapperLocations" value="classpath:com/ascent/mybatis/mapping/*.xml"></property>
</bean>
<!-- 配置自动扫描mapper接口 -->
<bean class="org.mybatis.spring.mapper.MapperScannerConfigurer">
   <property name="basePackage"  value="com.ascent.mybatis.mapper"/>
   <property name="sqlSessionFactoryBeanName" value="sqlSessionFactory"></property>
</bean>
```

图 10-31　修改 applicationContext.xml

修改 applicationContext.xml，添加 SqlSessionFactory，自动扫描 mapper 接口。

```
<!--配置 MyBatis SqlSessionFactory -->
  <bean id="sqlSessionFactory"
    class="org.mybatis.spring.SqlSessionFactoryBean" >
      <property name="dataSource" ref="dataSource"></property>
  <property
      name="mapperLocations" value="classpath:com/ascent/mybatis/mapping/*.
      xml">
    </property>
    </bean>

<!--配置自动扫描 mapper 接口 -->
<bean class="org.mybatis.spring.mapper.MapperScannerConfigurer">
  <property name="basePackage" value="com.ascent.mybatis.mapper"/>
  <property name="sqlSessionFactoryBeanName"
    value="sqlSessionFactory"></property>
</bean>
```

(12) 编写 Java POJO 类，建立 MyBatis Mapping 映射文件。

右击 com.ascent.mybatis.po 包，从弹出的快捷菜单中选择 New→Class，出现图 10-32。

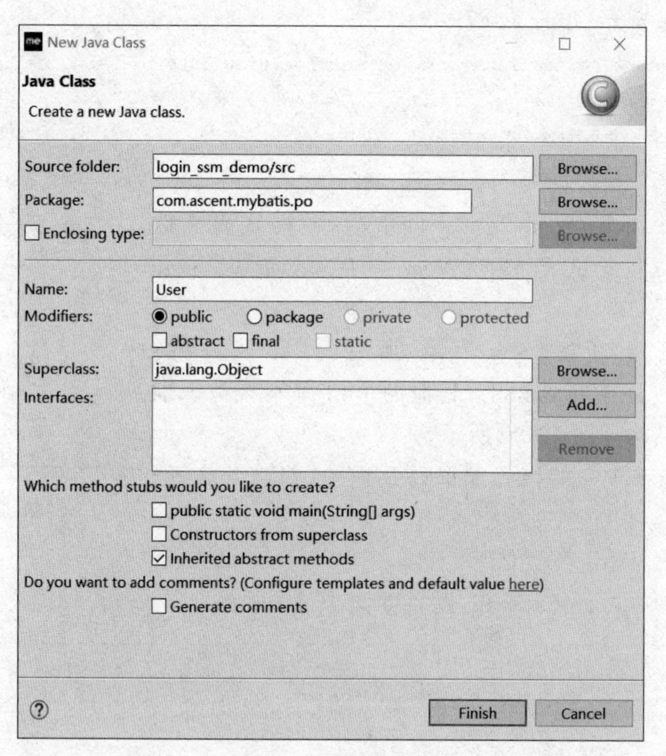

图 10-32　新建 Class

编写 User 类，包含 id、username、password 3 个属性以及 getter/setter 方法。

```
package com.ascent.mybatis.po;

public class User{
    private Integer id;
    private String username;
    private String password;
    public Integer getId() {
        return id;
    }
    public void setId(Integer id) {
        this.id=id;
    }
    public String getUsername() {
        return username;
    }
    public void setUsername(String username) {
        this.username=username;
    }
    public String getPassword() {
        return password;
    }
        public void setPassword(String password) {
        this.password=password;
    }
}
```

右击 com.ascent.mybatis.mapper，从弹出的快捷菜单中选择 New→Interface，新建 SQL 访问接口 UserMapper.java，如图 10-33 所示。

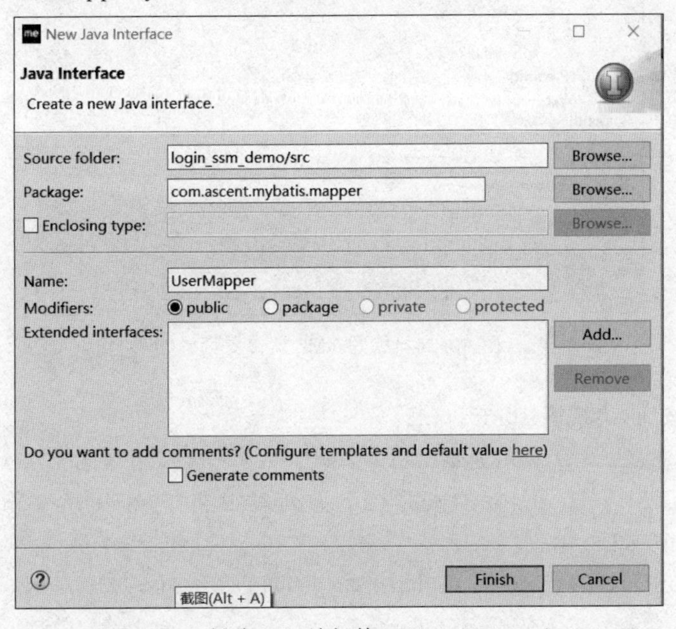

图 10-33 新建 SQL 访问接口 User Mapper.java

编辑 UserMapper 接口：

```
package com.ascent.mybatis.mapper;

import org.apache.ibatis.annotations.Param;

import com.ascent.mybatis.po.User;

public interface UserMapper {

public User userLogin(@Param(value="username") String username,
@Param(value="password") String password);
}
```

注意：@Param 是用来定义方法中 username 这个参数的，对应 mapping 文件中 SQL 语句中的相应参数，从而达到传递参数到 SQL 语句的目的。

右击 com.ascent.mybatis.mapping，从弹出的快捷菜单中选择 New→XML 文件，出现图 10-34。

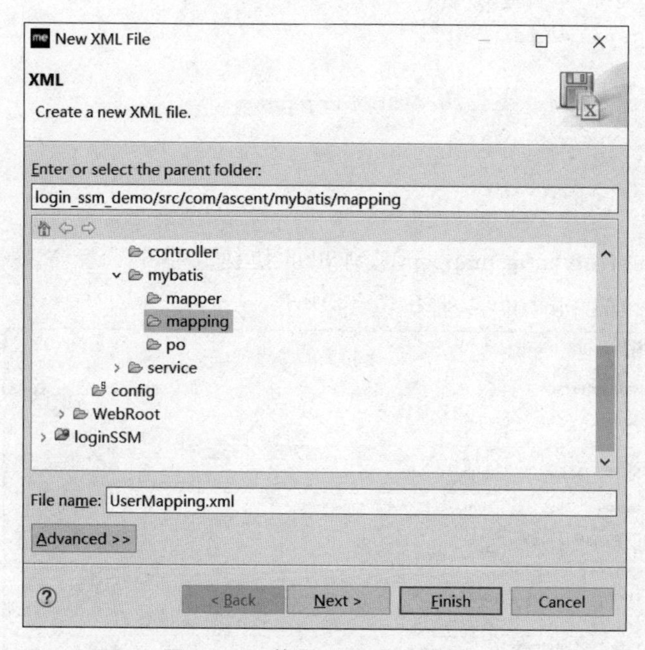

图 10-34　编写 XML 文件界面 1

单击 Next 按钮，出现图 10-35。

勾选 Create XML file from a DTD file，单击 Next 按钮，出现图 10-36。

选中 Select XML Catalog entry，并在下方列表中选择 mybatis-3-mapper.dtd，单击 Next 按钮，如图 10-37 所示。

注意：在 Public ID 处填写-//mybatis.org//DTD Mapper 3.0//EN，之后单击 Finish 按钮。

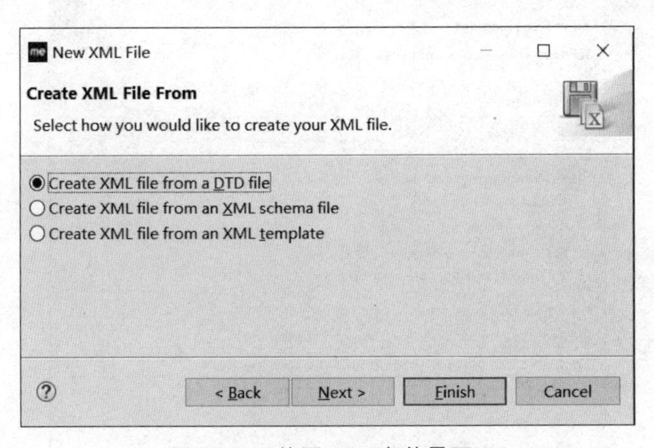

图 10-35　编写 XML 文件界面 2

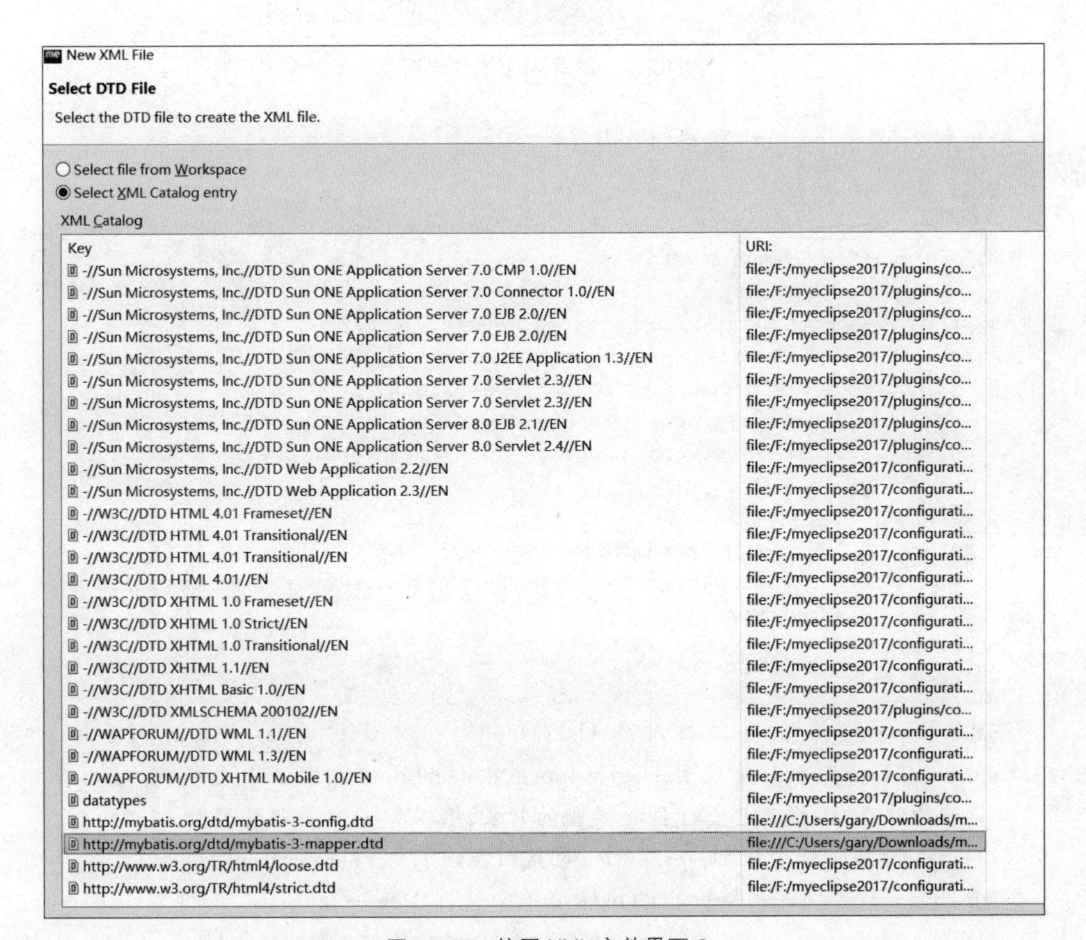

图 10-36　编写 XML 文件界面 3

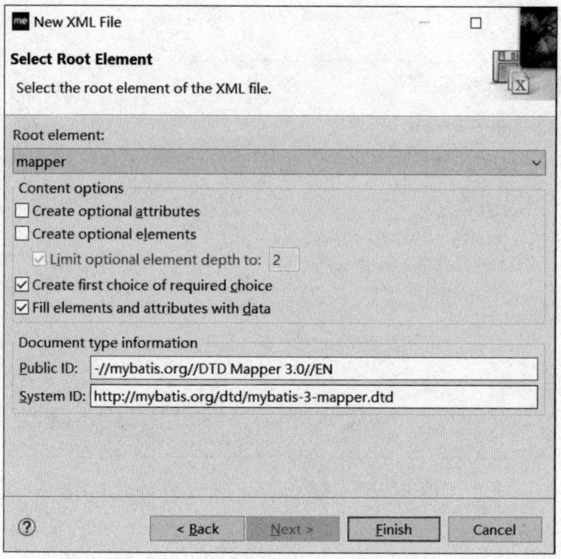

图 10-37　编写 XML 文件界面 4

编写 UserMapping.xml，填写如下内容：

```
<?xml version="1.0" encoding="UTF-8"?>
<!DOCTYPE mapper PUBLIC "-//mybatis.org//DTD Mapper 3.0//EN" "http://mybatis.
org/dtd/mybatis-3-mapper.dtd" >

<mapper namespace="com.ascent.mybatis.mapper.UserMapper">
  <resultMap type="com.ascent.mybatis.po.User" id="userPo">
    <id column="id" property="id"jdbcType="BIGINT"/>
    <result column="username" property="username"jdbcType="VARCHAR"/>
    <result column="password" property="password"jdbcType="VARCHAR"/>
  </resultMap>

  <select id="userLogin" resultMap="userPo">
    select id,username,password fromusr where username=#{username} and password=
             #{password}
  </select>
</mapper>
```

注意：select 标签的 id 属性对应 UserMapper()方法中的 userLogin()方法；#{username}、#{password}是映射 UserMapper 中 userLogin()方法中的参数；resultMap 是查询返回的结果，是之前定义的 Java 类型的封装，根据 id 进行关联。

(13) 编写登录业务接口 IUserService 和接口实现 IUserServiceImpl。

右击 com.ascent.service，从弹出的快捷菜单中选择 New→Interface，新建 IUserService 接口，如图 10-38 所示。

编写 IUserService 代码如下：

```
package com.ascent.service;

import com.ascent.mybatis.po.User;
```

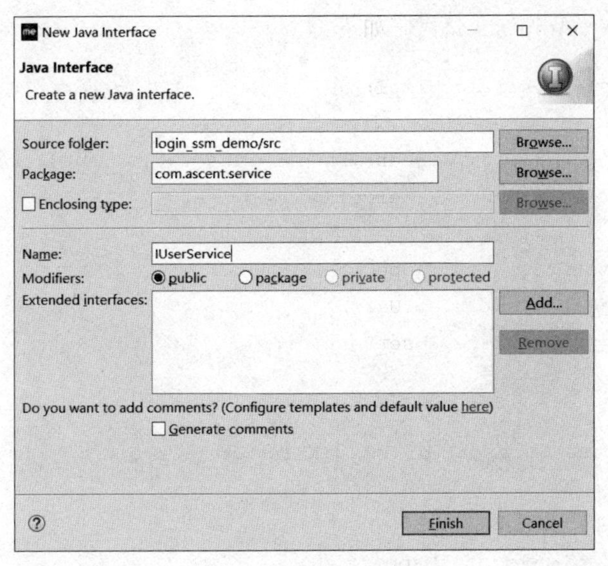

图 10-38　编写业务接口

```
public interface IUserService {

    public User login(Stringusername,String password);

}
```

右击 com.ascent.service.impl，从弹出的快捷菜单中选择 New→Class，实现 IUserService 接口，如图 10-39 所示。

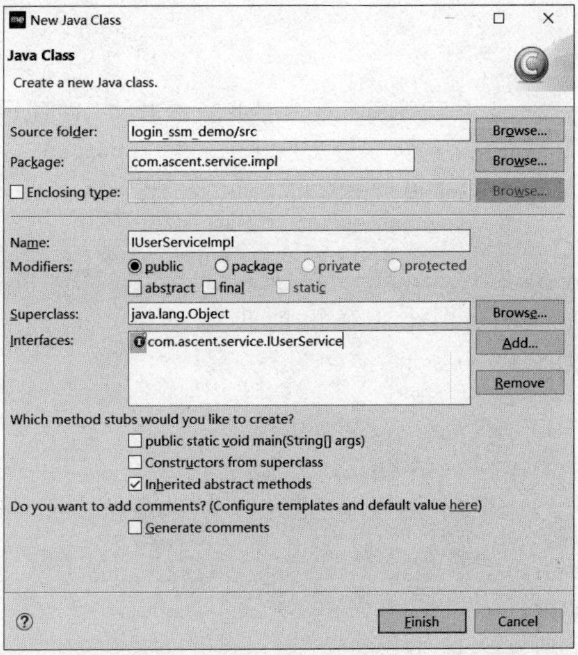

图 10-39　实现业务接口

实现 IUserServiceImpl.java，代码如下：

```
package com.ascent.service.impl;

import org.springframework.beans.factory.annotation.Autowired;
import org.springframework.stereotype.Service;

import com.ascent.mybatis.mapper.UserMapper;
import com.ascent.mybatis.po.User;
import com.ascent.service.IUserService;

@Service
public class IUserServiceImpl implements IUserService {

    @Autowired
    private UserMapper userMapper;

    @Override
    bpublic User login(Stringusername, String password) {

        //使用 UserMapper 接口,完成 SQL 查询操作
        //应该返回 User POJO 对象 可以为 null
        User user=userMapper.userLogin(username, password);

        return user;
    }

}
```

注意：必须加上@Service，否则 Spring 容器找不到 IUserService 接口的实现类，不能完成自动注入功能。@Autowired 表示该属性对象在运行时由 Spring 自动注入，这样用户就省去了获取 SqlSessionFactory 和 SqlSession 的过程，只需要从容器中获取想要的 Java API 接口 Mapper。

(14) 编写 Spring MVC Controller 对象，处理用户登录的登录请求。

右击 com.ascent.controller，从弹出的快捷菜单中选择 New → Class，新建 LoginController 类，如图 10-40 所示。

编写如下代码：

```
package com.ascent.controller;
import org.springframework.beans.factory.annotation.Autowired;
import org.springframework.stereotype.Controller;
import org.springframework.web.bind.annotation.RequestMapping;
import org.springframework.web.servlet.ModelAndView;
import com.ascent.service.IUserService;
@Controller
```

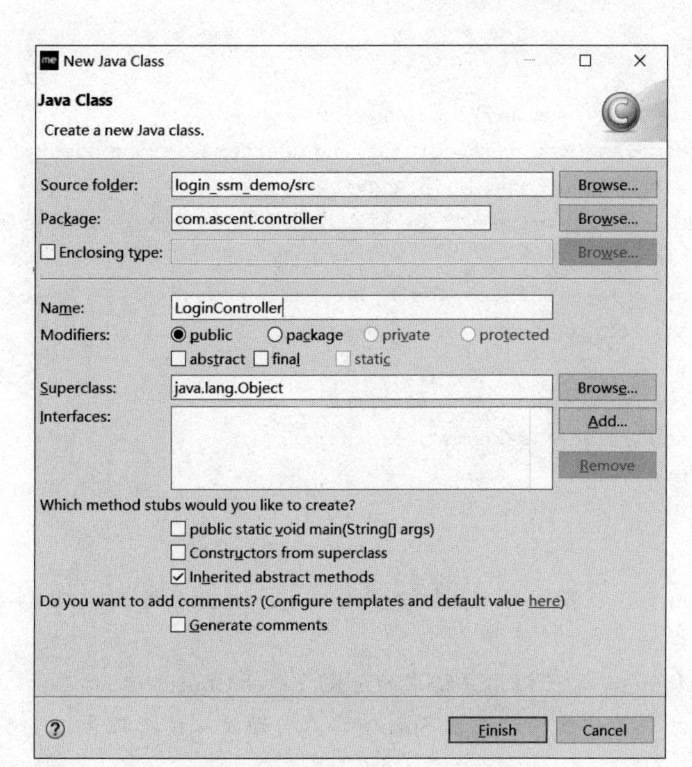

图 10-40 新建 LoginController 类

```
@RequestMapping(value="/")
public class LoginController {

    @Autowired
    private IUserService userService;

    /**
     * 设置默认请求首页的 url,跳转到 /WEB-INF/view/index.jsp 页面
     * 请求方式为 http://localhost:8080/login_ssm_demo/
     * 因为在 web.xml 中配置了 welcome-file index.action,所以不用再写 index.action
       后缀了
     * @return
     */
    @RequestMapping("index")
    public ModelAndView index(){
        return new ModelAndView("index");
    }

    /**
     * 用于处理 login.action 请求的控制器方法,读取用户名、密码,使用自动实例化的 IUserService
     * 接口的实现类 IUserServiceImpl 完成登录验证
     * @param username
     * @param password
```

```
     * @return
     * /
    @RequestMapping(value="login")
    public ModelAndView userLogin(String username,String password){
        ModelAndView mv=new ModelAndView();
        //使用 IUserService 获取登录信息,如果成功,则返回到 success 视图;如果失败,则
          返回到 fail 视图
        if(null!=userService.login(username, password))
            mv.setViewName("jsp/success");
        else
            mv.setViewName("jsp/fail");
        mv.addObject("username", username);
    return mv;
    }
}
```

注意:@Controller 必须标记,表示 DispatcherServlet 用来处理 URL 映射的一个 Controller 控制类;

@RequestMapping 是标记用户请求的 URL,和该 Controller 对应;

@Autowired 表示该属性对象由 Spring 容器自动注入被标记为该类型下@Service 的对象。

(15) 建立 3 个页面 index.jsp、jsp/success.jsp、jsp/fail.jsp。

右击 WEB-INF/view,从弹出的快捷菜单中选择 New→Folder,并命名为 jsp,如图 10-41 所示。

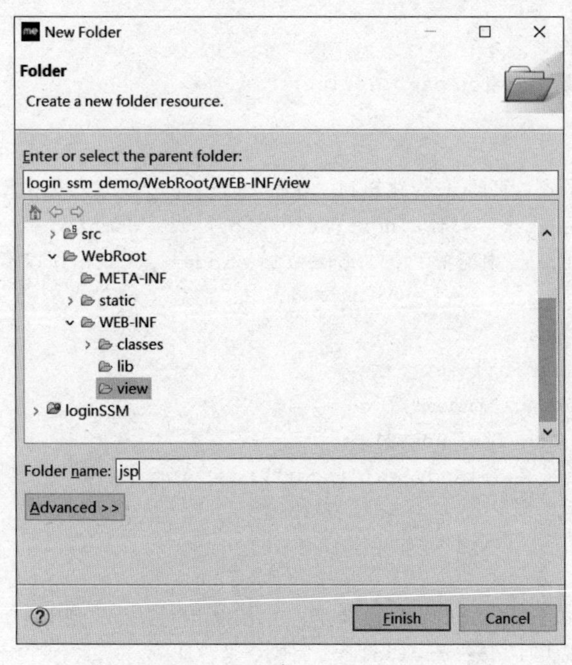

图 10-41　建立 jsp Folder

之后单击 Finish 按钮。

右击 WEB-INF/view,从弹出的快捷菜单中选择 New→JSP,新建 index.jsp,如图 10-42 所示。

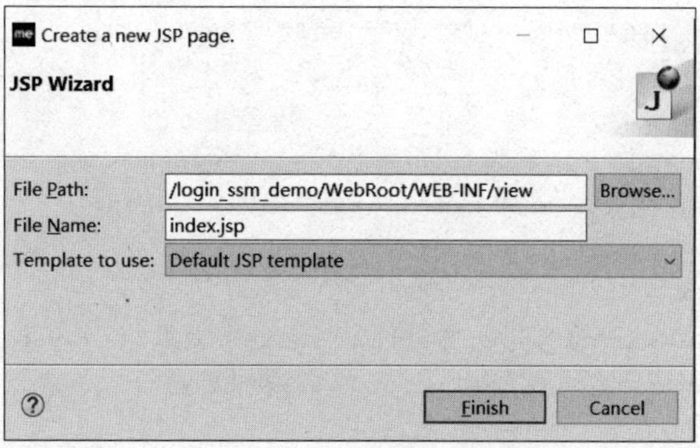

图 10-42　新建 index.jsp

编写 index.jsp,用户登录表单,代码如下:

```
<%@ page language="java" import="java.util.* "pageEncoding="utf-8"%>
<%@taglib uri="http://java.sun.com/jsp/jstl/core" prefix="c"%>
<%
String path=request.getContextPath();
StringbasePath= request. getScheme ( ) +"://" + request. getServerName ( ) +":" +
request.getServerPort()+path+"/";
%>

<!DOCTYPE HTML PUBLIC "-//W3C//DTD HTML 4.01 Transitional//EN">
<html>
  <head>
    <base href="<%=basePath%>">

    <title>登录页面</title>
    <meta http-equiv="pragma" content="no-cache">
    <meta http-equiv="cache-control" content="no-cache">
    <meta http-equiv="expires" content="0">
    <meta http-equiv="keywords" content="keyword1,keyword2,keyword3">
    <meta http-equiv="description" content="This is my page">
    <!--
    <link rel="stylesheet" type="text/css" href="styles.css">
    -->
  </head>

  <body>

    <div>
```

```
        <form action="<%=basePath %>/login.action" method="post" >
          <div>
            <span><label>账号</label></span><span><input type="text"
                id="username" name="username" ></span>
            </div>
            <div>
              <span><label>密码</label></span><span><input type="password"
                  id="password" name="password" ></span>
            </div>
              <div>
              <span><input type="submit" value="登录"></span>
              <span><input type="reset" value="重置" ></span>
            </div>
          </form>
        </div>
      </body>
    </html>
```

编写登录成功的页面，右击 WEB-INF/view/jsp，从弹出的快捷菜单中选择 New→JSP，新建 success.jsp，如图 10-43 所示。

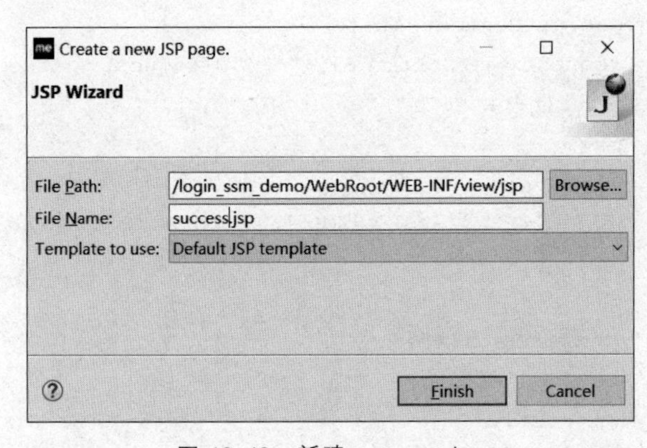

图 10-43 新建 success.jsp

编写 success.jsp，内容如下：

```
<%@ page language="java" import="java.util. * "pageEncoding="UTF-8"%>
<%@taglib uri="http://java.sun.com/jsp/jstl/core" prefix="c"%>

<%
String path=request.getContextPath();
StringbasePath=request.getScheme()+"://"+request.getServerName()+":"+
                request.getServerPort()+path+"/";
%>

<!DOCTYPE HTML PUBLIC "-//W3C//DTD HTML 4.01 Transitional//EN">
<html>
```

```
  <head>
    <base href="<%=basePath%>">
    <title>登录成功</title>
    <meta http-equiv="pragma" content="no-cache">
    <meta http-equiv="cache-control" content="no-cache">
    <meta http-equiv="expires" content="0">
    <meta http-equiv="keywords" content="keyword1,keyword2,keyword3">
    <meta http-equiv="description" content="This is my page">
    <!--
    <link rel="stylesheet" type="text/css" href="styles.css">
    -->
  </head>
    <body>
      <c:if test="${emptyrequestScope.username}">
        <a href="<%=basePath%>/">返回首页</a>
      </c:if>
        <c:if test="${not emptyrequestScope.username}">
          <h3>${requestScope.username}</h3><label>登录成功</label>
      </c:if>
    </body>
</html>
```

注意：如果登录成功，则使用 JSTL 标签和 EL 表达式获取 request 对象保存的属性 username，即 Spring MVC 模型中保存的数据，等于 jsp 中的 request.getAttribute。

编写登录失败的页面，右击 WEB-INF/view/jsp，从弹出的快捷菜单中选择 New→JSP，新建 fail.jsp，如图 10-44 所示。

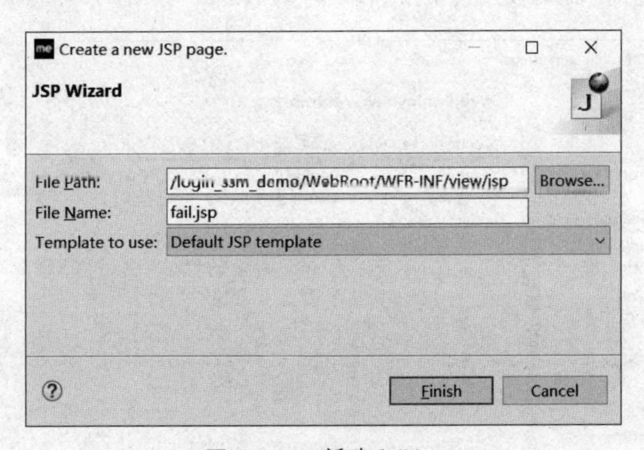

图 10-44 新建 fail.jsp

编写 fail.jsp，内容如下：

```
<%@ page language="java" import="java.util.*"pageEncoding="UTF-8"%>
<%@taglib uri="http://java.sun.com/jsp/jstl/core" prefix="c"%>
<%
String path=request.getContextPath();
StringbasePath=request.getScheme()+"://"+request.getServerName()+":"+
            request.getServerPort()+path+"/";
```

```
%>
<!DOCTYPE HTML PUBLIC "-//W3C//DTD HTML 4.01 Transitional//EN">
<html>
  <head>
    <base href="<%=basePath%>">
    <title>登录失败</title>
    <meta http-equiv="pragma" content="no-cache">
    <meta http-equiv="cache-control" content="no-cache">
    <meta http-equiv="expires" content="0">
    <meta http-equiv="keywords" content="keyword1,keyword2,keyword3">
    <meta http-equiv="description" content="This is my page">
    <!--
    <link rel="stylesheet" type="text/css" href="styles.css">
    -->
  </head>
  <body>
     <div>登录失败</div>
     <a href="<%=basePath%>/">重新登录</a>
  </body>
</html>
```

3. 部署与调试

先删除 WebRoot 下系统默认生成的 index.jsp，因为需要 Controller 自动导航到 WEB-INF/view/index.jsp。

(1) 在部署打包前，需要设置导出的依赖库，如图 10-45 所示。

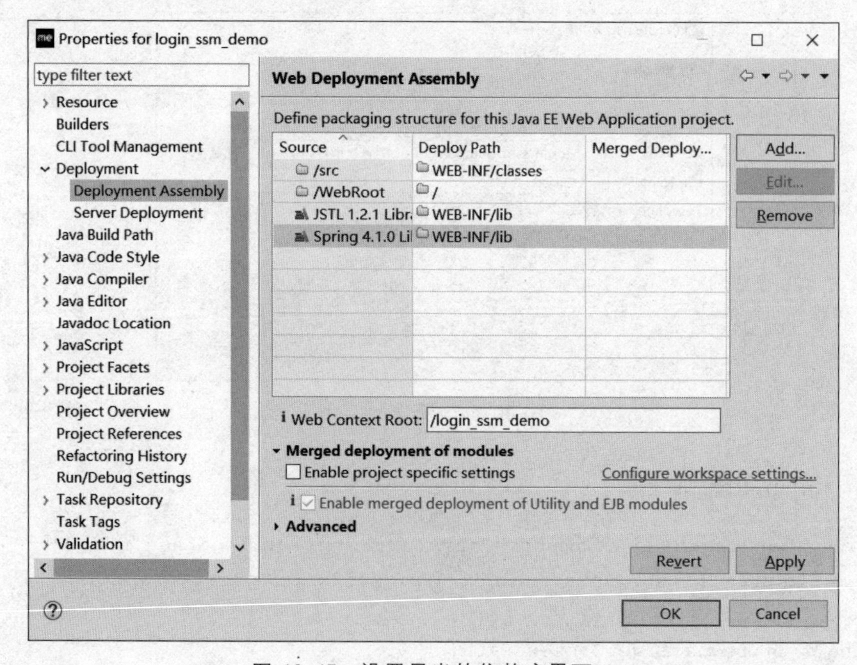

图 10-45　设置导出的依赖库界面 1

单击 Add 按钮，出现图 10-46。

图 10-46 设置导出的依赖库界面 2

选择 Java Build Path Entries，之后单击 Next 按钮，出现图 10-47。

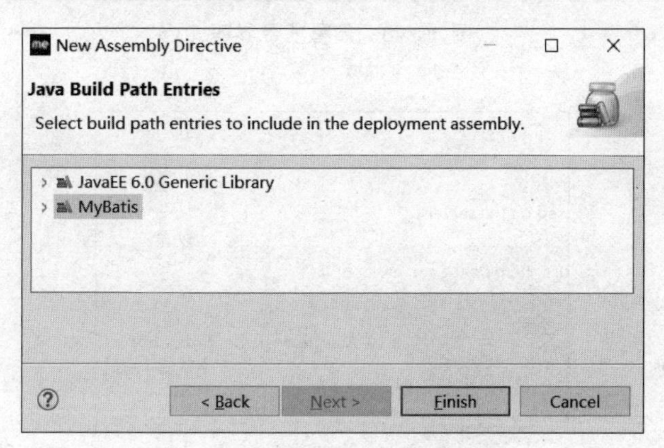

图 10-47 设置导出的依赖库界面 3

选择所定义的 MyBatis 库，之后单击 Finish 按钮。

在 login_ssm_demo 项目的 Build Path 中，配置 Order and Export，勾选 Spring 4.1.0 Libraries 以及 MyBatis 依赖库。

修改 web.xml，并修改默认的 welcome-file：

```
<welcome-file-list>
  <welcome-file>index.jsp</welcome-file>
    <welcome-file>index.action</welcome-file>
</welcome-file-list>
```

（2）部署 login_ssm_demo。

单击图标，出现图 10-48。

选择 MyEclipse Tomcat v8.5，单击 Next 按钮，出现图 10-49。

Deploy modules.

Deploy modules.

Deploy Java EE modules.

How do you want to select the server?

Choose an existing server　Manually define a new server

Select the server that you want to use:

type filter text

Server	State
MyEclipse Tomcat v8.5	Stopped

MyEclipse Tomcat v8.5 supports J2EE 1.2, 1.3, 1.4, and Java EE 5, 6, and 7 Web modules.　Columns...

Always use this server when running this project

< Back　Next >　Finish　Cancel

图 10-48　部署项目界面 1

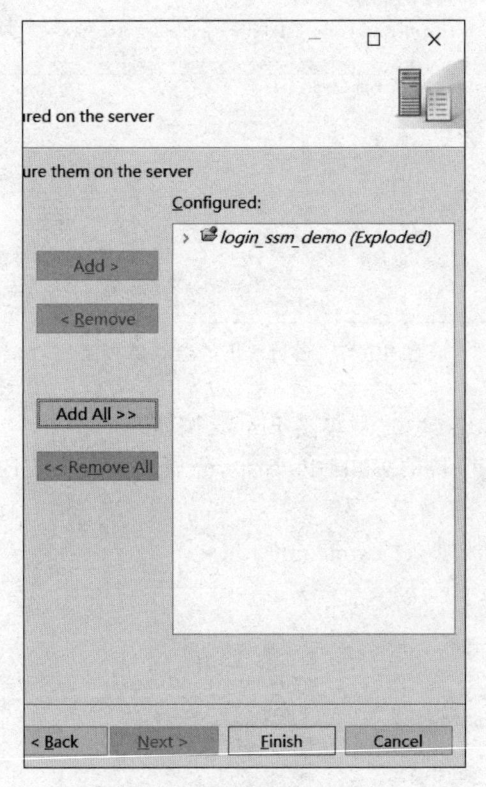

图 10-49　部署项目界面 2

单击 Finish 按钮，出现图 10-50。

Manage Deployments

Manage Deployments

Deploy and undeploy Java EE modules.

Module: login_ssm_demo

Server	Status	Mode	Location
MyEclipse Tomcat v8.5	Stopped		
login_ssm_demo		Exploded	
spring-web-4.1.0.RELEASE.jar		Packaged	

Add... Remove Publish Browse...

OK Cancel

图 10-50 部署项目界面 3

最后单击 OK 按钮。

(3) 运行项目。

单击 MyEclipse Tomcat v8.5 Start 启动项目，如图 10-51 所示。

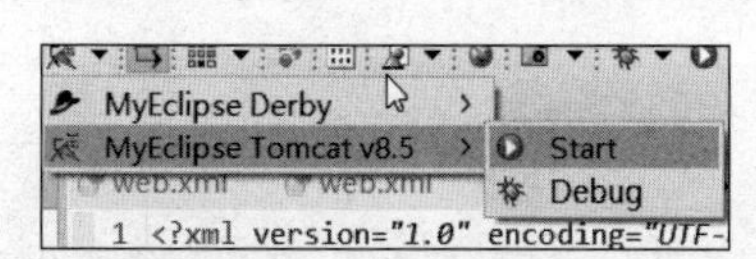

图 10-51 启动项目

观察 Console，保证启动没有错误，如图 10-52 所示。

```
INFO [localhost-startStop-1] - Refreshing WebApplicationContext for namespace 'springMvcServlet-servlet': startup date [Sat S
INFO [localhost-startStop-1] - Loading XML bean definitions from class path resource [config/springmvc.xml]
INFO [localhost-startStop-1] - Mapped "{[//index],methods=[],params=[],headers=[],consumes=[],produces=[],custom=[]}" onto pu
INFO [localhost-startStop-1] - Mapped "{[//login],methods=[],params=[],headers=[],consumes=[],produces=[],custom=[]}" onto pu
INFO [localhost-startStop-1] - Looking for @ControllerAdvice: WebApplicationContext for namespace 'springMvcServlet-servlet':
INFO [localhost-startStop-1] - Looking for @ControllerAdvice: WebApplicationContext for namespace 'springMvcServlet-servlet':
INFO [localhost-startStop-1] - Mapped URL path [/css/**] onto handler 'org.springframework.web.servlet.resource.ResourceHttpR
INFO [localhost-startStop-1] - FrameworkServlet 'springMvcServlet': initialization completed in 902 ms
九月 07, 2019 4:40:12 下午 org.apache.catalina.startup.HostConfig deployDirectory
信息: Deployment of web application directory H:\myworkspace\.metadata\.me_tcat85\webapps\login_ssm_demo has finished in 8,956
九月 07, 2019 4:40:12 下午 org.apache.catalina.startup.HostConfig deployDirectory
信息: Deploying web application directory H:\myworkspace\.metadata\.me_tcat85\webapps\manager
九月 07, 2019 4:40:12 下午 org.apache.catalina.startup.HostConfig deployDirectory
信息: Deployment of web application directory H:\myworkspace\.metadata\.me_tcat85\webapps\manager has finished in 31 ms
九月 07, 2019 4:40:12 下午 org.apache.catalina.startup.HostConfig deployDirectory
信息: Deploying web application directory H:\myworkspace\.metadata\.me_tcat85\webapps\ROOT
九月 07, 2019 4:40:14 下午 org.apache.jasper.servlet.TldScanner scanJars
信息: At least one JAR was scanned for TLDs yet contained no TLDs. Enable debug logging for this logger for a complete list of
九月 07, 2019 4:40:14 下午 org.apache.catalina.startup.HostConfig deployDirectory
信息: Deployment of web application directory H:\myworkspace\.metadata\.me_tcat85\webapps\ROOT has finished in 2,466 ms
九月 07, 2019 4:40:14 下午 org.apache.coyote.AbstractProtocol start
信息: Starting ProtocolHandler [http-nio-8080]
九月 07, 2019 4:40:14 下午 org.apache.coyote.AbstractProtocol start
信息: Starting ProtocolHandler [ajp-nio-8009]
九月 07, 2019 4:40:14 下午 org.apache.catalina.startup.Catalina start
信息: Server startup in 18177 ms
```

图 10-52 观察 Console

单击 ，打开浏览器，输入 http://localhost:8080/login_ssm_demo/，如图 10-53 所示。

图 10-53　输入 URL

输入账号 admin,密码 123456,登录成功,如图 10-54 所示。

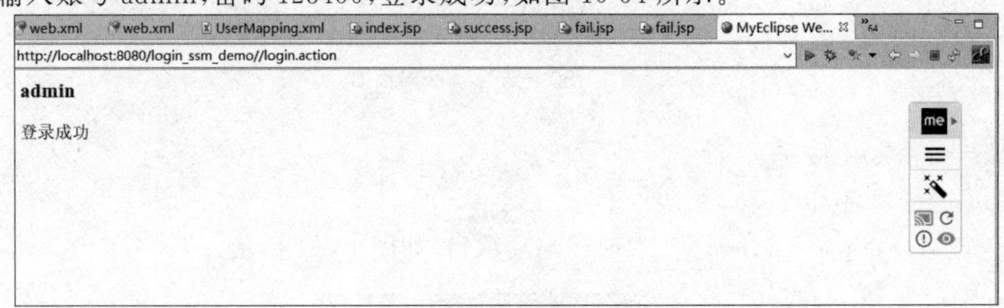

图 10-54　登录成功

输入账号 aaaa,密码 bbbb,系统中不存在此账号,登录失败,如图 10-55 所示。

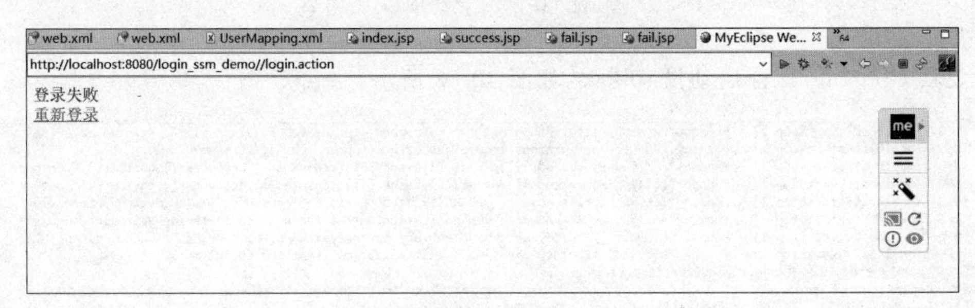

图 10-55　登录失败

至此,SSM 框架整合成功。

10.2　项目案例

10.2.1　学习目标

本章学习了使用 Spring MVC 框架接收用户请求,使用 MyBatis 查询和管理信息,使用 Spring 对整体业务中需要的对象进行配置的过程。

10.2.2　案例描述

目前我们的 electrone 项目已经配置了 Spring MVC、Spring DI/IoC 框架、MyBatis 框

架以及插件，因此，本案例进行 Spring MVC、MyBatis 与 Spring 框架的整合，完成一般用户浏览中头版头条新闻的展示、综合新闻的展示、发布新闻功能的整合。

10.2.3 案例要点

将查找头版头条新闻、查找综合新闻、发布新闻功能中我们需要的组件全部交给 Spring 托管和注入，在控制器 IndexController 和业务类 IAuthorizationServiceImpl、INewsServiceImpl 中仅声明需要使用的对象即可。这些对象之前的依赖关系全部交给 Spring 管理。

为了快速配置，本案例采用注解方式声明组件的依赖关系，使用 Spring 的 context-component 标签，按组件类型自动装配。

取消 MyBatis 的 mybatis-config 配置，交给 Spring 创建。

Spring MVC 控制器交给 Spring 托管并注入依赖对象。

因此，进行整合时，分两大步骤：一是简化 Spring MVC 框架和 MyBatis 配置并集中配置 Spring 的 applicationContext；二是修改已有的功能代码。

10.2.4 案例实施

(1) 导入 Mybatis-Spring 包，如图 10-56 所示。

图 10-56 导入 Mybatis-Spring 包

设置该项目的 Deployment 属性，部署时要同时部署 Mybatis-Spring 包，如图 10-57 所示。

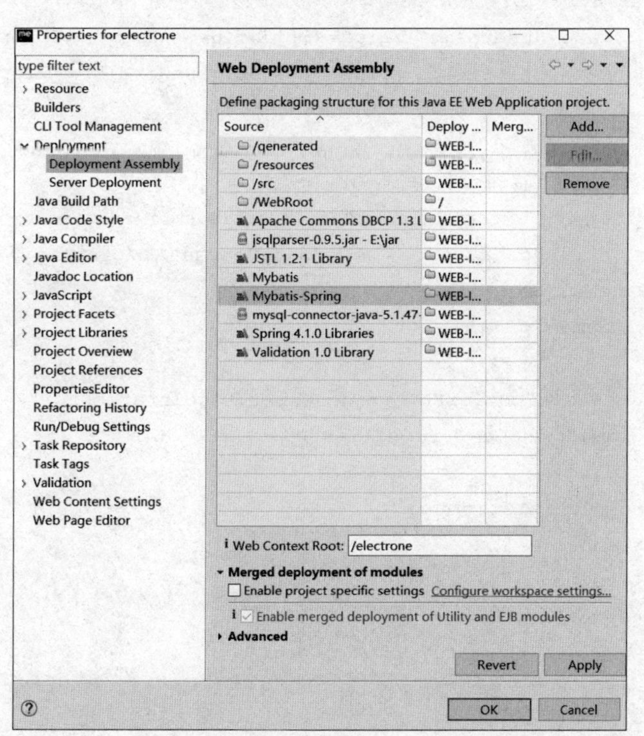

图 10-57 设置项目的 Deployment 属性

（2）在 config 目录下新建 db.conf 文本文件。

```
db.conf:
jdbc.url=jdbc:mysql://localhost:3306/electrones?useUnicode=true&characterEncoding=
utf-8&autoReconnect=true
jdbc.driver=com.mysql.jdbc.Driver
jdbc.username=root
jdbc.password=root
```

注意：username、password 前面有 jdbc.作为前缀，因为直接使用 username、password 会导致 Spring 读取信息失败。

（3）删除 mybatis-config.xml 文件，使用 Spring 的 applicationContext 配置 MyBatis 的 sqlSessionFactory 以及 PageHelper 插件，自动扫描 com.ascent.mybatis.mapping 文件和 Java Mapper 接口。

```
applicationContext.xml:
<?xml version="1.0" encoding="UTF-8"?>
<beans …这里省略头信息>

  <!--读取外部数据源配置文件 -->
  <context:property-placeholder location="classpath:config/db.conf"/>

  <!--配置数据源 ${jdbc.*} 获取 db.conf 中配置的条目-->
  <bean id="dataSource"
class="org.springframework.jdbc.datasource.DriverManagerDataSource">
     <property name="driverClassName" value="${jdbc.driver}"></property>
      <property name="url" value="${jdbc.url}"/>
     <property name="username" value="${jdbc.username}"></property>
     <property name="password" value="${jdbc.password}"></property>
  </bean>

  <!--配置 MyBatis 的 sqlSessionFactory -->
  <bean id="sqlSessionFactory" class="org.mybatis.spring.SqlSessionFactoryBean" >
     <property name="dataSource" ref="dataSource"></property>
  <!--配置 PageHelper -->插件
     <property name="plugins">
        <array>
          <bean class="com.github.pagehelper.PageInterceptor">
             <property name="properties">
               <value>
               helperDialect=mysql
               resonable=true
               </value>
               </property>
          </bean>
        </array>
     </property>
```

```
  <!--配置 sqlSessionFactory 需要自动托管的 mybatis-mapper 映射配置文件,这里为自动
托管某目录下所有的 xml 文件->
    <property name="mapperLocations"
      value="classpath:com/ascent/mybatis/mapping/*.xml"></property>
    </bean>

<!--使用 mybatis-spring 包中的 MapperScannerConfigure 自动扫描 mapper 接口 -->
<!--本案例配置扫描 com.ascent.mybatis.mapper 下所有的 Java Mapper 接口类 -->
<!--目的是可以 sqlSession,可以获取并调用这些接口-->
<bean class="org.mybatis.spring.mapper.MapperScannerConfigurer">
  <property name="basePackage"  value="com.ascent.mybatis.mapper"/>
  <property name="sqlSessionFactoryBeanName" value="sqlSessionFactory"></
property>
</bean>

<!--配置自动扫描 Service,注意,要扫描到 controller 目录的上一层-->
<context:component-scan base-package="com.ascent">
  <!--避免与 Spring MVC 配置产生冲突 -->
    <!--制定扫包规则,扫描除使用@Controller 注解的 Java 类 -->
  <!--Spring 在实例化包下的 class 时,会根据下面的过滤器(如标记了@Controller 的类,则
不托管)交由 Spring MVC 框架处理@Controller 语义 -->
      <context:exclude-filter type="annotation"
          expression="org.springframework.stereotype.Controller" />
</context:component-scan>

<!--配置业务事务处理 AOP -->
<bean id="txManager"
class="org.springframework.jdbc.datasource.DataSourceTransactionManager">
  <property name="dataSource" ref="dataSource"></property>
</bean>

    <!--事务 Advice 配置 -->
<tx:advice id="txAdvice" transaction-manager="txManager">
        <!--事务内容属性定义什么样的方法使用什么策略 -->
<tx:attributes>
      <!--所有用 find 开头的方法都是只读的,并且事务隔离特性使用避免脏读策略 -->
  <tx:method name="find*" read-only="true" isolation="READ_COMMITTED" />
  <!--所有以 insert 开头的方法都是具有事务的,事务隔离特性使用避免脏读策略 -->
  <!--rollback-for 是指该方法抛出什么异常时回滚 -->
<tx:method name="insert*" isolation="READ_COMMITTED"
    rollback-for="java.sql.SQLException" />
        <!--其他方法使用默认的事务配置 -->
  <tx:method name="*" />
</tx:attributes>
</tx:advice>
```

```
        <!--使得上面的事务配置对 Service 接口的所有操作有效 -->
        <!--配置 advice 将作用于哪些符合切点表达式的方法上 -->
<aop:config>
        <aop:pointcut expression="execution(* com.ascent.service.*.*(..))"
              id="pointCut" />
            <aop:advisor advice-ref="txAdvice" pointcut-ref="pointCut" />
</aop:config>
</beans>
```

Spring applicationContext 中包含了 MyBatis 基础配置，因此不需要单独配置 mybatis-config.xml。

(4) config/spring-mvc.xml 保持不变。

(5) 修改 web.xml，配置 ContextConfigListener。

```
<listener>
    <listener-class>org.springframework.web.context.ContextLoaderListener</
listener-class>
  </listener>
  <context-param>
    <param-name>contextConfigLocation</param-name>
    <param-value>classpath:config/applicationContext.xml</param-value>
  </context-param>
```

保持其他内容不变，可以将 applicationContext.xml 放入 config 目录，如图 10-58 所示。

图 10-58　将 applicationContext.xml 放入 config 目录

(6) 修改用于查询头版头条新闻和综合新闻的 IAuthorizationServiceImpl.java。

```
package com.ascent.service.impl;

import java.util.ArrayList;
import java.util.HashMap;
import java.util.List;

import org.springframework.beans.factory.annotation.Autowired;
import org.springframework.stereotype.Service;

import com.ascent.mybatis.mapper.NewsMapper;
import com.ascent.mybatis.utils.MybatisSqlSessionFactoryUtil;
import com.ascent.po.News;
import com.ascent.service.IAuthorizationService;
import com.github.pagehelper.PageHelper;
```

```
//需要使用@Service注解标记这是@Controller控制类所需要自动注入的业务对象
@Service
public class IAuthorizationServiceImpl implements IAuthorizationService {
    //告知 Spring 自动注入 NewsMapper 对象
@Autowired
private NewsMapper newsMapper;


@Override
public HashMap<String, List<News>>findHeaderNews() {
    //查询头版头条新闻
    News headerNews=newsMapper.findHeaderNews();
//构造空的 List<News>集合对象,集合成员必须为 News 对象
    List<News>headerNewsList=new ArrayList<News>(1);
    //将查询得到的头版头条新闻保存到 List 中
    headerNewsList.add(headerNews);
    //针对查询使用 PageHelper 进行分页,第一个参数 1 表示第一页,第二个参数 6 表示有 6 条
      记录
    PageHelper.startPage(1, 6);
    List<News>indexNews=newsMapper.findIndexNews();

    //定义并初始化 HashMap<String,List<News>>对象
    HashMap<String, List<News>>headerAndIndexNews=new HashMap<String, List<
News>>(2);

    //将头版头条新闻保存在 Map 中
    headerAndIndexNews.put("headerNews", headerNewsList);
    //将综合新闻保存在 Map 中
    headerAndIndexNews.put("indexNews", indexNews);

    return headerAndIndexNews;
    }
}
```

(7) 修改 INewsServiceImpl.java。

```
package com.ascent.service.impl;

import java.util.Date;

import org.springframework.beans.factory.annotation.Autowired;
import org.springframework.stereotype.Service;

import com.ascent.mybatis.mapper.NewsMapper;
import com.ascent.po.News;
import com.ascent.service.INewsService;
```

```
@Service
public class INewsServiceImpl implements INewsService {

    @Autowired
    private NewsMapper newsMapper;
    /**
        *定义保存新闻的业务逻辑
        *如果类型为头版头条新闻,则设置不跨栏目
        *假设用户 id 为 9,部门 id 为 1
     */

    @Override
    public boolean addNews(News news) {

        Integer deptid=1;
        Integer usrid=9;
        boolean isSuccess=false;

        try {
            //创建新闻对象
            Date dt=new Date();
            news.setAuthor("测试 3");
            news.setPublishtime(dt);
            news.setCheckstatus(0);
            if ("1".equals(news.getType())) {
                news.setCrossstatus(1);
            }
            news.setStatus(0);
            //目前静态设置,实际项目需要从登录用户信息里获取 deptid 和 usrid
            news.setUserid(usrid);
            news.setDeptid(deptid);

            newsMapper.insert(news);
            isSuccess=true;

        } catch (NumberFormatException e) {
            //TODO Auto-generated catch block
            e.printStackTrace();
            isSuccess=false;

        }

        return isSuccess;

    }
}
```

(8) 修改 IndexController.java。

```
package com.ascent.controller;

import java.util.HashMap;
import java.util.List;

import javax.servlet.http.HttpServletRequest;
import javax.servlet.http.HttpServletResponse;
import javax.servlet.http.HttpSession;

import org.springframework.beans.factory.annotation.Autowired;
import org.springframework.context.ApplicationContext;
import org.springframework.context.support.ClassPathXmlApplicationContext;
import org.springframework.stereotype.Controller;
import org.springframework.validation.BindingResult;
import org.springframework.validation.annotation.Validated;
import org.springframework.web.bind.WebDataBinder;
import org.springframework.web.bind.annotation.InitBinder;
import org.springframework.web.bind.annotation.RequestMapping;
import org.springframework.web.servlet.ModelAndView;

import com.ascent.mvc.validator.NewsValidator;
import com.ascent.po.News;
import com.ascent.service.IAuthorizationService;
import com.ascent.service.INewsService;
import com.ascent.service.impl.INewsServiceImpl;

/**
 * 所有在 URL/下发出的请求,都由该 index 控制器处理,Class 相当于包含了命名空间的对象,
可以有多个不同的 URL 映射
 * @author gary
 *
 */
@Controller
@RequestMapping(value="/")
public class IndexController {

    @Autowired
    private IAuthorizationService authService;

    @Autowired
    private INewsService newsService;

    /**
     * 所有进入 IndexController 控制的请求,都先经过 initBinder 的校验
```

```
     * @param dataBinder
     */
    @InitBinder
    public void initBinder(WebDataBinder dataBinder){

    dataBinder.setValidator(new NewsValidator());

}

    @RequestMapping(value="addNewsnewsAction")
    public ModelAndView addNews(@Validated News news,BindingResult
        bindingResult){

        ModelAndView mv=new ModelAndView();;
        if(bindingResult.hasErrors()){

            //mv=new ModelAndView("jsp/issue","typ",news.getType());
            mv.addObject("typ",news.getType());
            mv.setViewName("jsp/issue");
        return mv;
    }
    //直接调用 newsService
    boolean isSuccess=newsService.addNews(news);

    if(isSuccess){
    //保存成功,回到管理界面 selectIssue.jsp
        mv.setViewName("jsp/selectIssue");
    }
    else{
            //保存失败,退回到新闻录入界面
            mv.addObject("typ",news.getType());
            mv.setViewName("jsp/issue");
        }
        return mv;
    }

    @RequestMapping(value="indexauthAction")
    public void index(HttpServletRequest request,HttpServletResponse
    response){
        //获取 HttpSession
        HttpSession session=request.getSession(true);

        //直接调用 authService
        HashMap<String, List<News>>indexMap=authService.findHeaderNews();
```

```
        List<News>headerNews=indexMap.get("headerNews");

        if(headerNews.size()>0){
            News news=(News) headerNews.get(0);
            session.setAttribute("typ1",news);
        } else {
            session.setAttribute("typ1",null);
        }

        List<News>indexNews=indexMap.get("indexNews");
        session.setAttribute("typ2", indexNews);

        //mv.setViewName("");

        //return null;
    }

}
```

取消了获取 ApplicationContext、IAuthorizationServiceImpl 的过程,并取消了创建 INewsServiceImpl 对象的过程,直接使用声明@Autowired 的对象。

(9) 部署运行。

测试头版头条新闻和综合新闻,输入 http://localhost:8080/electrone/,如图 10-59 所示,表示 SSM 整合头版头条新闻查询和综合新闻查询成功。

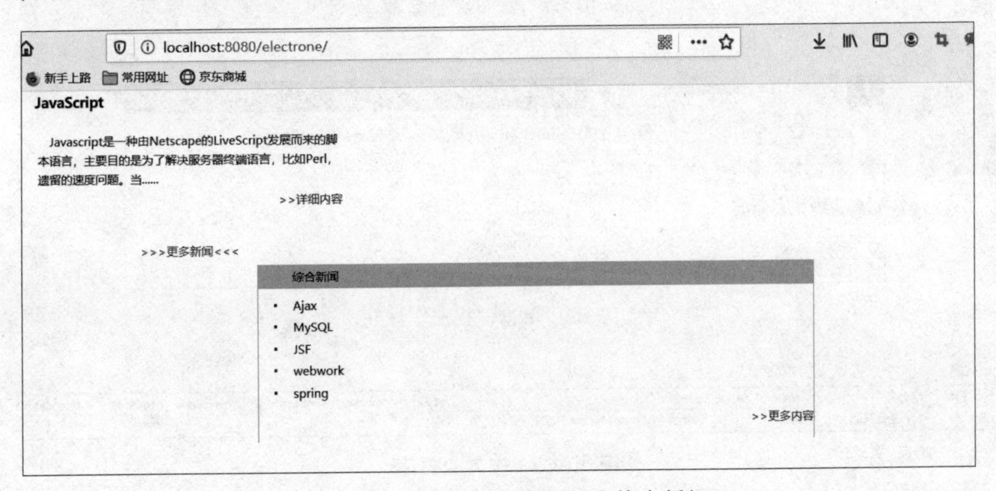

图 10-59 测试头版头条新闻和综合新闻

测试发布新闻,输入 http://localhost:8080/electrone/jsp/selectIssue.jsp,如图 10-60 所示。

单击"发布综合新闻"链接,出现图 10-61 所示。

单击"提交"按钮,如图 10-62 所示。

查看数据库,如图 10-63 所示。

图 10-60 测试发布新闻

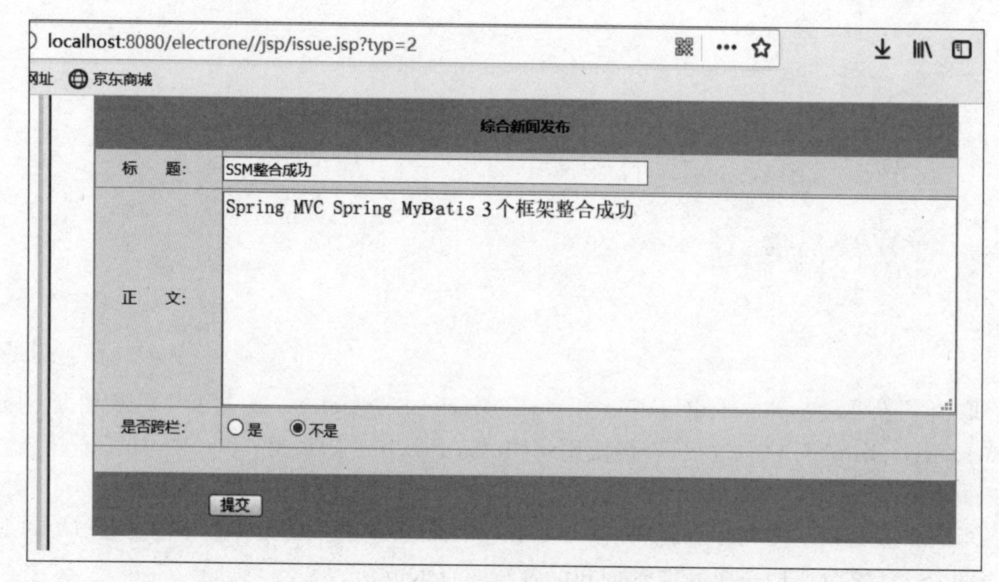

图 10-61 发布综合新闻

图 10-62 单击"提交"按钮

id	title	author	deptid	content	type	checkopinion	checkopinion2	checkstatus	crosscolumn	crossstatus
40	SSM整合成功	测试3	1	Spring MVC Spring MyBatis 3个框架整合成功	2	NULL	NULL	0	NULL	1

图 10-63 查看数据库

至此，三大框架整合成功。

10.2.5 特别提示

Spring MVC、Spring 以及 MyBatis 都使用日志 Log4J 框架输出程序中的执行信息，因此需要在整合后配置 log4j.properties。

在 electrone/src 下新建 log4j.properties 文件，如图 10-64 所示。

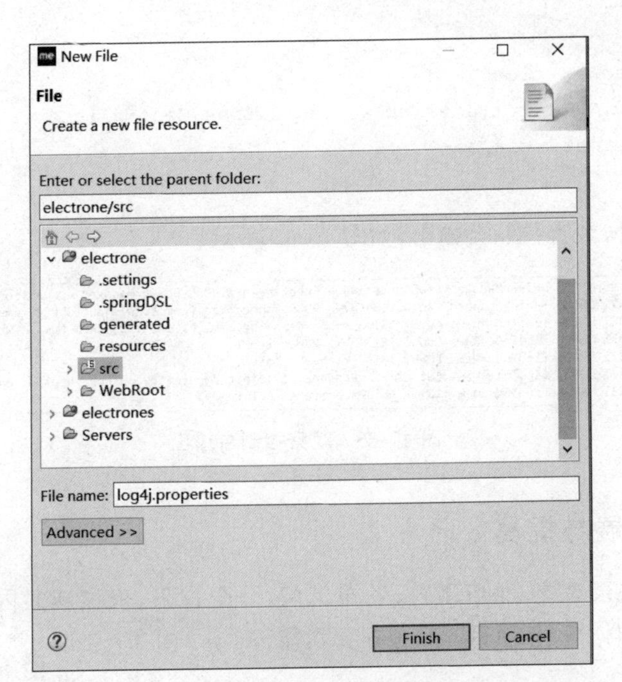

图 10-64　新建 log4j.properties 文件

编写 log4j.properties。

```
#定义 stdout 为控制台信息
#必须定义一个自定义日志级别,如 myinfo
log4j.rootLogger=myinfo,stdout

#输出调试信息 debug 到控制台
log4j.appender.stdout=org.apache.log4j.ConsoleAppender
log4j.appender.stdout.Target=System.out
log4j.appender.stdout.layout=org.apache.log4j.PatternLayout
log4j.appender.stdout.Threshold=debug
log4j.appender.stdout.layout.ConversionPattern=%d %-5p %c{1}:%L- %m%n

#必须配置 org.mybatis=信息级别,与 Threshold 相同
log4j.logger.org.mybatis=DEBUG
log4j.logger.com.ascent.mybatis.mapper=DEBUG
```

该配置配置了控制台输出 MyBatis 的 SQL 语句信息。

最后修改 web.xml。

在配置 contextConfigLocation 前配置 log4J。

```
<context-param>
        <param-name>log4jConfigLocation</param-name>
        <param-value>classpath:log4j.properties</param-value>
    </context-param>
<!--定义 Log4J 监听器 -->
```

```
<listener>
    <listener-class>
        org.springframework.web.util.Log4jConfigListener
    </listener-class>
</listener>
```

重新部署项目,观察控制台输出,如图 10-65 所示。

```
020-04-15 13:13:55,305 DEBUG DriverManagerDataSource:142 - Creating new JDBC DriverManager Connection to [jdbc:mysql://loca
020-04-15 13:13:55,321 DEBUG SpringManagedTransaction:87 - JDBC Connection [com.mysql.jdbc.JDBC4Connection@4d6b3733] will n
020-04-15 13:13:55,321 DEBUG findIndexNews_COUNT:181 - ==>  Preparing: SELECT count(0) FROM News n WHERE n.status = 1 AND n
020-04-15 13:13:55,321 DEBUG findIndexNews_COUNT:181 - ==> Parameters: 
020-04-15 13:13:55,321 DEBUG findIndexNews_COUNT:181 - <==      Total: 1
020-04-15 13:13:55,321 DEBUG findIndexNews:181 - ==>  Preparing: select id, title, author, deptid, content, type, checkopin
020-04-15 13:13:55,321 DEBUG findIndexNews:181 - ==> Parameters: 6(Integer)
```

图 10-65 观察控制台输出

10.2.6 拓展与提高

请独立完成登录系统,注册用户,发布新闻,分配权限,审核新闻内容,查看发布的新闻、更多的新闻等电子政务项目的其他业务功能。

习题

1. Spring＋Spring MVC＋MyBatis 集成需要哪些包?
2. Spring＋Spring MVC＋MyBatis 集成的步骤是什么?
3. 如何在 applicationContext.xml 中编写 MyBatis 配置信息?

图书资源支持

感谢您一直以来对清华版图书的支持和爱护。为了配合本书的使用，本书提供配套的资源，有需求的读者请扫描下方的“书圈”微信公众号二维码，在图书专区下载，也可以拨打电话或发送电子邮件咨询。

如果您在使用本书的过程中遇到了什么问题，或者有相关图书出版计划，也请您发邮件告诉我们，以便我们更好地为您服务。

我们的联系方式：

地　　址：北京市海淀区双清路学研大厦 A 座 714

邮　　编：100084

电　　话：010-83470236　010-83470237

客服邮箱：2301891038@qq.com

QQ：2301891038（请写明您的单位和姓名）

资源下载：关注公众号“书圈”下载配套资源。

书圈

获取最新书目

观看课程直播